THE POTTERS AND POTTERY OF MIRAVET

Anthropology Matters : Scholarship on Demand

A new series focusing on significant contributions that demonstrate the scholarship and depth of the traditional anthropological monograph, perhaps without wide commercial appeal, but with unquestionable academic merit. It will include important collections that open up new themes, and will publish research reports that deal with specific development or policy related issues, and are major interventions in current affairs and decision-making. All of these exemplify the way 'Anthropology Matters'.

Volume 1: Of Alien Kings and Perpetual King: Contradiction and Ambiguity in Ruwund (Lunda) Symbolic Thought
Manuela Palmeirim

Volume 2: Living with Things: Ridding, Accommodation, Dwelling
Nicky Gregson

Volume 3: The Potters and Pottery of Miravet: Production, Marketing and Consumption of Pottery in Catalonia
Rob van Veggel

Volume 4: Sin, Sex and Stigma: A Pacific response to HIV and AIDS
Lawrence James Hammar

Volume 5: Mutuality and Empathy: Self and Other in the Ethnographic Encounter
edited by Anne Sigfrid Grønsethe and Dona Lee Davis

The Potters and Pottery of Miravet

Production, Marketing and Consumption of Pottery in Catalonia

Rob van Veggel

Anthropology Matters, Volume 3

Sean Kingston Publishing
www.seankingston.co.uk
Wantage

First published in 2009 by

Sean Kingston Publishing
www.seankingston.co.uk
Wantage

British Library Cataloguing in Publication Data
A catalogue record for this book is available from the British Library.

Printed by Lightning Source

ISBN 978-0-9556400-2-5 hardback

Contents

List of illustrations

Maps

Plates

Tables

Acknowledgements

Many people and institutions have supported, encouraged and assisted me in the research and writing of this book and the dissertation upon which it is based.

The Progama D'Estudis Catalans 'Joan Maregall', the Fundació Caixa Barcelona and the Program for Cultural Cooperation between Spain's Ministry of Culture and US Universities deserve a very special thank you for their financial support of the research. In addition, the writing of the dissertation was financially supported by a position at the Nederlandse Organisatie voor Wetenschappelijk Onderzoek. The position was obtained with the help of Dr Johannes Fabian and Dr Jojada Verrips to whom I am especially grateful. A generous gift from my aunt Nelly van Leeuwen is also greatly appreciated.

While in the field, discussions of my work with anthropologists at several Catalan universities provided invaluable ethnographic feedback on my research I would especially like to thank Dr Silvia Carrasco, Dr Mercè Fernández Martorell and Dr Joan Prat i Carós. The information and feedback received from the pottery afficionado Santiago Albertí has also been most helpful, for which I am very grateful.

My professors at the University of Chicago who trained me and supervised my research and dissertation deserve special acknowledgement and thanks: Dr Jean Comaroff, Dr James W. Fernandez, Dr Nancy D. Munn and Dr Marshall Sahlins, and at the University of Amsterdam, Dr Johannes Fabian. In addition, I am grateful to Dr Leonora Auslander for her pointed feedback on an early draft and to Dr Enrique Luque who, as a visiting professor, read and commented on an early draft.

My gratitude also goes to the people of Miravet: they not only provided me with the data for this study, but also made the years of my fieldwork some of the most enjoyable of my life. I would like to mention Carme de Bertó who cared so well for me as owner of the local hotel and restaurant; Ingrid and Ulli Voss who let me stay in their house; Diana Campos, Meritxell Segarra and Manolita, who improved my skills in Catalan and Castillian; and Vicent Benaiges, Tomás Borell, Susanna Bros, Quim Marsal and Paco Navarro, who included me in their circle of friends and with whom I spent many pleasant hours. Numerous people have shared generously of their lives, and while there is not room to acknowledge everyone, Joan Avante, Santiago Benaiges, Pepe Jornet, Pepe and Josep Papaseit, Dominica Fabregat, Josep and Antoni Pedrola, Vicen Sales and Milio Treig Jr. deserve special mention and my personal gratitude. Two people who not only were important participants in this study, but who supported and encouraged me in many ways and without whom this research would have been virtually impossible, are Ferran Segarra and Teresa Pedrola, to whom I extend special thanks.

There are also a number of friends who have given their time and energy and deserve special thanks. In Amsterdam Jaap Vinken and Lucette ter Borg helped sharpen my thinking with their probing questions, and also helped me on special occasions, as in writing grant applications. In Chicago my classmates Michael W Scott and Judy A Boruchoff supported me by critiquing ideas I developed in the preparation of my fieldwork and during the writing of the dissertation. They also supported me with their sympathy as fellow sufferers in the process of obtaining a PhD at the University of Chicago. Furthermore I wish to thank Donald Whitfield and my spouse Allan Segall for helping me with my English, critically commenting on the text and probing me to formulate my ideas with greater clarity. To turn the manuscript into something readable, the editing by Ruth Higgins was essential.

Finally, I would like to thank and dedicate this book to my parents Pieter and Lily van Veggel. Though anthropology is something they don't quite understand, they have unfailingly trusted that I would accomplish this study and were perhaps even happier than I with its completion. Unfortunately, my mother passed away before she could see this book, but we were able to celebrate together the news that it would be published.

Chapter 1

Introduction

This is a study of the potters of Miravet and their pottery, a study of the dialectic between the social relationships which these potters constituted through the production, marketing and consumption of pottery, and the meaning their pottery acquired in this production, marketing and consumption. Based mainly on interviews with the potters and fellow villagers, I reconstruct the history of these people in a southern Catalan village as they produced and marketed pottery and learned about the consumption of their products. The period examined runs from the beginning of the twentieth century, when their production and marketing of pottery was within a local economy based on both subsistence and market-oriented agriculture, to the early 1990s, as this production and marketing became integrated into a modern international economy. By juxtaposing the local history of this small group of producers (numbering no more than 50 people), to macro histories, I localize the European transition to one single integrated market economy.

The context of the fieldwork

My interest in ceramics goes back to 1968, when I was 14 and in Catalonia for the second time. My parents, brothers and I were vacationing at the Costa Brava campsite. To get from the campsite to the beach, one had to pass through some dunes. Here I often encountered a young man who, accompanied by his colourfully adorned donkey, was selling pottery. He used the donkey to transport the pottery but also to attract the attention of tourists. We became friends. I let him ride my bike and in exchange he gave me pieces that were chipped and could not be sold. I brought these pieces back to the Netherlands and displayed them in my bedroom. I can still see them: they were small, about 20 to 30 centimetres high, like vases – one with the head and tail of a cock – of red unglazed earthenware decorated with scrolls in a different gloss. Now I know that it was pottery made in Salvatierra de los Barros. Over time I have lost these pieces, but I continue to collect pottery and ceramics.

These interests were further developed when training as an art/art history teacher. I became familiar with the history of Western art, including pottery and ceramics. Later on, during my training as an anthropologist, I took courses in Far Eastern, South Asian and Islamic art, courses in which I concentrated on ceramics and pottery. Complementing this academically acquired knowledge was more practical information gained from my aunt Greet Coerwinkel, a ceramist. She explained techniques and taught me to appreciate ceramics and pottery from a more technical angle. I became fascinated by pottery and ceramics produced by different people working in different traditions, by the diversity of objects produced by one of the oldest techniques known to mankind. I began to appreciate the plasticity of clay, which has been exploited by virtually every culture since the Neolithic era. I also learned that different cultures and social groups – sometimes even individual potters – are clearly identifiable by their pottery and ceramics. Indeed, Ruth Benedict's metaphor of a 'cup of clay [from which each] people [drink] their life' (Benedict 1989: 22) often used to represent the diversity and uniqueness of each culture, could be an appropriate epigram for this investigation.

In the mid 1970s, my parents bought an apartment for their retirement in a tourist resort on the Costa Brava. Their house was located in a neighbourhood like so many that had sprouted since the 1960s in Spanish tourist resorts and on the outskirts of Spanish cities, towns and villages. These were neighbourhoods consisting of what the Catalans call *xalets*: whitewashed houses of one or two storeys with many decorative windows, roofs, arches, balconies and terraces with surrounding gardens. On these terraces and balconies and in these gardens, one could cultivate flowers in a wide variety of pottery planters.

In 1986 I visited my parents with a friend, Erik Pol, who shortly thereafter decided to import such planters to the Netherlands. These planters with their blond terracotta colour, their simple but lively shapes and 'natural' look appealed to the current interior and exterior decorating fashions. The planters appeared in glossy magazines and within one year they appeared in every upscale florist and interior decoration shop. A shop opened in Amsterdam that carried only this pottery. In subsequent years these Spanish planters became so popular that even the cheapest department stores stocked them.

I helped my friend set up his business. Directed by a 'guide to potteries of Spain and Portugal' (Sempere 1982a), we visited potters in many parts of the Iberian Peninsula. I was struck by the great variety of potters' production, repertoire of products and approach to wholesalers like us. On the one hand, we visited workshops that had almost fully automated assembly lines with jiggers for moulding and a conveyor belt moving pieces through glazing booths and kilns. On the other hand, we also visited a workshop which could not be reached with our van because the access route pre-dated motor vehicles. We came across pottery that ranged from pots for storing olive oil that evoked a rural/traditional lifestyle, to planters with garlands and lion heads that evoked aristocratic gardens. Sometimes we were received by people specially appointed for welcoming clients. These 'greeters' spoke foreign languages, and offered colour-photo catalogues and assistance with export regulations. We were also met by potters who were suspicious of

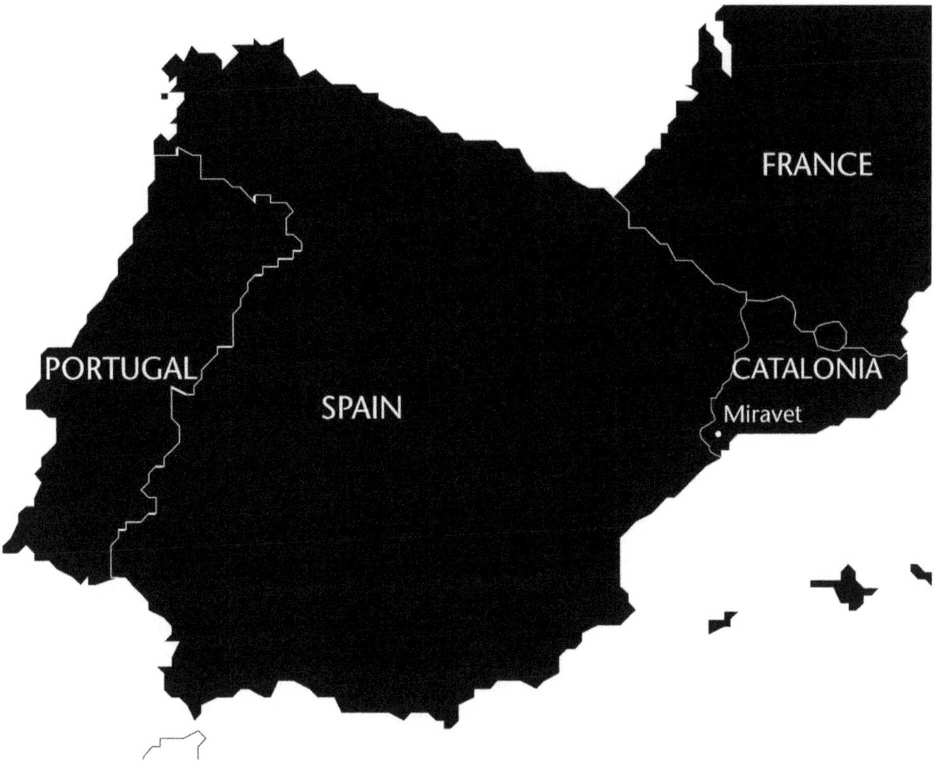

Map 1 *Map of Spain.*

strangers and who initially refused our orders, only accepting them if we paid in advance.

Although the ethnographic literature on Spain is rich, there is hardly anything published on pottery.[1] Several monographs on rural Spain and agricultural production have been published, many dealing with modernization and in particular the transition from a combined market-oriented and subsistence agriculture to a solely market-oriented one, but none has appeared on the rural industry of pottery production. No ethnographer has yet studied the historical process of the transition to a capitalist economy and commercialized society from the perspective of a specialized and inherently market-oriented form of production. Thus, I decided that this should be the topic for my dissertation research which is the basis of this book.

On one of my trips I visited Miravet and met Ferran Segarra, a very friendly potter who spoke English. We talked about pottery and he showed me his collection of pottery and books on pottery. After explaining to him my intended study, he responded enthusiastically and offered his help with my project. In Miravet, there were eight potters' workshops, all apparently thriving. These workshops showed interesting variations in production, repertoires of products and marketing practices. Since the literature could

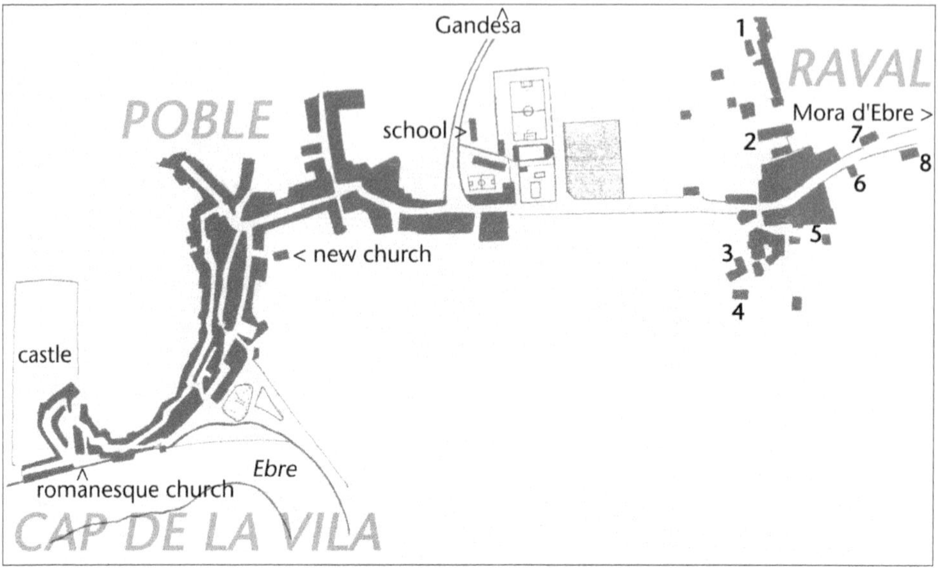

Map 1 ***Map of Miravet.*** The Poble consists of the area from the Romanesque church to the school. The Raval is the area to the right, where the potters' workshops are. The potters' workshops are: 1 – Jornet, 2 – Ventura, 3 – Papaseit, 4 – Treig, 5 – Societat Fabregat-Sales, 6 – Pedrola, 7 – Segarra, and 8 – Avante (on the road to Mora d'Ebre). Based on *Miravet,* a handout published by the Municipality of Miravet, n.d..

not give me much guidance for selecting a field site, these variations and the warm encouragement helped me choose Miravet.

Miravet is a small village, no more than 800 inhabitants at the time of the fieldwork, and it is located in the southern Catalan province of Tarragona, in the *comarca* ('county') of La Ribera d'Ebre, 'the Bank of the Ebre'. Close to the Mediterranean, the climate is mild with average temperatures between 2 and 32° C (Anon., *La Ribera d'Ebre* 1989:14). In the heart of this county is a plain and Miravet sits at the point the Ebre leaves this plain and enters a mountain range. The old part of the village – locally called the Cap de la Vila – is located high above the river. At the centre of this area is a Romanesque church. The Cap de la Vila was once enclosed by ramparts, still visible. Just outside these and a bit higher up the mountain are Iberian and Moorish ruins. (Someone once showed me a shard of pottery she had found at that site after heavy rainfall. It had a glaze with a metal reflection, a Moorish glaze in a now almost extinct technique.) Just above the Cap de la Vila and on top of a huge rock overlooking the river and the plain, the Templars built a castle. When protection by the ramparts was no longer needed and when the annual flooding of the Ebre came under control, the villagers began to build their houses on lower terrains closer to the river, and Miravet descended. This lower part of the settlemet is now the core of the village, housing most of its inhabitants. It contains the new church, shops, bars, the marketplace, the agricultural co-operative and the school. The villagers call the new and the old part together 'the Poble' as opposed to 'the Raval' – the

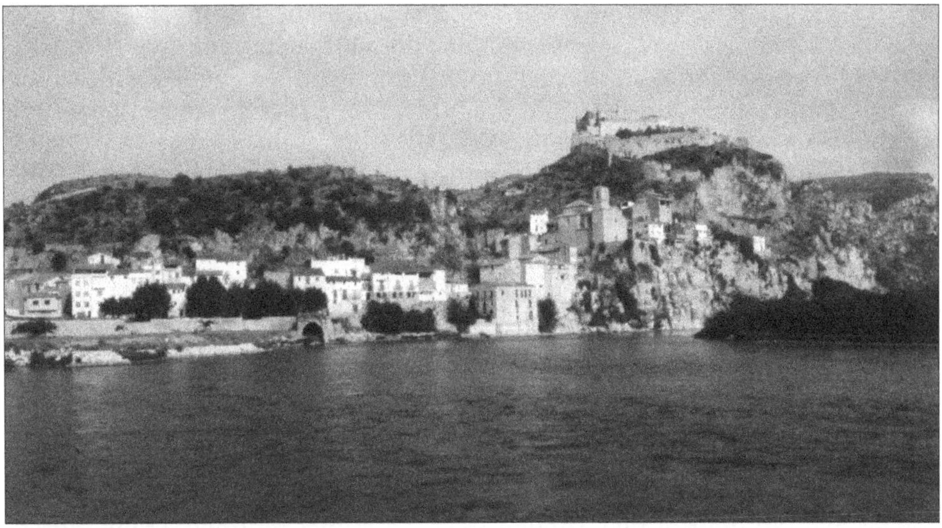

Plate 1.1 ***Miravet, view from the river.*** On the left sits the caste sits with the Romanesque church and the Cap de la Vila underneath it. The houses on the right form the Poble. Author photo, 1991.

suburb – which is a small conglomeration of houses and workshops some one hundred metres from the Poble. Although there are still some shops in the Raval, it is, and always has been, very dependent on the Poble, which almost every inhabitant visits daily. It is in the Raval that the potters of Miravet live and have their workshops. This part of Miravet has been officially called, in different periods, Terriseries and Alfarerías, Catalan and Castillian, respectively, for 'potter's workshops'.[2]

On 7 March 1990, I arrived in Miravet to start my fieldwork. I had prepared myself by learning Castillian and, to a lesser extent, Catalan. Castillian, also known as Spanish, has been imposed on the Catalans for centuries. Though most Catalans have perfect command of Castillian, they have continued to express themselves on many occasions in their own language; a language with a longer written history than Castillian, and with its own distinctive grammar, vocabulary and orthography (see Woolard 1989). Upon arrival I went back to see Ferran Segarra and consulted with him about beginning my fieldwork by working in a potter's workshop in order to observe the activities in these workspaces. He introduced me to his brothers-in-law, who had a large workshop where I could work. The next day I began by doing all kinds of odd jobs. Within a few weeks the Pedrola brothers discovered that I knew how to paint houses, which was a good thing, as they needed someone to paint their recent workshop extension, as all the painters had left the village to work in Barcelona for the upcoming Olympics. So, for the duration of my fieldwork, I became a part-time painter of workshops, houses and furniture, mostly for potters. This allowed me to supplement my income but also provided a practical way to integrate into the village community.

While working in potters' workshops, I noticed the regularity with which friends, family and tourists visited the potters for a friendly chat. Depending on the task a potter was engaged in at the time, these chats formed a welcome distraction, as they kept the potter informed of what was going on in the village. I started to make similar visits to the potters' workshops, introducing myself as someone who was writing a book in order to acquire an academic degree.

To my surprise, none of the potters seemed to wonder why I visited them to enquire about their craft. They perceived me as one of the many usual visitors, which included foreigners, Spaniards and Catalans. They seemed unfazed by my writing a book about them. Other authors, like the one who had written a guidebook on the potters, had also visited to collect data for their books. Though most potters had only skimmed through these books, they owned several of them. Visitors, including authors, did not usually stay longer than an hour or two, though they sometimes returned. Many potters were surprised, however, that I stayed longer and enquired about other matters such as kinship relations, which were never investigated by other visitors. 'What does that have to do with pottery?' was a typical response. Eventually, they got used to my odd questions, and after two years I had established myself with the potters as a professional academic researcher, an ethnographer.

When I began visiting workshops, I casually talked with potters about their work, usually what they were doing at that particular moment, or about the pottery and equipment around the workshop. I gradually directed my questions towards the topics that emerged from my theoretical framework. After a few visits, when I felt more confident, I introduced my tape recorder. Afterwards I would transcribe the tapes and index the data. Eventually I arrived at workshops with a printout list of questions. I quickly learned what time to arrive for the best the response and during which tasks I could interview potters best, and I kept track of the stage in the production process for each potter.

When I arrived at a workshop I began with small talk. Sometimes these conversations evolved into talking about the craft, at which point I asked if I could tape the conversation. Sometimes, however, the participant suggested I should take out the tape recorder as he or she thought that the conversation might contain interesting information: the tape recorder became the explicit marker that our conversations were part of my research. If during this initial conversation, I felt that the moment was inappropriate for interviewing, I would leave. At other times, I would ask if I could ask questions, a request that was hardly ever denied. I also interviewed visitors to workshops. For instance, the Pedrola brothers had a retired uncle who visited them daily, and I interviewed him extensively. Later I also made house visits to formally interview villagers, but often in bars and other public places interviews were more informal.

Since my active command of Catalan was not strong, I interviewed mostly in Castillian, though interviewees replied in either Castillian or

Catalan, and often in a mixture of the two. Indeed, it was not unusual to hear conversations in which one person spoke Castillian (always a native Castillian-speaker married into the village) while the other replied in Catalan. Many of these native Castillian speakers had been communicating with their fellow villagers in this bilingual way for decades. For example, I met an Aragonese woman, in her sixties or seventies, who had come to live in Miravet when she married a half century earlier. She still communicated in Castillian, even with her family. My interviewees did not expect otherwise from me, though they appreciated my attempts to speak Catalan. In this book I use Catalan words and expressions.

The recorded interviews make up the bulk of my data in this study. The interviews concentrated on life histories of potters and other people working in potters' workshops, on histories of potters' workshops, and on the different pieces of pottery. With respect to life histories, I gathered data on an individual's kinship and friendship relationships, his/her skills and responsibilities, idiosyncrasies and anecdotes of his/her life. With respect to the workshop histories, data was collected on the personnel, equipment, work space, location, repertoires of products, labour exchanges, marketing strategies and clients. Regarding the pottery itself, I concentrated on the names and descriptions of the different pieces, how each type was produced and marketed, and how and by whom each piece was consumed. The transcriptions of these interviews were organized by participant. During transcription, I also filed data under 'life histories', 'history of workshop' and 'types of pottery', according to the participant. While working on specific topics, I organized data specifically for them. For instance, when working on marketing to wholesalers and shop-owners, the data was organized on clients according to each workshop, and histories of the clients they had served were constructed for each workshop. Furthermore, while transcribing, I entered keywords on which I could do computer searches. As my fieldwork developed, I added new keywords as I deemed necessary. These files and keywords enabled me to make cross-references, detect gaps in my data and to formulate new questions.

The participants in my study generally appreciated these interviews (and I took care to keep them pleasant, as in most cases they were best perceived as 'chats'). Most participants found themselves pleasantly distracted from their jobs. Some participants even complained when I had not visited them in a while. They were usually flattered by my interest in their craft. As I soon discovered, the potter's craft had long been looked down upon, but had begun to be more appreciated since the 1960s. Someone coming all the way from Chicago to learn about their work added to this appreciation. My participants were offered an opportunity to talk about things that they seemingly never discussed. For example, one participant, Josep Papaseit, was in his mid twenties and had lived most of his life in his parental home, working in the family workshop. At one point, his father and I were talking about certain marketing practices and when we were finished, Josep remarked that he had never heard about this and found it very interesting. His mother, also always working in the workshop, confirmed that she had not known about marketing practices either, and that she, too, had enjoyed listening to it. In this way, I generated unexpected conversations.

Along with this systematically collected material, I was also able to gather data serendipitously. The fact that I lived in Minaret for two years provided me with numerous opportunities to observe the fabric of the society wherein this pottery was produced. I shopped, frequented bars, tutored English and was tutored in Castillian and Catalan, received invitations for trips and dinner parties, watched television at friends' homes, and went with friends to rock concerts as well as the Barcelona opera. Whenever I observed or heard something interesting, I made a note of it. These notes were diary entries made once or twice a day.

I complemented the gathering of this oral data by archival research. Potters let me consult their account books as well as those of their co-operative. (Since household economics are considered a very private matter, I only consulted these books when they were offered. I never requested them.) The municipality had an archive that went back to the Civil War (1936–9): records from before the Civil War had been burnt by anarchists during the conflict. The parish church kept a much older archive, one of the oldest in the region. The provincial archive, the *Arxiú Historic*, was incomplete. According to my sources, truckloads of the information stored in this archive disappeared when the Socialists came to power in 1981–2. Moreover, at the time of my fieldwork, this archive was in the process of indexing its holdings. On the provincial level, I also consulted the Camara de Comercio, the Delegación de Industría and the provincial branch of the Instituto Nacional de Estadistica.

The key participants

During my fieldwork, I made contact with many people but, of course, the key participants were all potters.

I met Ferran Segarra (born 1947) on my first visit to Miravet, and went to see him on my return. Throughout my fieldwork he and his wife Teresa Pedrola (born 1952) were very supportive. They introduced me to other participants and to friends of theirs, had me over for dinner parties and picnics, arranged a house for me, and even invited me several times to stay in their guest apartment. This support was an extension of previous similar support: Josep Maria Saez, the parish priest, had written a short book on the pottery of Miravet (Saez 1992) and Ferran had proofread the manuscript (Ferran himself had published a series of short articles on pottery in a local magazine, Segarra 1990a, b and c, and 1993). Moreover, Ferran and Teresa had a comprehensive collection of antique Spanish pottery – a passion of theirs – about which they taught me much. If we were not discussing food, the favourite topic of conversation in Miravet, but especially with the Segarras since Teresa was a fine cook and Ferran a gourmet, we were talking about pottery. We became good friends.

As a potter, Ferran Segarra was highly exceptional in that he had grown up outside Miravet and had come to Miravet to learn the craft and to begin his own workshop. He had been born in Miravet, but at an early age he migrated with his parents to Barcelona. During summer vacations he returned to the village and started courting Teresa Pedrola, the daughter of a local potter. After finishing school he worked as a waiter at the airport and had to learn foreign languages, especially French and English. He married and after a few years decided to return to Miravet and to learn, at the age of 30, the craft of the potter from his father-in-law. After a few years, he and Teresa began their own workshop. Almost from the beginning, they had combined their pottery production and marketing with trading of antique pottery and, to a lesser extent, furniture.

After we became better acquainted he told me that he had brought, in his words, 'the mentality of the city to the village'. When he decided to become a potter he had perceived the commercial potential of a new, urban clientele then coming into existence. At the time of my fieldwork, he appeared to be the Miravet potter with the greatest economic success. On another occasion he explained to me that though he and his wife lived in a village, they considered themselves to have an urban lifestyle. In contrast to most of his fellow villagers, he had an extended social network that encompassed even foreigners. For instance, he had close friends in Paris, Rome and Amsterdam in whose homes he had stayed. (Though his school education was limited, he had become familiar with art through his travels and through literature.) He had a similarly broad approach to his craft. He read books and magazines on pottery and ceramics and was up-to-date on publications on Catalan and Spanish pottery. He visited the national annual pottery fair in Zamora and in 1993 an international conference on pottery and ceramics in Istanbul.

On that first day of fieldwork, Segarra introduced me to his brothers-in-law, Antoni (born 1944) and Josep Pedrola (born 1947). Their father Antoni Pedrola Senr. (born 1917) was in ill health and had retired; he only occasionally swept the workshop. The two brothers, Antoni and Josep, now ran the business. Antoni was the more outgoing one of the two: he usually addressed visitors and dealt with clients, and in our conversations he always clearly expressed his emotions. Josep was more introverted and was the one concerned with the potter's technique. Since their early youth, Josep and his brother-in-law Ferran had experimented with glazes, discussed technical problems and exchanged technical knowledge. Josep had a great interest in improving production methods and products. He had designed an ingenious trolley for use in their workshop and had modified a machine for pottery production. Josep took pride in the green colour of the glaze that they improved upon during this fieldwork. I discussed issues of technique with him. Next to working as potters, Antoni and Josep had both pursued other interests as well. Antoni played the clarinet in one of the two bands of Miravet and performed at dances and festivals. Josep modelled reliefs, among them an impressive one, more than one metre in length, that featured the typical view of Miravet with the castle overlooking the river. After the brothers Pedrola had taken over the workshop from their father they had made important investments in its modernization. Now they owned the workshop with the largest production in Miravet, and employed three full-time permanent workers. All this made for a very lively workshop: there were often several simultaneous conversations and, on top of that, two dogs added to the bustling atmosphere.

When I had learned that one could make an informal visit to a potter's workshop, I tried my luck with the Pedrola's neighbours. Vicen Sales (born 1934) and his assistant Santiago Benaiges (born 1927) worked there. Vicen Sales Senr. (1914–85), with his brother-in-law Josep Fabregat (born 1914), had founded a society, the Societat Fabregat-Sales. There had been many people working in this workshop, but at the time of my fieldwork Vicen Sales Senr. had passed away, Josep Fabregat had retired, other employees had left, and only Vicen Jr. and Santiago remained. Both were looking forward to their imminent retirement, when this workshop would close.

Vicen had worked all his life in this workshop, with only a short break for military service. Santiago came from a potter's family, but had not taken over his father's workshop. Instead he worked for a wage in potters' workshops or on the land. When I began fieldwork in 1990 he had been working at Societat Fabregat-Sales about 12 years. Vicen and Santiago had known each other their whole lives. Unlike the Pedrola workshop, this was a quiet workshop, a bit out of the way, with few visitors. However, especially in winter and during bad weather, it was regularly visited by nearby farmers. Both Vicen and Santiago had endless patience and I could return to them again and again to obtain more details on aspects of their craft. If one could not answer my question, then the other could. Also, Vicen and Santiago tended to elaborate each other's stories.

On my first visit to Miravet I also met Pepe Jornet (born 1939). At that time I hardly spoke Castilian, and no Catalan at all, so we could not communicate well, but he showed me his yard where he kept a variety of poultry, which he enjoyed eating, and his dogs for hunting, a favourite pastime. On my return to Miravet I went to see him again. At the beginning of the fieldwork his son was fulfilling his military service, and Pepe was occasionally assisted by his wife or by a distant family member. Later, his son worked with him regularly. His workshop was at the edge of the Raval and hardly anyone visited: we were often alone. In this tranquillity Pepe gladly answered my questions. Other times he invited me for brunch, *botifarra* with *allioli*, and we talked about potters and pottery.

As I had done with Vicen Sales and Santiago Benaiges, I introduced myself to the Papaseit family. To my astonishment they had heard about me and were already anticipating my visit! Pepe Papaseit (born 1934), his wife Dominica Fabregat (born 1937) and their son Josep (born 1970) created a family atmosphere in their workshop. They were often joined by their daughter, Pepe's mother and Dominica's father. Pepe possessed an encyclopaedic knowledge of the history of his village and his craft. He proved to be, for me, the ultimate resource when it came to unravelling problems in my understanding. Furthermore, he also attempted at times to carefully explain social processes (he often commented that he himself could not understand how he and his fellow villagers had been living a few decades ago). As I made progress with my research, and began to perceive patterns in social and historical data, I often discussed my findings with him.

His son Josep had learned the potter's craft from his father and his grandfather, but Josep had also attended the art school in Tarragona. Josep had a similar interest in pottery to Ferran. He too collected books on Spanish pottery, visited potters in other villages to talk with them about their craft and products, collected antique pottery, and had also written short articles about pottery. But besides pottery, he was also interested in ceramics as an artistic expression. He had been trained as a ceramist, and in his free time produced ceramics. His plan was to focus more on this, though at the time pottery proved to be more lucrative. It was Dominica who mostly dealt with collectors who visited their workshops: she remembered them all, and even kept an address list of the most loyal ones, and hence was a good source of information on them.

Shortly after I began my fieldwork, my friend who imported pottery to the Netherlands visited Miravet and introduced me to Joan Avante (born 1940). The Avante workshop was on the road between Miravet and Benissanet. Here Joan Avante was working together with his sons Jordi, who had also attended the art school in Tarragona, and Joan Jr., who had begun university but then decided to return to work with his father. Both sons were in their twenties. Also there was Joan Senr.'s nephew, Pepito Pedrola, a man in his forties. I talked mostly with Joan Avante Senr.: he was usually throwing and hence easy to talk to in contrast to the other less sedentary people at the workshop. Moreover, the others felt that, because of their youth, they could not adequately inform me.

The last key participant was Milio Treig Jr. (born 1955). I presented myself at his workshop, where he worked with his father, Milio Treig Senr. (1926), usually assisted by his mother, Teresa Adell, and on occasion by his wife. Unfortunately, his father had a hearing problem and I could only communicate with him through his son or his wife.

The theoretical framework: encompassment in the social relationships constituted through objects

The theoretical framework of this study consists of three interrelated premises:

1) the analysis of social relationships that are constituted in interaction with objects, has to be complemented by the analysis of these objects;

2) the social relationships constituted in interaction with objects have to be analysed by using the mutually encompassing concepts of production, exchange and consumption. As these social relationships are constituted through objects, the third premise follows;

3) objects have to be analysed as being differentiated in categories according to how people interact with them and constitute social relationships through them.

* * *

Objects do not have an inherent meaning for subjects or people. Their objective characteristics become meaningful in people's interaction with them and their meanings are in relations between objects and people. Yet these relations between objects and people are aspects of the relationships subjects have with each other, of social relationships. Objects and people constitute a semiotic space with three dimensions;

1) one dimension is constituted by the relations between objects, i.e. their physical similarities and dissimilarities;
2) another dimension is constituted by the relations between people, by their social relationships;
3) the third dimension is constituted by the relations between objects and people.

Hence the two complementary inquiries. To study social relationships, one can study meanings that objects acquire in people's interaction with them. To study the meaning of objects, one has to study the relationships between subjects as articulated through objects.

This premise is worked out in a methodology drawn from Arjun Appadurai's assertion: one must complement the 'theoretical' perspective that 'things have no meanings apart from those [meanings] that human transactions, attributions and motivations endow them with', with the 'methodological' perspective that 'things-in-motion ... illuminate their human and social context' (Appadurai 1986: 5). Therefore, I gathered data on the histories of the individual potters and of the social entities (i.e. workshops) in which they produced and marketed their products, and also gathered data on the pottery: data on its physical shape in relation to its production, marketing and consumption. Furthermore, I specified data on potters (and the social entities they constituted, such as workshops) with respect to the kinds of pottery, and I specified data on pottery with respect to the kinds of social interaction involving potters.

By describing and analysing the social relationships that the potters of Miravet had through their pottery and, at the same time, the meanings of the pottery produced in Miravet as constituted in these social relationships, I further developed this approach. This study is divided into a social history of the potters and a history of their pottery: both histories complement each other as one cannot understand one without the other. Thus this study attempts to correlate changes in the social relationships in which potters were involved, with changes in the physical shape and the meanings of pottery, and vice versa. I show how the social history of the potters is inherently linked with the material aspects of this pottery, and which changes over time have resulted from and affected their role in human relations. I complement the 'social life of things' (Appadurai 1986) with the material life of people.

* * *

To analyse social relationships as constituted and articulated through objects, I use a set of three concepts that together encompass them: production, exchange and consumption.

Production is the process of making objects. People do this by employing 'means of production' (i.e. materials, tools and techniques) to which they have socially structured access (i.e. property and use rights), and by employing their own and other's labour (i.e. in social 'units of production' and by hiring labour). Consumption is the process of putting a product to use – of satisfying a need with it – in order to reproduce life or a way of life (see also Miller 1987, 1995a, b and d). These two processes of producing and consuming objects are interrelated.

Arguing against analyses in which production and consumption are perceived as separate or consecutive processes, Marx analyses how production and consumption mutually encompass each other in three ways. First, production and consumption encompass each other 'immediately' in that a producer consumes materials, tools and labour, and in that a consumer produces him/herself. The latter Marx exemplifies with nourishment, the consumption of food that at the same time produces the body. Second, production and consumption 'mediate' each other. 'Production mediates consumption; it creates the latter's material; without it, consumption would lack an object. However, consumption also mediates production in that it alone creates for the products the subject for whom they are products' (Marx 1972: 229). Marx elaborates on this form of mutual encompassment:

> Production also gives consumption its specificity, its character, its finish. Just as consumption gave the product its finish as product, so does production give finish to consumption. Firstly, the object is not an object in general, but a specific object that must be consumed in a specific manner, to be mediated in its turn by production itself. Hunger is hunger, but the hunger gratified by cooked meat eaten with a knife and fork is a different hunger from that which bolts down raw meat with the aid of hand, nail and tooth. (Marx 1972: 230).

The third way in which production and consumption mutually encompass each other is that 'each of them ... creates the other in complementing itself, and creates itself as the other' (Marx 1972: 231). With this, Marx refers to production 'creating' the need or 'motive' (ibid.: 230) for consumption, and consumption as 'creating' the 'producer's inclination' (ibid.: 230): a connection which is usually called 'the relation of demand and supply' (ibid.: 231).

The mutual encompassment of production and consumption implies exchange, or in Marx's words 'distribution' (Marx 1978: esp. 232). Exchange is the act of handing over an object or product, from one subject to another. Exchange constitutes a social relationship between two partners and such a social relationship expresses itself in exchange. Exchange takes place when producer and consumer are not the same person. That is, a person produces an object not for his/her own consumption, but rather to exchange this object with another person for another object that he/she needs for personal consumption and for reproducing his/her lifestyle. Exchange is not necessarily limited to taking place between producer and consumer. Middlemen are possible within this process. Hence, exchange is a function of division of labour, and in that sense is itself encompassing of the interrelation of production and consumption.

These concepts of production, exchange and consumption as mutually encompassing, cover all social relationships constituted and articulated through objects, all people's interaction with objects, and hence provide a thorough analysis of meanings of objects constituted in people's interaction with those objects. Because these three concepts make it possible to focus on a particular aspect of the complex formed by people's interaction with objects, the concept of encompassment was crucial to this study and provided five points of focus for my study.

First, I look at the relationships potters constitute by the consumption of materials, tools and labour in their production. That is, I study which raw materials and tools or equipment they need, in which social relationships they acquire and own these materials and tools, and how they organize their labour in order to produce specific pottery. Second, I look at the social relationships potters constitute by the exchange or distribution of their products, relationships constituted in the marketing to middlemen and to consumers of the pottery, and again specified by the kinds of pottery. Third, I look at the relations in which both production and consumption become specified. These are the social relationships potters constitute with consumers of their products in the form of their knowledge of the consumption and consumers. Fourth, I study the relations in which production creates the motive for consumption, as consumption also creates the producer's inclinations. I study these by looking at the interrelation of forms of production, marketing and consumption as an aspect of social relationships potters constitute with middlemen and with consumers. That is, how the first three points of focus are mutually conjoined, especially in reference to specific pottery. The fifth point of focus relates to 'potters' themselves. As potters are engaged in all these social relationships, they also constitute relationships amongst themselves. I look at these mutual relationships with respect to their production, marketing and consumption of their products.

With these five points of focus, I study the social relations potters constituted in my period of research through production, exchange and consumption of pottery. In order to understand the dynamic of these social relationships, I also analyse these points as taking place within social structures of a much wider extent. The potters formed a small group, numbering not more than 50 people, who through production, exchange and marketing of pottery had been influenced by, and reacted to, developments in the Spanish and wider European political economies. Therefore I juxtapose the local histories, even life histories, of the potters with these macro histories. (The most individual level is indicated by italics. The texts in italics are close renditions of parts of conversations with participants on their lives, opinions and accounts of events.) This spatial dimension is complemented by a temporal one – social history – based mainly on oral data. Hence the period covered is the one from the 1920s and 1930s until the 'present' – the ethnographic present of the early 1990s.[3]

* * *

To analyse the physical characteristics of the pottery as becoming meaningful in human interaction I use the concepts of differentiation and categorization.

Physical characteristics of objects become meaningful only when people interact with them. This interaction is specific. A person interacts in a specific way with a certain object because of that object's specific physical characteristics: the meanings of an object are these specificities. Different forms of interaction correspond with physical differences or dissimilarities; similar forms of interaction correspond with physical similarities. Hence, in their interaction with objects, people differentiate and categorize objects, but they accomplish this differentiation and categorization as aspects of social relationships. That is, the differentiation and categorization of objects has to be situated in specific social relationships.[4]

The two complementing concepts of differentiation and categorization provide two further points of focus for my study. First, I look – in the most literal sense of this verb, including also other senses like touch – at the pottery, as its physical characteristics become distinctive for differentiation into categories of pottery. I study which differentiations and categorizations are made, by whom, in which interaction with the pottery, and in which social relationships. Second, I examine how pottery, as one category and as several subcategories, becomes differentiated from, and categorized, with others kinds of objects. I study in which wider categories of objects pottery is included, who makes these wider categorizations in which interaction with these objects and in which social relationships they do so.

Summary of the book

The structure of this study is dually based on the progression of time from the beginning of the twentieth century to the time of the fieldwork (the early 1990s), and on the course pottery undergoes from production through marketing to consumption. The division in chapters is based on essential differences between marketing relationships and concomitant forms of consumption as they occurred during this time period.

* * *

During the first half of the twentieth century, the potters of Miravet sold their products to muleteers, fellow villagers who peddled the pottery to the southern part of Catalonia, the northern part of Valencia, and to more distant Aragon. The relationships potters formed through the exchange of pottery with muleteers and the differentiation and categorization of pottery in these exchanges is the topic of Chapter 2, 'Potters and muleteers'.

Throughout this period, Spain was basically a rural country with a combined market-oriented and subsistence agriculture. The village of Miravet formed a close knit, relatively independent community. The people of Miravet judged productive activities and subsequent livelihoods by the indigenous values of 'money,' 'food ' and sociability. In addition they were circumscribed in their production by the socially distributed access to the means of production. People without access to sufficient land could complement their own agricultural production with the second most desirable means of access to food, peddling. These were the muleteers. People entirely without access to land could work for a wage, or could produce pottery. As two different parties in trade, potters and muleteers were opposed to each other. Their respective locations in the local political economy implied a power imbalance in their exchanges and potters were relegated to the weaker position.

Pottery acquired meanings in these exchanges. Potters and muleteers reached agreements on aspects of these meanings but they often contested other aspects, most importantly price calculations. Within Miravet, potters and muleteers fell into two political-economic categories but were related to each other also as fellow villagers, that is as friends and kin. On this personal level potters and muleteers varied their relationships by mitigating or sharpening their contestations of the meanings of the pottery. Because of these opposing levels, their relationships remained ambiguous. This ambiguity was dialectically related to the agreed and contested meanings of the pottery: as mediating between two political economic categories, the meaning agreed was expressed in money, but as mediating between persons, the meaning with regard to the price calculation was agreed and contested in relation to the nature of the personal relationship.

* * *

Potters sold their products to muleteers, who in turn sold them to consumers of the pottery. Through muleteers, potters related to consumers: these relationships are the topic of Chapter 3, 'Potters and consumers during the period of the muleteers'.

During this period, with the exception of the Civil War, Spain constituted one homogenous and free market for pottery. The market area of the pottery of Miravet varied widely. It encompassed villages for which pottery was one of the few products acquired from the outside, and towns in which people acquired the latest European consumer goods.

Within this context, potters constituted relationships with consumers of their products by a particular meaning of the pottery, its 'social use-value' or simply 'use-value' (Marx 1984: esp. 43–53). 'Exchange-value' (ibid.) was created in, while itself also creating, the marketing relationships between potters and muleteers. Use-value was created in the consumption of this pottery and thus created the relationships between potters and consumers. Here I combine Marx's conception of use-value with Baudrillard's insight of how people differentiate themselves from each other by consuming objects that are at the same time also differentiated from each other (Baudrillard 1981). The knowledge potters have of the consumption of their products is thus inherently linked with their perception of the consumers and, by this knowledge, potters locate themselves in a social space.

Potters knew about the consumption and consumers of their products through different channels. This knowledge was more or less detailed with respect to the several forms of consumption and to the several consumers, depending on a potter's individual location in the network of potters and muleteers. Furthermore, potters combined this knowledge with knowledge acquired by observing consumption and consumers in their own village.

Since the consumption and consumers within this period varied little, one could perceive this knowledge as it pertained to space, that is, as an actual and as a semiotic space. The potters' knowledge of the consumption and the consumers formed a 'geography of consumption'.

* * *

During the 1940s two vanguard potters began to explore a new practice by marketing their products to wholesalers and shop-owners. Other potters hesitated to follow. During the 1950s, however, especially at the end of this decade, potters were increasingly forced to shift to marketing their products to wholesalers and shop-owners. By the early 1960s, this practice became universal among the potters. This shift is the topic of Chapter 4, 'Selling to wholesalers and shop-owners'. To analyse this shift in the potters' relationships with middlemen and in their mutual relationships, I apply Sahlins' analysis of historical 'events' (Sahlins 1991), observing how the marketing to wholesalers and shop-owners was 'continuous' with certain socio-cultural structures, how it was discontinuous with other structures, and how the continuities and discontinuities formed a 'structure of conjunction'.

Already by the 1930s the monopoly of muleteers on transportation had been undermined with the development of modern means of transportation: train, bus and truck. During the 1940s Franco, the victor of the Civil War, pursued a policy that resulted in an extensive black market. Through their different locations in the political economy, potters and muleteers were affected differently by the black market and the mutual tension between them increased.

Incited by the increased tension between potters and muleteers, the two aforementioned potters formed marketing relationships of a very different social nature than the previous ones with muleteers. Due to this different nature, other potters, as previously mentioned, hesitated to follow these two in their new marketing practice. But in 1948 potters organized themselves in a co-operative in order to strengthen their negotiating position against muleteers, and to explore collectively the new marketing practice. Using Sahlins' distinction of principles in 'economic transactions' as being 'reciprocity' or 'pooling' (Sahlins 1972: 188), I analyse several forms of co-operation between potters, why the potters chose a co-operative for their organization, and why this co-operative was eventually to be short lived. During the 1950s muleteers disappeared and all potters were then forced to market their products to wholesalers and shop-owners.

In these new marketing relationships, pottery acquired new meanings, especially as the prices were calculated differently and sizes became standardized.

Throughout the 1940s and 1950s the consumption and consumers of this pottery remained the same, but the channels through which potters learned about them changed as their marketing practices changed

* * *

The 1960s were a pivotal period for the potters: simultaneously, they began to change their productions, repertoires of products and clients. The 'synergetic effect' of these changes was that they individualized their production, repertoires and clients. This is the topic of Chapter 5, 'New production techniques, new products, new clients, new consumers'.

During the 1960s Spain went through tremendous changes. It became an industrialized and urban country, it developed tourism, and consequently experienced tremendous economic growth. Of significance for the potters was that the demand for what had previously been their main types of pottery declined. Fortunately, the ever-adaptable potters responded by taking new types of pottery into production to meet a newly developed demand. Also significantly, rural wages rose and potters responded to this by investing in labour-saving, capital-intensive equipment.

As the potters responded to these changes, they made small shifts, but as these shifts were conjoined they had the 'effect of a structural synergy' (Sahlins 1991: 68). Here I look at the changing productions of pottery, the changing relationships between potters and clients, the changing relationships between potters, and the changing physical appearance of pottery.

The economically stronger potters bought machinery, increased their unskilled labour force and produced flowerpots, a new type of pottery. The economically weaker potters shifted to the production of decorative pottery, though not all the potters producing decorative pottery produced the same repertoire. All potters acquired clients in accordance with their repertoire of products, and clients themselves also requested new types of pottery. Thus, in these new marketing relations, the types of pottery became a variable and became identified with that relation: potters could recognize a type of pottery, by whom it was produced, and often also for whom. The synergetic effect of all these shifts resulted in potters becoming individualized – or differentiated from each other – by their production, their repertoire of products and their clients.

Concomitant with these shifts, potters changed their relations with the consumers of their products: as potters individualized their repertoires of products, and subsequently their marketing relationships, their knowledge of consumption and consumers also became individualized. Increasingly, each potter only knew where, how and by whom his products were consumed. Now he hardly knew anything of his colleagues' products. The changes in consumption and consumers coincided with a geographical shift from consumers in rural regions of eastern Catalonia and Aragon to consumers along the Catalan tourist coast and to consumers in cities who represented a new group of people with a lifestyle and purchasing power different from that of the potters.

* * *

Chapter 6, 'Potters and their clients – wholesalers and shop-owners', combines the type of analysis of Chapter 2 on potters and muleteers (i.e. of the relationships potters formed through the exchange of pottery with their clients and the categorization and differentiation of pottery in these exchanges, but now during the 1970s and 1980s) with the type of analysis of the chapter on the pivotal 1960s changes – i.e. the synergetic effect of shifts in marketing relationships, productions and repertoires.

Despite the recession between 1974 and the mid 1980s, the Spanish economy continued to grow remarkably. When Franco died in 1975, Spain underwent a political restructuring by becoming a democracy. In 1986 Spain joined the European Community. The main result of these changes for the potters was a growing, internationally expanding demand for their products.

Looking again at the local political economy (like I did in Chapter 2), people of Miravet now judged productive activities and resulting livelihoods basically by the money one made. Because of the growing demand for pottery, potters became prosperous and pottery making was no longer a deprecated economic activity but was valued with regard to the money one made. When selling their pottery to muleteers, potters had been in the weaker position, but now in a sellers' market, their clients were.

Potters individualized their forms of production and their repertoires of products in relation to their different clients. Pottery became categorized in relation to different potters. Simultaneously, potters continued perfecting the finish of their products, though in different degrees, and charged prices accordingly. As potters differed in the quality of their products and the prices they charged, pottery also became differentiated in price and quality in relation to its producer. Potters also differentiated their marketing relationships with respect to their repertoire of products and as to whether clients were Catalans and Spaniards, or foreigners. Potters sold all types of pottery to Catalan and Spanish clients. These clients tended to be shopowners who regularly visited the potters, with whom potters did business over longer periods, and with whom potters developed cordial relationships. Potters sold only planters to foreign clients. These clients were invariably wholesalers with whom potters had only occasional and restricted personal contact over relatively short periods.

* * *

Potters marketed the bulk of their production to middlemen, hence they were economically dependent on these exchanges. But at the same time, they also marketed pottery to people coming to their workshops. Though these exchanges were not economically lucrative, the relationships constituted by them were socially important. These relationships are the topic of Chapter 6, 'The exchanges at the workshops'.

With visitors, potters had two very different kinds of relationships. The first was the relationships with locals, members of the same community, with whom potters had kin- and/or friendship relations in which the exchanges of pottery, as well as other objects and services, were embedded. The second relationship was with visitors, people coming from afar to visit potter's workshops with whom the potters had relationships based on the trade of pottery. These two kinds of relationships entailed different forms of exchange, different knowledge of consumers and consumption of pottery, and different pottery.

The motivations of these visitors coming from afar related to two historical developments: tourism and the search for Catalan cultural identity. Potters understood these motivations differently and hence marketed and related to visitors differently. Applying Benjamin's concept of 'aura', I show that although exchanges with visitors were not especially profitable for potters, they were important as potters derived self-esteem from them, a self-esteem particular to pottery.

* * *

The relationships between producers and consumers as constituted in the potters' knowledge of consumers and consumption of their products are the topic of Chapter 3. In the following chapters, I trace the implications of changes in production and marketing with respect to these relationships and this knowledge. These relationships and this knowledge at the time of the fieldwork are the topic of Chapter 8, 'Potters and consumers in the early 1990s'.

To highlight the difference between the beginning of the twentieth century and the early 1990s, I apply the same analysis as I did in Chapter 3. In the early 1990s, the political-economic context of the potters' relationships with consumers of the pottery was formed by a capitalist international market for goods being transported across borders and people travelling abroad. To analyse these relationships, I again apply conceptions developed by Marx and Baudrillard.

Previously, potters had acquired knowledge of consumers and consumption through their relationships with muleteers. Potters shared common knowledge, though one might know more detail about certain types of consumption and types of consumers. In the 1990s, potters learned about consumers and consumption through their direct contact with visitors in their workshops, through their contacts with distributors by way of orders placed, and as potters themselves delivered orders to distributors' warehouses and shops. They acquired knowledge by their differentiated productions, marketing and self-identifications. Their knowledge of consumers and consumption had become specialized. Hence, this knowledge no longer forms one 'geography of consumption' but 'geographies of consumption'. In this seventh chapter, I present three separate geographies.

1. By American and British ethnographers, only one article on Spanish pottery is published. In 1962, Curtis published a short general description of pottery production in Bailén, an Andalucian village. In the 1970s and early 1980s two German ethnographers, Vossen and Köpke, published a few surveys of production techniques and repertoires (Köpke 1974 and 1985; Vossen 1972a and b; Vossen, Seseña and Köpke 1980); Köpke published one article – also a survey – on religious beliefs and practices associated with pottery production (Köpke 1986); and Vossen published two articles on systems of valuing and labeling pottery – again surveys (Vossen 1974, 1984). A very early publication by another German ethnographer, also focusing on production techniques, and worthwhile mentioning is Bierhenke 1932.

Of course in Spain itself, a great number of publications on pottery have appeared (see for instance Guerrero Martín 1988 for a very extensive bibliography of these publications). These publications are in general surveys of production techniques and types of pottery (see for instance surveys on the level of the Spanish state: Guerrero Martín 1988, Llorens Artigas and Corredor Mateos 1979, Sempere 1982a, Seseña 1976, Torres 1982; and on a smaller level, Alvaro Zamora 1980, Cabézon et al. 1984, Carretero et al. 1984, Feito 1985, García Alén 1983, Gonzalez Antón 1987, Pérez Vidal 1983, Ramos Perez 1976, Schútz n.d., Sempere 1982b, 1985 and Subirats 1988). Many of these publications contain interesting ethnographic and historical data, but lack a systematic ethnographic approach. They are clearly not written by ethnographers for professional readers of anthropology and ethnography. (The only exception that I know of is Cabezón Cuélar et al. 1984, published by the Instituto Aragonés de Antropologia.) See also Prat et al. 1985/86. This is to date the most comprehensive bibliography of ethnographic and anthropological literature on Spain published between the mid-1950s and 1985 and does not contain one single entry explicitly dealing with pottery, except the afore-mentioned Curtis article.

2. My participants could recall only one potter's workshop in the Poble. It closed in the late 1940s.

3. Two historians, Pasqual Ortega and A. Bladé Desumvila, have been working specifically on the history of Miravet (see for instance Ortega 1986 and Bladé Desumvila 1966). They mention potters incidentally, but offer too little information about them to reconstruct the history of pottery production, marketing and consumption before the period of which oral data is available.

4. See van Veggel 1991. In this article, I critiqued ethno-science studies using Peirce's conception of semiosis. Crucial in Peirce's conception is the notion that meaning always exists in relation to an 'interpreter', something absent in ethno-science studies. Here I argue for addressing the concerns of ethno-science – i.e. categorization and differentiation – but avoid the reification of the signs, that is, the pottery as being meaningful in and of itself.

Chapter 2

Potters and muleteers

The national and international context of the first part of the twentieth century

During the first half of the twentieth century Spain was economically isolated from other countries by high tariffs and low exports (Shubert 1990: 22; Carr 1982: 398)[1] and had one of the lowest living standards in Europe (Carr 1982: 400), remaining an economic 'laggard' (Shubert 1990: 11). In and around Barcelona the textile industry, especially cotton (and to a lesser extent wool and leather), had developed, but Cantabrian heavy industry was in decline (Carr 1982: 406). On the whole, Spanish industry lagged behind international competition by remaining small scale (Shubert 1990: 15–16; Carr 1982: 405; Tamames 1986: 133–4).

Spain was still an agricultural country, and oranges were the main export product (Carr 1982: 401; Shubert 1990: 22–3), but the cultivation of oranges was exceptional. Agriculture was mainly dry agriculture, that is the cultivation of wine, olives and cereals, especially wheat (Carr 1982: 402). Although agricultural productivity was increasing (Shubert 1990: 13 and 15), in general, agriculture remained rather primitive, with labour intensive techniques 'not much better than those used in Roman times' (Lison-Tolosana 1983: 120; see also Carr 1982: 403) and agricultural production vacillated between market-oriented and subsistence agriculture (Harding 1984: 30–1, 37 and 54). Agricultural production varied from the huge southern land holdings, worked by wage labourers cultivating cash crops, to the small northern parcels, usually worked by members of a household primarily providing for their own consumption.[2] Though many people moved to the industrial cities (Barcelona, for instance, doubled its size), Spain remained a rural country with a mixed capitalist and pre-capitalist agriculture (Carr 1982: 413–17; Shubert 1990: 45).

The main economic activity in La Ribera d'Ebre was agriculture, and, as was most common in Spain, dry cultivation. People here had relatively equal access to land: in the province of Tarragona, where this county is situated, the land distribution was one of the most equal in Spain.[3] Miravet is only 125 kilometres from Barcelona, though hardly anyone migrated there, nor to any other industrial centre, nor abroad.[4] On the contrary, Miravet's relative prosperity was manifested in that boys from the poorer Aragon came to work there as *mozos* – live-in workers.

Though agriculture was the main economic activity, there was some industry in the county of La Ribera d'Ebre. For example, there were some very small industries based on the processing of agricultural products: flour, pastry, soap, wine, alcohol, ice cream and soda drinks. Some small industries produced construction materials such as bricks, cement and lime, and there was one mattress factory. The largest regional factory was a chemical plant employing about 400 workers (Anon., *La Ribera d'Ebre* 1989: 366–7). Other economic activity included small-scale mining of lead and minerals used in producing construction materials, and there was also handicraft production: baskets and other palm-fibre objects produced in Rasquera; brooms and reed objects made in the villages along the river like Miravet and Benissanet; and pottery made in Benissanet and Ginestar, but mainly in Miravet.

Several characteristics of this context affected the relationships between potters and muleteers. Though industry was emergent, agriculture was the dominant mode of production in Spain. In the region of Miravet, landed property was relatively equally distributed, hence agriculture was conducted in a combination of market-oriented and self-subsistence agriculture. Though naturally there were economic differences between people, these were not as harsh as elsewhere and the region was in general prosperous compared to other rural regions in Spain.

Potters and muleteers as members of their village community

As people form social structures they have different access to means of production – i.e. power differentials – in such forms as property and use rights. This social distribution of means of production circumscribes a person in his economic activity, but a person is also motivated in his economic activity by the cultural values he shares with members of his society or community (see Sahlins 1976: 206 and Roseberry 1994). Which means of production did the potters and muleteers have access to within this context, and what were the cultural values that motivated them in their respective economic activities?

The potters' and muleteers' social context was primarily the village of Miravet. In the first half of the twentieth century, Miravet formed a relatively secluded, close-knit community, with a population of between 1,400 and 1,500 – the population of an average Spanish village. The inhabitants had their own pastor, their own primary school (hardly anyone ever went to high school), and their own doctor and veterinarian. They could find all their daily goods in the local shops and market square. There was hardly any advanced means of transportation: one participant explained how the bicycle he and his brother had bought in the early 1940s was considered a novelty in the village. The male villagers socialized with each other on a daily basis during their recreational time in the local cafes and taverns; the female villagers socialized during domestic chores like shopping, fetching water at the fountain or river, or doing laundry. Though Miravet was a part of a larger social whole – an administrative unit, for instance, within the Spanish state and within the Catholic Church – it was primarily within this small and dense social context that people reproduced their lives, and wherein potters and muleteers derived the values that motivated them in their respective economic activities.[5]

When recalling the period of the muleteers, the participants judged economic activities and resulting living standards according to two separate values: *diner,* 'money', and *dinada,* 'food' (or synonymous expressions such as 'to have to eat'). For example, when I asked Joan Avante about economic activity at that time, he replied: 'At that time, neither the muleteer, nor the cutter of branches, nor the potter, nor the peasant were getting rich. All were poor. Really all. The peasants did not have a nickel either, they had food because of what they harvested. But with respect to money, nothing.' Joan Avante typically explained by stating that these livelihoods were equal, as opposed to the 'rich', in that they did not have 'money'. At the same time the poor were differentiated with respect to 'food' – specifically their access to food.

By 'money', the participants referred, of course, to currency, but in this context they referred especially to wealth. The rich people in Miravet had, above all, landed property; but they also owned houses, agricultural equipment (like olive mills), equipment for the pottery production (like kilns) and merchandise.[6]

By using the two values, 'money' and 'food', the participants recognized that people who lacked wealth were not necessarily equal: economic activity with its resultant living standard was also differentiated with respect to 'food'. In the aforementioned quote, Joan Avante equated 'muleteers', 'cutter of branches', 'potters' and 'peasants' with respect to their lack of 'money' or wealth, but he differentiated the 'peasants' from the rest by their direct access to 'food'. People in Miravet were predominantly engaged in agriculture, mainly the cultivation of olives, wine, almonds, carob and cereals on dry land, horticulture on irrigated land, and some breeding of sheep, goats and pigs. The land was worked in two ways. First, small parcels were either owned, rented or sharecropped by peasants who produced primarily for their own consumption. These peasants sold or bartered only a small surplus to complement their own production with other produce, with artisanal products, and with products from emergent industry. Second, large areas of land were owned by the 'rich' minority who employed labourers to supply the market, as well as themselves, with food. Everyone who had land, ate from it, and had direct access to food (see also Harding 1984: 30–58 and 81–112).[7]

Land and the direct access to food were part of each villager's identity. For the people of Miravet, and Spain at large, *casa* connoted a complex of persons and things. The casa was the unit of production as well as consumption, and as such it encompassed a household (at least one nuclear family which often extended to two or three generations), the house in which these people lived, and the means of production of that household and its wealth, which was most importantly land (see Jociles Rubio 1989: 29–30 on the province of Tarragona, Barrera González 1990 on northern Catalonia, and Harding 1984: esp. 23 on a village in northern Aragon). People were identified by their casa; one referred for instance to someone as 'Josep of ca Pedrola'.[8] Parcels of land were also identified by their casa; one knew to which casa a parcel belonged. This identification of land and people stretched out over generations by primogeniture. The rich, landed people could attain this ideal of identification with land, the poor and landless could not.

By working or owning land one had the most direct access to food, but the next best access could be attained by peddling. Peddlers (i.e. muleteers) used mules and carts as their primary means of transport, and only used other means of transportation to vary and expand their range of action.[9] When they peddled nearby, they would make short trips, no longer than a week. When they peddled in distant regions, they could be away from home for up to three months at a time. For such trips, they preferred to ship the main part of their merchandise. When the Ebre was still navigable in the first part of the twentieth century,[10] they shipped their merchandise by barge upstream to Mequinença, or downstream to Tortosa or Amposta.[11] From the time the nearby station at Mora la Nova opened in 1891, muleteers hired wagons and shipped to places like Saragossa, Huesca, Fraga and Jaca. During the 1930s trucks became available and they could ship anywhere. If merchandise was shipped, the muleteer himself would travel by cart and try to reach the place of destination the same day as his shipment. There he had a depot, and from there he would make trips peddling with his mule and cart, from time to time returning to the depot to replenish his cartload. Hence, although a peddler might use other means of transportation to expand his range of action, the mule and cart was his basic means.

Supplementing the pottery produced in Miravet, muleteers peddled palm-leaf baskets – which also served as packing material – made either in the nearby village of Rasquera across the river from Miravet, or (with the finer baskets) in Regués, a village downstream in Valencia.[12] They sold or bartered these products to acquire produce such as cereals, flour, bread, dried beans, chickpeas, potatoes, onions, pig products like sausages, fodder and sheep and goat skins. They acquired such produce from peasants and farmers at the lowest seasonal and regional rates. All muleteers of course prioritized providing their household with food. A muleteer's grandson remembered how his grandfather returned from his trips with food for their household, and that they always had plenty to eat. However, they might also trade this produce for a profit.[13]

A mule and cart was something worthwhile, provided one could use it intensively. People with a mule and cart were people who worked the land and/or transported things locally and regionally. Therefore, people who worked the land could complement that activity with peddling. In such a case, the economic importance of peddling for a particular household differed in inverse ratio to that of working the land, i.e. the quantity and quality of the land a household held. For instance, one muleteer, Tapit, owned a sizeable portion of land. With a single trip per year Tapit complemented his agricultural production sufficiently. In contrast, another muleteer, Llop, had little land – only some dry land with olives and almonds – and therefore was away from home nine to ten months a year. He was almost totally dependent on

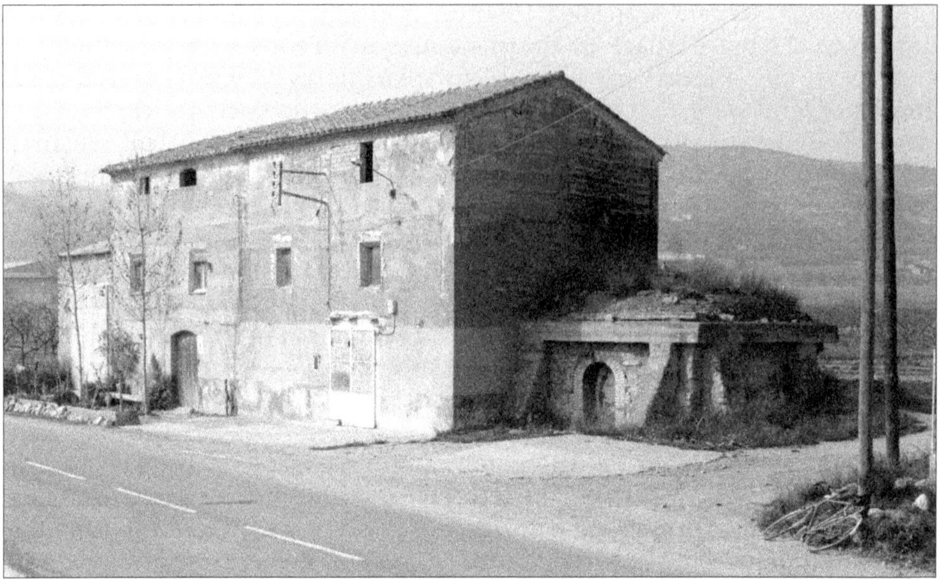

Plate 2.1 ***Two houses with an adjacent kiln, built to be rented out to potters.*** Note that the kiln had not been used for about 20 years when the photograph was taken. Author photo, 1991.

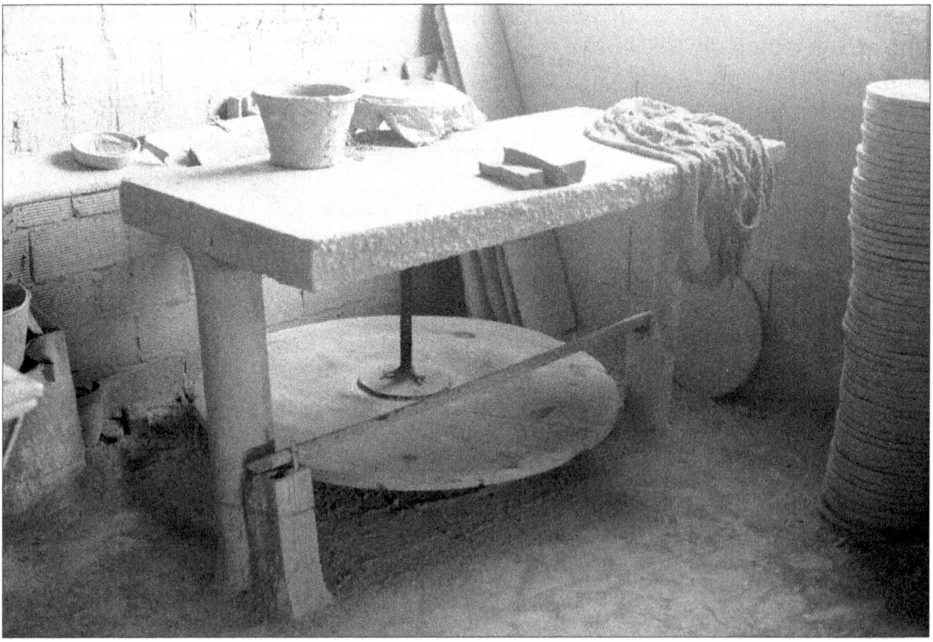

Plate 2.2 ***Kick-wheel in the Avante workshop.*** This wheel is the same or very similar to potters' wheels in use until the mid 1970s. Author photo, 1992.

peddling to obtain produce for his own consumption, as well as on trading to make some money. Some muleteers did not have any land: they might still have a mule and cart because they transported things and peddled year round. These muleteers usually cut, or co-operated with men who cut, the shrub growing in the mountains surrounding Miravet. With their mules, muleteers brought the branches down to the potters who used them for firing their kilns. They bartered the firewood for pottery that they peddled. Hence, since peddling had different economic importance for different muleteers, muleteers differed in their turnover and in the regularity of peddling. There had been only one potter who had a bit of land that was (just) enough to permit him to own a mule and cart with which he could peddle his own production.[14]

Compared with farmers, peasants and muleteers, potters had a less desirable access to food. Indeed, if they had land, potters would most certainly work it. For instance, the Pedrolas inherited land and first worked that, and only as an addition worked in their pottery workshop. Potters were forced into an economic activity that did not require any property rights over land, and which required only low start-up costs: a potter's wheel and other simple equipment. Furthermore, potters had free access to clay and access to necessary equipment – a kiln – that could be acquired in exchange for their products.[15] Since they did not own land, potters could not produce their own food, nor could they maintain a mule and cart to peddle their own products.[16] Hence potters had to buy their food in the local stores and from local peasants: they had access to food only through money, the money they made by selling their products. This access was problematic since potters hardly made any money to begin with.

The potters were poor so they needed to rely on currency as little as possible, by limiting – or if necessary, postponing – payments. Therefore, they settled their rents with pottery. They gave owners of kilns pottery in exchange for the use of the kiln. They bartered pottery for lead, the main component of glaze and the only raw material that had to be purchased. They bartered pottery for firewood, and, occasionally, they bartered pottery for food with a peasant or farmer. These forms of barter between potters and people connected with pottery production and trade seemed, in general, to be accepted matter-of-factly (though a member of the economically strongest potter household in the pre-war period boasted that they always paid for their firewood). Since the potters had no money, they had to buy everything on tab. For many potter households, this mode of purchase was perpetual: when they received money, they could only pay off their debts, and then start buying on tab all over again. By allowing potters to buy in this manner, the shop-owners clearly granted the potters a favour – one a household could only partly reciprocate by its patronage. Since potters could reciprocate only in part, they stayed in debt. Indeed, a household buying on tab compared unfavourably with other households that could pay immediately. But one potter, Pepe Papaseit, describing this situation, added that he was not at all embarrassed that his mother could not pay immediately when she went shopping: none of the potters' wives could.

The humiliation arising from difficult access to food is reflected in a saying in Miravet: girls were warned against marrying a potter, for if she did, she would have to eat grit. This grit was the waste from riddling *terra refractaria*, a main component of the potters' clay, i.e. this was what the potters were 'producing' instead of something edible. This was not a mere saying: at least one potter admitted the problem he had had with his parents-in-law, who initially refused to let their daughter marry him. This saying was the background against which the brothers Pedrola could contrast their household, boasting that they had always had plenty of food, but then this household was exceptional in that it had a decent amount of land and their father peddled his own pottery production.[17]

Potters might have been worse off than people working the land and muleteers, but potters fared better than *jornalers*, 'day labourers' (e.g. the 'cutter of branches' in Joan Avante's comparison). None of the potters worked for a wage – with the rare exception of temporary work during the wine harvest, or odd jobs for young men making pocket money. Potters at times employed wage labourers, usually for specific occasional tasks, like the unpleasant job of mixing clay and water with their bare feet. Very occasionally some workshops even employed wage labourers full time, often *mozos*, boys from poor families – usually Aragonese – who lived in the potter's household and worked for boarding and pocketmoney.

Both potters and workers were in the same position of having access to food through money (though both did not earn much money), but potters were obviously better off, earning marginally more than the workers, and they were better off by earning on a regular basis with independence from the seasonal labour market. Though the potters had no direct access to food, none of them were forced to become a peasant on a rent or sharecrop arrangement, which must have been even less attractive than being a potter. (Note that the peasants and wage workers differed from each other only in degree: some people had no land at all and were full-time wage workers, others complemented working their own land with working for a wage. Moreover, because of the natural cycles of households, a household might have excess labour at one point which could be employed for a wage. As a result of this labour fluctuation, people usually preferred to apply their agricultural skills in working their own or others' land accordingly, instead of acquiring new skills like those of the potters.)

The values of 'money' and 'food' explain how in the social distribution of means of production (especially that of landed property) people working the land, muleteers, potters and wage-labourers, reproduced their positions in the local political economy. Yet these values do not explain why potters did not abandon their craft to become full-time muleteers (they would have better access to food and would be able to use a mule and cart intensively). By examining instances of potters who became muleteers and muleteers who became potters, one can discern another value – a weaker and more ambiguous one – for valuing economic activities: the value of sociability.

To become a muleteer and peddle, one needed a mule and cart, and the knowledge necessary for peddling. Both were accessible to potters. However, according to participants' recollections, only two potters had become muleteers. Both of them did so in advanced age in order to peddle the production of their sons, who were working in their workshop and living in the same household. Several of the potters' sons also became muleteers for a similar reason (that is, to peddle the family pottery production). One such person, Santiago Benaiges, peddled only until his marriage, not wishing to leave the village and his wife for an extended period of time. He preferred the sociability he found among his fellow villagers, especially with his new wife, above living among strangers.

As one participant put it, 'to have to make one's living outside the village was *misèria*', 'misery'. A muleteer often had to sleep in lodgings, and since he did not have much to spend, this usually meant sleeping in the stable with his mule. Also, he often had to miss out on one of the most important enjoyments in life, good meals, and instead contended with a piece of bread, some ham and a can of sardines. (Though one former muleteer said that in his youth he had loved dancing, and as a peddler he often had the opportunity to visit fairs where he danced.) Then, too, life on the road could also be humiliating, as the following anecdote illustrates.

One muleteer's son, Josep Fabregat, became a potter. (Note that this move is not the same as a potter becoming a muleteer: the craft of pottery had to be learned at a young age and required that the apprentice commit himself for a long period. A muleteer did not require this degree of skill, and, indeed, people have become muleteers at relatively advanced ages). Vicen Sales, his nephew, explained to me by an often-told story why his uncle did so. Josep Fabregat and his father had shipped pottery to Barbastro and were peddling in its surroundings. Upon arrival in these tiny villages, both would carry some pots on their back and together they would walk through the village attracting customers. One late afternoon they arrived in a village and were taking their turn through it when they came across a sewing circle. They encountered seven or eight women sewing. The elder Fabregat asked a lady if she wished to buy something. After a short deliberation with her daughter, she did so. This daughter, the same age as Josep, was pretty. When they delivered the pottery to this lady's house, Josep's father asked the daughter for a bit of ham and bread since both had not eaten that day and there were no shops in those villages where one could buy food. Josep was so embarrassed and angered by his father's request that he told him that he would never again peddle. He considered it begging for alms and therefore humiliating; he never made another trip to peddle.[18]

Since food and lodgings were not commercially provided in many small villages, most muleteers had developed networks of friends in whose houses they could stay and eat, favours they reciprocated with gifts of food and pottery. Such networks sometimes passed from father to son, and former muleteers recounted them with pride. Muleteers would tell me about the good friends they had during their peddling trips – friends who received them as family members – and how they still maintained these friendships with occasional visits.

By necessity, muleteers missed much of the social life in their village, but potters were among its most active participants. Potters worked mainly in their workshops, hence within the village, and were often visited by their fellow villagers. Moreover, after brunch, potters went to the daily *tertúlia*, the 'social gathering', in the centre of the Raval, and after the midday meal, to the tavern for a game of cards. Indeed the potters were stereotypically seen as tavern-loafers. Potters participated even more in the social life than people working the land: the latter could not always return home for the midday meal when they had to work fields far away from the village.[19]

The structure of the values of 'money', 'food' and sociability together with the social distribution of means of production, especially land, resulted in sons who usually followed their fathers' vocations. The potters and the former muleteers whom I interviewed had been producing pottery or peddling respectively for several generations, often even for as long as they could trace their family history. They had become potter or muleteer through routine. One interviewee asked me why one should not learn a profession when one had it 'at home'? A boy gradually became integrated in the workforce of his household and thus gradually learned the profession of his father. The few sons of potters who became muleteers – exceptions – did this in order to peddle the production of their households.[20]

The exchanges between potters and muleteers

Potters sold their products to muleteers. Both parties had opposing interests and therefore contested the price of the pottery: potters naturally wanted a higher price, muleteers a lower one. Potters had no land but knew a craft and owned or rented their means of production. As such they were independent producers, as opposed to wage labourers, but they were dependent on the barter or sale of their products to muleteers. Muleteers varied in the amount of land they worked, to the extent they cut and/or transported firewood, and hence in their dependence on peddling. In their negotiation of the price of the pottery, both parties had unequal positions: all potters were dependent on the muleteers, while, conversely, muleteers were in varying degrees dependent on potters.

To strengthen their structurally weaker position against muleteers in the price negotiations, potters assessed collective prices, *el preu de Miravet*, 'the price of Miravet' as the interviewees referred to these prices.[21] These represented the minimum prices a potter should receive for his products. Yet, as production was organized in *cases*, potters assessed these collective prices only in an ad hoc manner and did not organize themselves to back these prices up. There was no means of enforcing these prices on each and every potter, and the worst that could happen to a potter who sold under these prices was that he got a bad name.[22]

In addition to receiving monetary prices, potters also exchanged their products. Potters frequently bartered their products for the use of the kiln (one muleteer owned a kiln), for lead (muleteers often acquired lead on their trips) for firewood, or to settle rents. In these exchanges, one balanced the bartered goods according to customary and stable standards that were not expressed in money, and hence were never negotiated like monetary prices. For instance, a potter could use a kiln in exchange for a certain number of pottery pieces. The participants did not remember any changes in these arrangements so long as they were a form of direct barter. When the owner of the kiln was no longer interested in pottery, he calculated and asked for the market value of those pieces. From the moment the rent was expressed in money, it increased, just as all monetary prices usually do. (Occasionally, potters bartered their products with muleteers for food. They balanced these exchanges by expressing the values of the bartered goods in money, which was negotiated.)

The collectively assessed minimum prices were an aspect of the dynamic between potters and muleteers as economic actors. In order to understand the dynamic between them as individuals, the price calculation will be explained in detail.

Pottery as marketed to muleteers

Potters exchanged pottery with muleteers. Such an exchange bridges two parties. There is a dialectic within an exchange: as an exchange bridges, it constitutes meanings by which two parties communicate, but as it bridges two parties, it also constitutes meanings that these parties can mutually contest. Potters and muleteers communicated with each other, that is, agreed upon certain aspects of the meanings of pottery as constituted in their exchanges. They agreed on a specific appreciation of the pottery, and on articulating their exchanges with money as a 'common denominator' (Marx 1981: 28). The appreciation encompassed qualities typical of pottery while the monetary denominator by its purely quantitative nature bridged these exchanges and made them compatible with other exchanges. At the same time that potters and muleteers were in agreement, they also contested other aspects of these meanings: as opposing trade parties, they contested the actual prices. How did potters and muleteers appreciate the pottery, and how did they agree and contest the price calculation?

Just as potters and muleteers knew each other as individuals, potters and muleteers knew the pottery on an 'individual' level, that is, by its producer. Both potters and muleteers could easily recognize a piece of pottery by the idiosyncrasies of its maker. For instance, Josep Papaseit showed me how his grandfather's storage vessels could be recognized at the underside of rims, where one could feel a shallow rim. One could recognize these storage vessels because Papaseit's grandfather had a certain idiosyncratic way of holding the piece of reed when finishing rims, which he did unintentionally. Other idiosyncrasies might be expressions of personal characteristics of a potter. For instance, Vicen Sales let a certain type of pot dry to become firmer and then placed it on the wheel again and defined its shape by going over it with a scraper. The Pedrola brothers, on the other hand, threw such a pot immediately into its final shape. Josep Pedrola explained that Vicen used dryer, firmer clay, which barely gave a shape when coiled. Vicen pushed the clay into a preliminary shape by sheer strength, and defined the shape later in a separate step. Josep used softer, more elastic clay, paid more attention to the shape when coiling, and threw it immediately into the final shape. Josep could not throw like Vicen, as he was not as strong as Vicen. Although the resulting pots of both potters were of the same type, someone intimately familiar with the Miravet pottery could recognize the different makers of these pots by the surface texture.[23]

Potters and muleteers appreciated the idiosyncrasies of the pottery according to several criteria. First, a pot should have a *linea bona*, 'good line'. This line is the contour that should never swerve and always be clear – difficult to achieve especially when one has to throw pieces in several steps. (When throwing in several steps, the clay differs in its wetness, i.e. percentage of shrinkage, and when this is not well anticipated, the finished product can have swellings.) Second, a pot should weigh relatively little for its size, and the clay should be 'well divided', i.e. the walls should be equally thick. Third, pots of a particular type should be equal in size and shape. For instance, a batch of medium-sized bowls should all be equal in shape and size. Fourth, a pot should be well glazed. The colour of the glaze was not important but its even distribution and its gloss was, since bowls easily showed flaws in the glaze. These criteria varied in application to the different types. For instance, a bowl should be thrown in exact measurements in order to nest. A storage vessel should have the clay well distributed: the walls should be thin, but the rim, which carried the weight, thick. The flaws in the 'line' of a storage vessel were more apparent than this type of flaw in other pieces.

Potters differed in their ability to produce according to the above criteria, as well as in the pursuit of these criteria: some focused on quality and made the best pottery possible, others focused on quantity and made as much pottery as possible. Of course, muleteers preferred the higher quality pottery, and the quality was an important factor in the relationships between individual potters and muleteers. However, muleteers did not simply pay more for this pottery. Potters and muleteers calculated prices according to two principles.

Potters and muleteers knew that consumers applied other, simpler, criteria to judge pottery than they did. Consumers only judged pottery by whether it was well fired, whether it was without cracks and holes, and whether it was entirely glazed in a shiny manner.[24] Hence, as a first principle of price calculation, muleteers valued pottery as either good or flawed. Good pottery was without cracks or holes, was well glazed and was valued at the full price. Pottery with cracks, holes or a bad glaze – *mal* or *de tara*, that is, 'flawed' pottery – was valued at half the price.[25] When it came to judge whether a particular piece was good or flawed, potters and muleteers might disagree. Yet potters invariably contested this principle of either good or bad and being paid accordingly full or half price, although they were forced to accept this calculation. When explaining this calculation principle to me, every potter immediately added that muleteers fixed the flawed pottery. Cracks and holes could easily be filled with a mixture of chalk and egg white, and muleteers usually sold the fixed pottery to consumers for regular prices. Fortunately for the muleteers, the consumers were unaware of this process. However, potters could not fix flawed pottery and charge muleteers regular prices since by their intimate knowledge of pottery, muleteers immediately recognized 'fixed' pottery. Understandably, potters contested this principle of being paid half for flawed pottery since ultimately muleteers sold the inferior pottery anyway for the full price. In the end, muleteers held the view that they did not need to pay the full price for something that they judged inferior.

The second principle of price calculation was very complex and implied a categorization of the pottery. In their exchanges, potters and muleteers labelled and organized the pottery. They distinguished glazed from unglazed pottery. Potters sold and muleteers bought glazed pottery in quantities called *peça*, 'piece', or *dotzena*, 'dozen'. (Interviewees used the words 'piece' and 'dozen' interchangeably: though 'dozen' of course in general refers to the quantity of twelve, it did not necessarily mean so in this context.[26]) One 'piece' consisted of a certain number of pieces of one particular size and type of pottery, and there were certain arithmetic, hierarchical orders between sizes and numbers of a type of pottery. (Only these 'pieces' appeared on sale bills. Pepe Papaseit described the bills' appearance as short and simple. They contained one line with the total number of 'pieces' of glazed pottery, without indication of the different sizes and types which formed these pieces, and a few lines with the numbers of unglazed pottery. Only the unglazed pottery was specified by size and type.) Many of the labels of the glazed pottery referred to this categorization.

The glazed pottery consisted of *ribells*[27] and *gerres*, the unglazed of *cántirs* and *cossis*.

The first type of glazed pottery is the ribell or 'bowl'. Bowls have different sizes and a few variants from the basic shape and decoration. Table 1 gives the different labels for the different sizes,[28] and how different bowls were decorated differently. Moreover, this table gives the relative values of these bowls by showing how different sizes formed 'pieces'.

Label of size	Appr. height	Appr. diameter	Shape/decoration	Relative value
mandonguer	25 cm	65 cm	linear design	2 'pieces'
mitjaner	18 cm	50 cm	half-circle design	1 'piece'
sangant	17 cm	39 cm	half-circle design	2 in 1 'piece'
quatreun	12 cm	25 cm	half-circle design	4 in 1 'piece'
grela	12 cm	25 cm	upstanding rim/ marbled or splashed design	4 in 1 'piece'
pitanca	10 cm	20 cm	upstanding rim/slightly rounded/marbled or splashed design	6 in 1 'piece'

Table 2.1 ***Ribell or bowl – arithmetic order.***

Most labels for the different sizes for all types of pottery were only used and known by potters and muleteers, as they referred to their organization of the different sizes.

The largest bowl is the *mandonguer*, a derivative from *mandongo*, the filling of sausages.[29]

The next bowl smaller in size is the *ribell de pieça* or the *mitjaner*: this size bowl was the 'piece': *ribell de pieça* means 'bowl of the piece'. Its other name, mitjaner, is a derivative of the adjective *mitjà*, which can be glossed as 'average' or 'standard'. The next smaller bowl from the mitjaner, de *sangant*, is also called *ribell de dos per un*, or shorted to *dos per un*; 'two for one'. Two sangants had the equivalent value of one 'piece'. The name sangant is derived from *sang*, 'blood'. The next smaller bowl is the *ribell de quatre per un*, or *de quatreun*, usually shortened to *quatreun*, the bowl of 'four for one'. These bowls, from mandonguer to quatreun, are supposed to nest in each other.

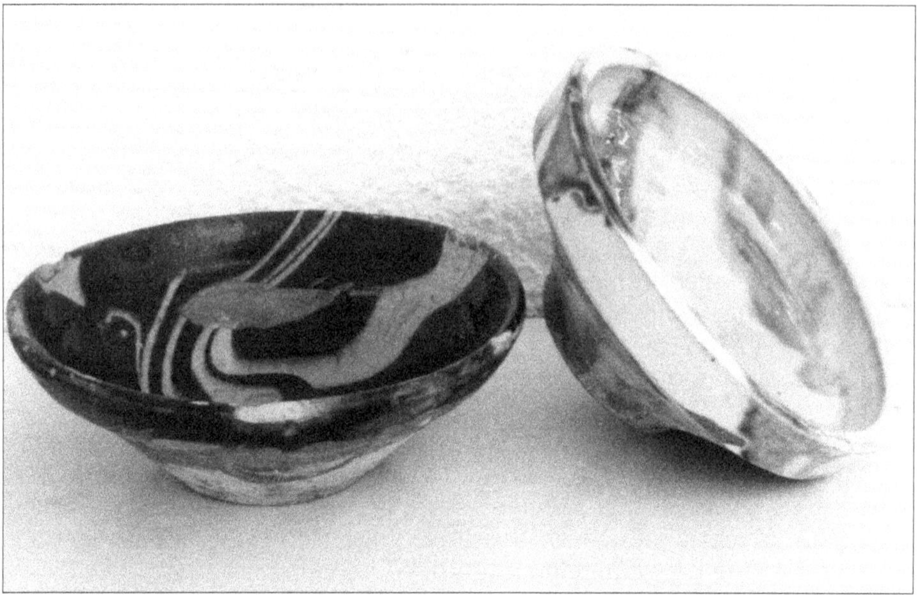

Plate 2.3 ***Two pitançes.*** The one on the left has a marbled design (note that the glaze is also chipped), and the one on the right has been splashed, which is made by first dipping the bowl in engobe so that the engobe covers the whole inside and rim. Next, one splashes engobe in a contrasting colour on it. Author photo, 1997.

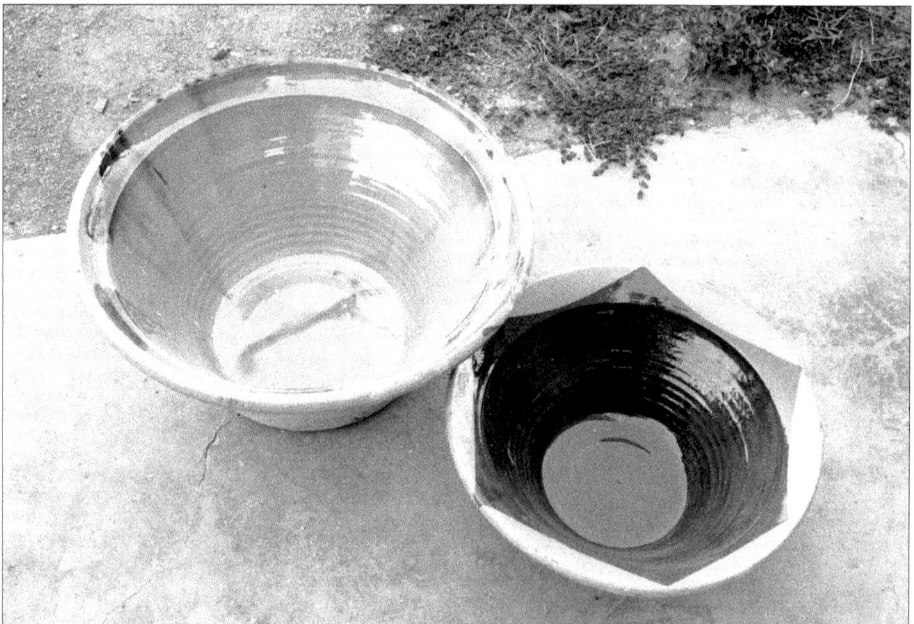

Plate 2.4 ***Two bowls or ribells.*** The one on the left is what is locally called Catalan bowl; the one on the right a regular bowl. The Catalan bowl is also called 'with the wide bottom', has a rounded rim, and a design of green splashes. The regular bowl has a conical shape – is therefore also called 'with the narrow bottom' – a flat, slanted rim, and the half-circle design. (Note that both specimens were made in 1992.) Author photo, 1992.

The *grela* is of the same size as the quatreun only it has an upstanding rim. A smaller bowl still is the *plateret*, or *ribell de sis per un*, the last shortened to *sisperun*. Six of these had the value of one 'piece'. There is also a version of this bowl with an upstanding rim like the rim of the grela. This bowl is called *pitança*, a word colloquially used for 'daily bread' and 'food' (Vilanova et al. 1988: 202). It differs from the other bowls in that it is slightly rounded, while the others have a straight cone shape.

There is another type of bowl within this category, the *ribell català*, or the 'Catalan bowl', which was probably introduced in Miravet in the early 1930s. The oldest participants, potters in their late 50s and early 60s, had always thrown this type but still referred to it as a *còpia*, 'copy' (as they used this particular word), of the bowl made in Esparreguera, a pottery centre near and west of Barcelona. It has a wider bottom than the other type, so it is called the 'bowl with the wide bottom', while the other is the one 'with the narrow bottom'. It is made in sizes comparable to the mitjaner, sangant and quatreun. It is completely red in colour or white with splashes of green, which was not used for other pottery at that time. The latter finish is a regular glaze, but with an added chemical containing copper sulphate and used in viticulture. One splashed this chemical on the still wet glaze, not the engobe, and in the finished product these splashes came out green.

The second type of glazed pottery is the *gerra*, the vessel. In this book, I use 'vessel' as a gloss for gerra, thus limiting the meaning of vessel to this storage pottery. Gerres are vessels of different sizes but with one particular shape. Table 2 gives the different labels and sizes. Apart from its approximate, non-fixed measurements, each size is marked by being thrown in three, two or one phases; the number of *anses*, i.e. 'handles' or 'grips'; the flat or rounded rim; and the technique of glazing. This table also contains the relative values of these bowls, their arithmetical organization into 'pieces'.

Label of size	Appr. capacity	Markers of size	Relative value
gerrot	200 litres	2 phases / 4 handles / grooved / splashed on glaze / flat rim	
gerreta	150 litres	3 phases / 3 handles / can be grooved / splashed on glaze / flat	
solsera gran	90 litres	2 phases / 4 handles / can be grooved / splashed on glaze/flat rim	together with *solsera petita* one 'pair'
solsera petita	75 litres	2 phases / 3 handles / can be grooved/ splashed on glaze / flat rim	
tafol	40 litres	2 phases / 4 handles / splashed on glaze / rounded rim	1 'piece'
solsero	10 litres	1 phase / 2 handles / dipped in glaze / rounded rim	12 in one 'piece'
petxin	5 litres	1 phase / 2 handles / dipped in glaze / rounded rim	24 in one 'piece'
saboner	1 litre	1 phase /2 handles/ dipped in glaze / rounded rim	48 in one 'piece'

Table 2.2 ***Gerra or vessel – arithmetic order.***

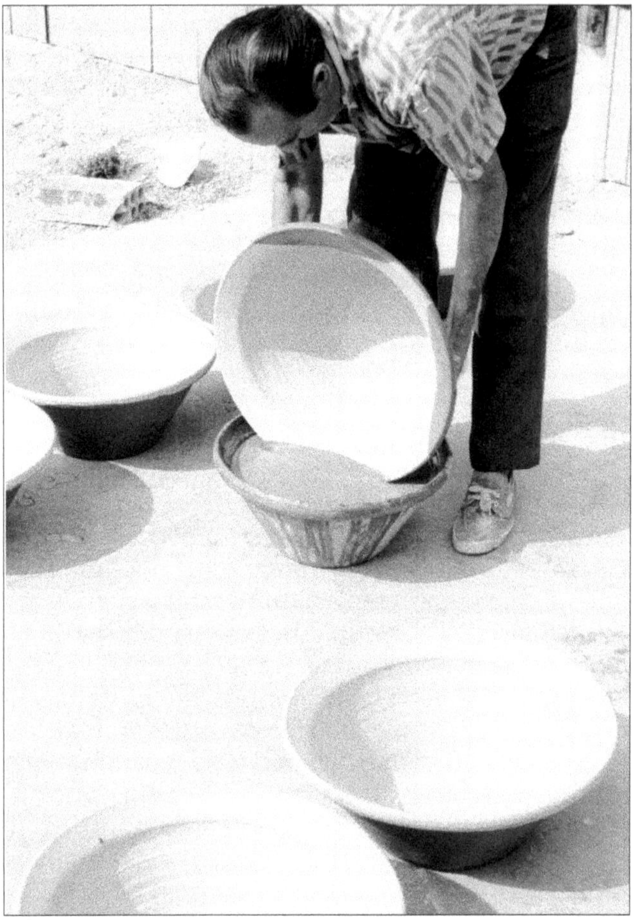

Plate 2.5 ***Pepe Jornet demonstrates how to make the half-circle design on bowls.*** Bowls in the sizes of mitjaner, dos per un and quatreun have a design that one makes by first dipping the bowl in red or white engobe, so that the engobe covers the whole inside and rim, and then dipping the bowl more or less perpendicularly in engobe of the opposite colour. The result is a half-circle on the rim. The dipping in engobe of the opposite colour is repeated several times while one turns the bowl and takes care that the half-circles do not overlap. Thus one creates a symmetrical pattern with several axes, a kind of star. Author photo, 1992.

The labels for vessels do not refer to uses, as do some labels of the bowls, nor to their organization, except for some labels being diminutives. Still, these labels were also exclusively used by potters and muleteers.

The largest vessel is also called *gerra*. It is exceptionally large, standing almost man-sized. The more common large gerra are referred to as *gerrot* or *gerreta*.[30] Gerrot is the masculine diminutive, gerreta the feminine of gerra. A gerrot is thrown in two phases. The first and lower part is left to dry, becoming firm so that it can support the second, upper part. The difference in dryness of the two parts makes them shrink at different rates. This is allowed for by the

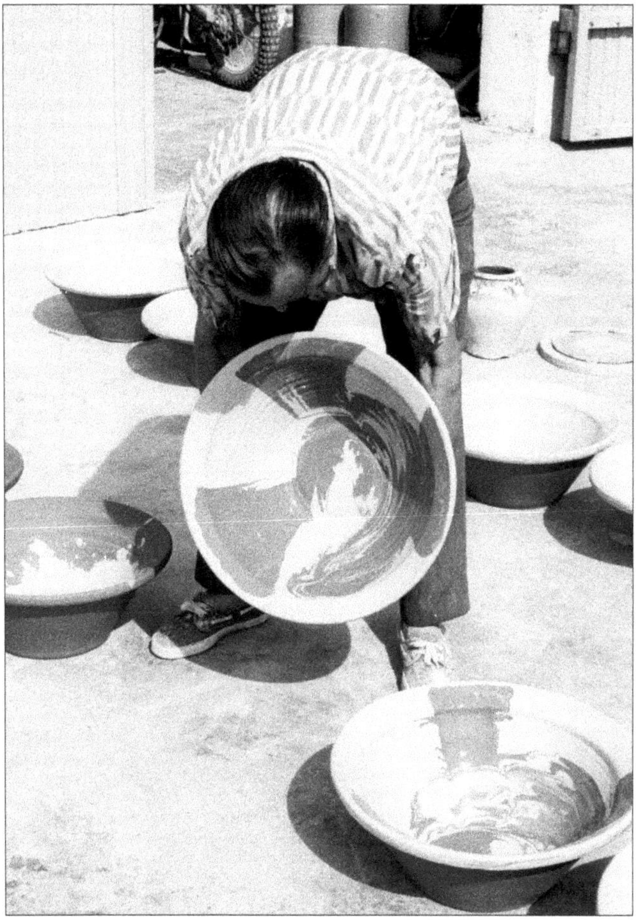

Plate 2.6 ***Pepe Jornet demonstrates how to make a marbled design.*** First one dips the bowl in the engobe. Next one splashes engobe in a contrasting colour on it, simultaneously giving the bowl a sharp twist. Author photo, 1992.

potter. Still, in the finished pot, the joint is easily discerned by a swelling. The gerreta is thrown in three phases, hence has two swellings. Moreover, the gerrot and the gerreta otherwise differ in their shapes. The gerrot's overall shape is sturdy and blunt, the gerreta's slender and lifted. In addition, the gerrot has four handles and the gerreta three. Both types can come in a large, *gran,* and a smaller, *petit,* version. The gerreta is the only vessel that is occasionally without glaze. The next smaller vessels are the *solsera gran,* and the *solsera petita.* A solsera gran and a solsera petita are called a *parella,* a 'pair'; and the potters sold them to the muleteers as such. The next smaller vessel is the *tafol.*[31] Smaller still is the *solseró,* the masculine diminutive of solsera. Then comes the *petxin.* The smallest is the *saboner,* whose name is derived from *sabó,* 'soap'.

The potters and muleteers categorized the several sizes of vessels according to their measurements hierarchically, but these measurements were not standardized. Over time, the potters tended to throw vessels that were larger

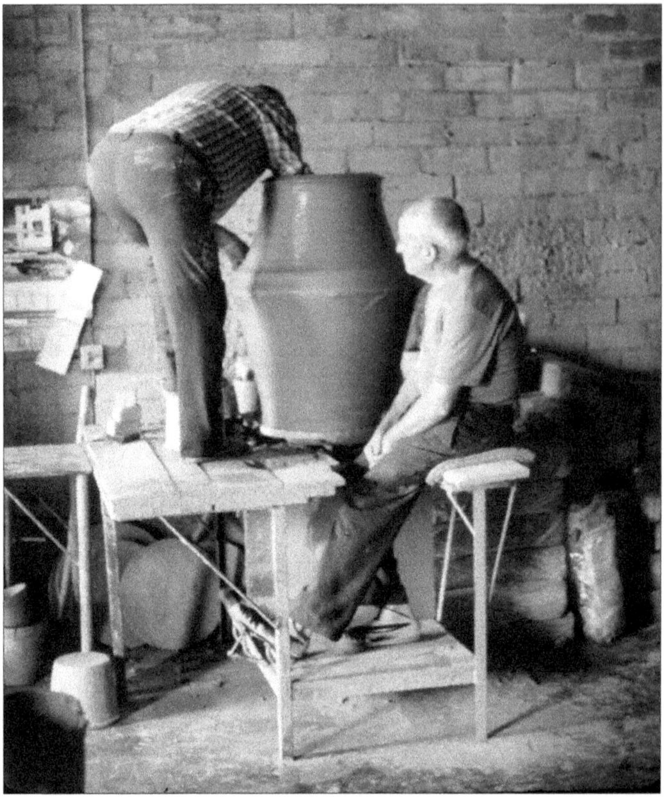

Plate 2.7 ***Throwing a gerrot.*** Vicen Sales throwing the second phase of a gerrot while Santiago Benaiges rotates the wheel. Author photo, 1991.

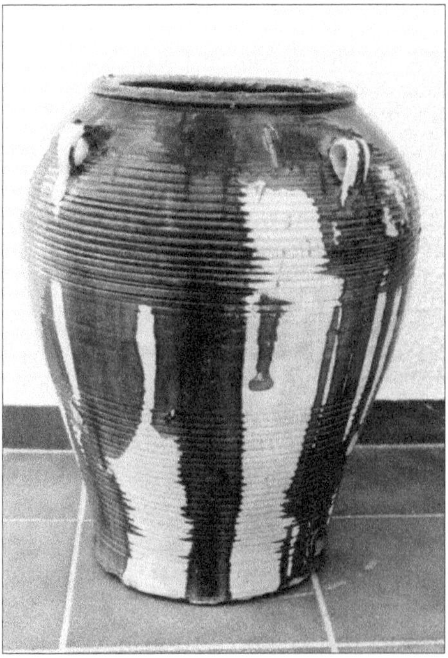

Plate 2.8 ***Tafol.*** The *gerrots* and the *gerretes* always have a shallow horizontal groove spiralling over their whole surface, making the pot look striped. The *solseres* and *tafols* might or might not have this groove. In the later case, this depends on the personal preference of the potter, though there has been a clear move to omit this groove. The *tafols*, as well as the *gerrots*, *gerretes* and *solseres*, have a flat rim. These pots are glazed by splashing the liquid glaze with a piece of cloth. Consequently, the glaze runs down from the rim. Height: 54 cm. Author photo, 1997.

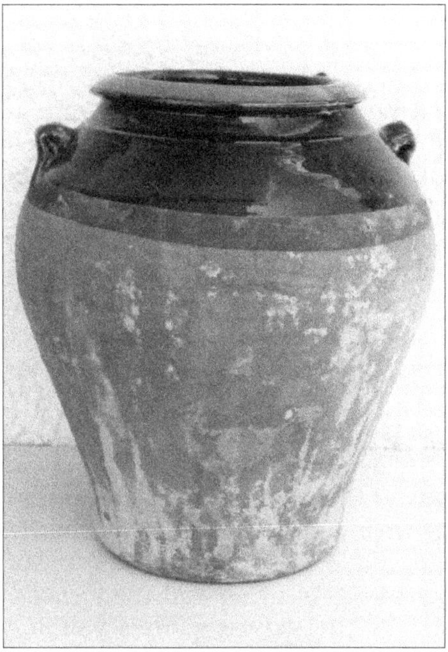

Plate 2.9 ***Petxin.*** This pot, as all the smaller ones, is thrown at once and can be dipped in engobe and glaze. Its outer part is glazed: the line between the glazed and the unglazed part is quite straight and almost horizontal. With the two smaller pots, the engobe on the inside and on the rim is often white. Height: 34 cm. Author photo, 1997.

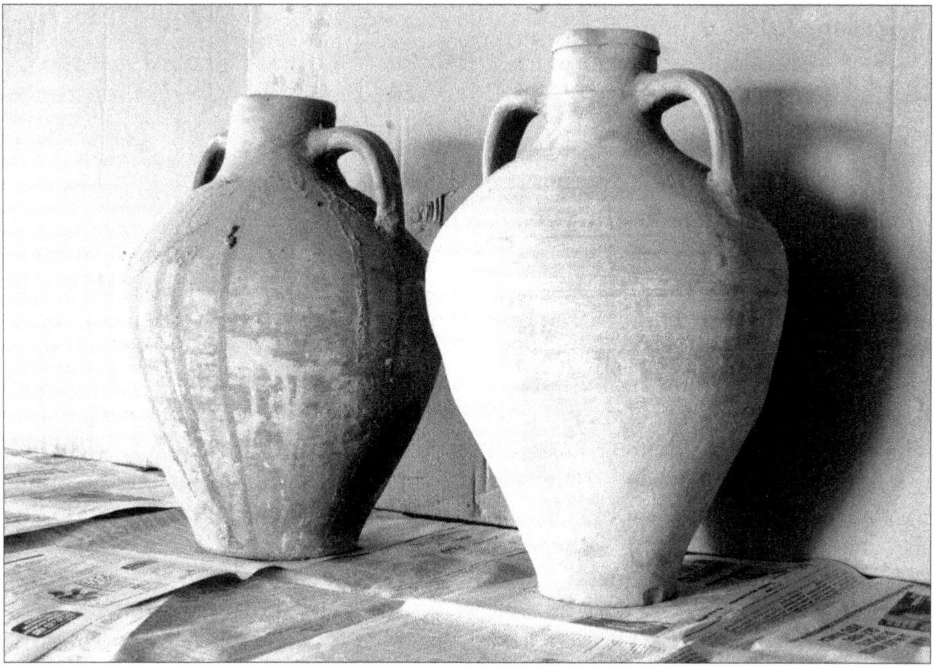

Plate 2.10 ***Cántírs.*** The one on the left has a wide bottom, the one on the right a narrow one. The cántírs with a wide bottom are thrown in one step, the ones with a narrow bottom in two, hence the bump in the curve. Author photo, 1992.

and larger. For instance, Pepe Papaseit had an old solsera, probably from the first half of the twentieth century. As he pointed out to me, because of its capacity, this vessel seemed to be a tafol, but it was not since it had a flat rim, i.e. a mark of a solsera gran. Its capacity was probably 80 litres and at the time this pot was made, the potters made solseres grandes with such capacity. Also, potters working concurrently differed among each other in these measurements. Josep Benaiges was known to throw vessels larger than those of his colleagues. He would make solseres with capacities of 115 litres, while others would only throw with capacities of 90 litres. (With the other main type of pottery, the bowls, such changes in measurements could not happen since the mitjaner, sangant and quatreun all had the same shape – they lacked markers of sizes like the vessels' different rims – and thus their labels referred to specific measurements.)

The first type of unglazed pottery was the *cántir*. There are two types of cántirs, the *cántir de cul ample*, one 'with a wide bottom' and the *cántir de cul estret*, one 'with a narrow bottom'. The body of the cántir with a wide bottom is fairly spherical, the body of the cántír with a narrow bottom is bluntly pointed at the bottom. These cántirs were made in two sizes, the *gran*, or 'large', also called *normal*, with a capacity of 12–16 litres, and the *aiguader* or *mitjà*, with a capacity of 9–10 litres. Like the vessels, potters working concurrently varied the capacities of the cántirs, but unlike the vessels, they did not increase these capacities over time. The *canteret*, a diminutive of cántir, was similar in shape to the cántir but smaller, with a capacity of 5–6 litres, with a strainer covering the opening, and a spout on the bulge between the two handles.

The second type of unglazed pottery was the *cossi*. The cossi is a tub of which there are two types. The *cossi calandrín*, or simply *calandrin*, measures 80 cm in height. The other type is usually referred to as cossi, *cossi ample*, 'wide tub', or *cossi de paret*. Cossi de paret, 'tub of the wall', refers to a particular technique of packing it in the kiln. It measures about 80 cm to 1 m in height with a diameter of 120 cm.[32]

As the potters and muleteers categorized the pottery in 'pieces', potters also contested this categorization, and as to the vessels, they contested the labelling. In their exchanges, the potters and muleteers priced a certain vessel as it was categorized and labelled by its size, but the potter participants were quick to point out that the muleteers charged their clients a price for a vessel based on its actual measurements. For instance, Josep Benaiges threw vessels, let's say a solsera, larger than any other potter at the time, but was still paid like every other potter. Yet, the muleteer who bought this larger than average solsera could then charge his client a price higher for that piece than for the smaller solseres thrown by other potters.

With regard to the glazed pottery, potters and muleteers contested the hierarchical organization of the several types of this pottery in 'pieces'. My potter participants were eager to demonstrate how the muleteers profited by selling in these units. The muleteers bought the pottery per 'piece', but sold these in single pieces. For instance, a muleteer bought one tafol for the same price as four petxins. But since he would sell these four petxins separately, he could make proportionally more profit on each petxin than he could make on one tafol. In contrast, as my potter participants argued, a potter invested more labour and raw material when he produced four petxins than when he produced one tafol, but he would get the same amount of money for these four petxins as for one tafol.[33]

The personal relationships between potters and muleteers

As economic actors, potters were united among themselves as were muleteers: each group was united by their opposing interests around price calculation. But apart from being economic actors, potters and muleteers were also members of the close-knit community of Miravet and related to each other on a personal level.

Always short of money, potters preferred muleteers who paid as soon as possible. Normally, the muleteers would pay when they came to pick up their merchandise, but they would also pay in advance, or during a trip by sending the money home, or when they returned. As muleteers differed in their turnover and in the regularity of peddling, potters also preferred to work for muleteers who had a high turnover rate and a reliable frequency of order placement.

Muleteers preferred potters whose vessels were relatively large, for instance potters who threw vessels labelled solseró that were larger than the average solseró as they could make a larger profit on these pieces. They preferred potters whose products were well finished, well fired and well glazed, products that consumers appreciated and that sold well. (Though some potters finished and glazed their products better then others, the firing was always a risk factor and therefore potters did not differ that much in this process.) Since muleteers had to plan their trips in advance, they preferred reliable potters, potters who finished the products on time.

Potters and muleteers competed among each other for preferred commercial partners and eventually preferred potters worked for preferred muleteers, and preferred muleteers bought from preferred potters.

As his son Santiago explained to me on several occasions, Josep Benaiges produced good pottery, and had the tendency to throw vessels that were larger than average. He was a popular potter who never had problems finding a buyer: the muleteers always came to him to ask if he would accept an order. He always sold all his products upon unpacking the kiln. Llop, as he was known, bought from him. Llop had a bit of dry land with olives and almonds and was peddling away from home nine to ten months a year. Therefore, he had considerable turnover and was a regular buyer. Moreover, he always paid Benaiges on time.

Potters and muleteers traded with the same partner for varying lengths of time: from longer than a generation, over the length of a lifetime, many years, down to one single occasion. The long-term marketing relationships were between trade partners who preferred to trade with each other: their length was based on each others' qualities as a preferred potter or muleteer, as well as the fact that the length itself developed these qualities vis-à-vis each other. Potters and muleteers chose attractive partners, but also took care to *conservar*, 'conserve' as one expressed it, a preferred muleteer or potter. Josep Benaiges produced good pottery and relatively large vessels: as a result he was attractive to Llop, and he took care that he kept on producing such pottery for Llop. Josep Benaiges supplied several muleteers, but he gave Llop priority over the others, hence was a very reliable supplier to Llop. On his end, Llop would take care to pay Josep Benaiges as promptly as possible. The interviewees referred to potters and muleteers who were long-term partners with possessives, for instance Llop was 'Josep Benaiges' muleteer' and Josep Benaiges' was 'Llop's potter', and with the verb 'to have', for instance Llop 'had' Josep Benaiges as a potter and Josep Benaiges 'had' Llop as a muleteer. In contrast, participants talked about occasional trade partners as chosen from sheer necessity.

Potters and muleteers related to each other through their exchanges, but as fellow villagers they also related through kin and friendship ties. Kinship was the basis for the units of production and consumption, the household or casa, made up of at least one nuclear family. Upon marriage, sons and daughters left the household, except, usually, for the last-to-marry son or daughter: he or she would stay with his or her nuclear family and in due time care for the ageing parents. Apart from invitations for family celebrations like first communions and weddings, sons and daughters who left their parental household did not have any obligation towards it in terms of providing it with goods or services, or of co-operating with it. Nor did parental households have any obligation except upon death of the parents. The sons and daughters who had left the parental household would inherit something, but the son or daughter who stayed usually inherited the house and possibly some land. Of course, with the death of one's household members a person might end up alone. Kin would feel the obligation to care for that person, and they usually did so by taking the lone relative in, thus making him or her a household member. Outside the household, kin did not have any obligation to each other in terms of providing goods and services, or co-operating with each other, including forming marketing relationships. However, kinship might be an aspect of a marketing relationship, but these relatives also had to be on friendly terms with each other.

Friendship was a fluid form of relationship. It was hardly linguistically differentiated: participants said that they were all *amics*, 'friends', of each other, while clearly when talking about certain events, they had not been treating each other equally well. The only differentiation that was linguistically made was the expression *de confiança*, defined as 'trusted, intimate, confidential': this represented the highest level of friendship. Friendship was associated with kinship in that friends emphasized possible kin ties. For instance, a participant had a good friend and often when referring to this friend he mentioned that he was also 'somehow' (if distantly) related to him. Quite probably he was related in some way to half of the village, but in the context of this friendship, he was aware of his kinship. Friends could even create a ritual form of kinship by making a friend the godfather or godmother of a child, and create a metaphorical form by encouraging their children to address their friends as *oncle*, 'uncle', or *tia*, 'aunt'.

Though hardly linguistically articulated, as previously mentioned, friendship could be articulated in the context of marketing relationships. Moreover, it could be so in association with kinship. Potters and muleteers preferred to trade with people they liked. They were not obliged to trade with kin or friends, but when they had to choose a trade partner, they clearly tended to look among kin and friends. As trade partners, they could strengthen these ties. As with preferences for certain potters and muleteers, these long-term relationships were based on, and created, friendships, as partners continued trading with each other, their relationships developed into one of de confiança 'friends'.[34] Thus, potters and muleteers who were long-term partners also tended to relate to each other as kin and/or good friends.

Llop was Josep Benaiges's most important buyer. Josep Benaiges and Llop were the same age: they had been to school together, had been quintos,[35] *and had served together in the army. During their lifetimes, they had been closest friends, though Llop was not related. Santiago, Josep Benaiges's son and another interviewee, grew up addressing him as 'uncle' and used this word in our conversations. Though both started to work before Santiago's birth, Santiago was adamant that they must have started to trade with each other from the beginning of their working lives. He was very sure of this: he could not imagine his father doing otherwise.*

In a few cases, such long-term relationships existed between brothers and had originated when they were still members of one household. Usually potters did not have a mule and cart, and had to sell their products to muleteers, yet some potters were exceptional in that their sons, while still a member of a potter household, decided to peddle their household production. Barceló was such an exception. He had two sons. When his sons were still living at home, one worked in the workshop while the other peddled the products of his father and his brother: this way they made the most out of their production. When this peddling son married and left the household, he kept peddling the products of his father and brother.

This relationship between the brothers Barceló originated when it became advantageous for one son to peddle the production of his father and brother. This advantage also worked in reverse, as when a potter married into a household of a muleteer. This happened to Josep Fabregat Senr., a muleteer, who married a daughter of a potter and moved in with that household. He subsequently peddled the products of his father-in-law.

Partners who preferred each other and were simultaneously de confiança had long-term relationships. Although it was not the case that all people that were de confiança were also trade partners, as they might never had had the need to trade with each other. They could, however, be partners for short-term relationships or even single occasions. For instance, if the regular muleteer of a potter had fallen ill and was unable to peddle during his illness, a friend and/or kin muleteer might offer to help that potter out by placing an order so that the potter would not be without buyers and income. But potters and muleteers might also be short-term exchange partners forced by circumstance, not because of mutual preferences or other social bonds. When forced, they might be rather distrustful and inimical of each other. The variation in the nature of these exchange relationships from being de confiança to being distrustful parallels a variation in the articulation of opposed trade interests: the long-term partners mitigated the opposition between their trade interests, the occasional partners who were not kin and/ or friends sharpened their opposition.

Sahlins perceives in 'primitive societies' (Sahlins 1972: 186) an 'interplay between reciprocity, social relations and material circumstances' (ibid.: 190). He argues that these three factors interplay and thus form a continuum. One extreme is 'generalized reciprocity' (ibid.: 193): 'transactions' or 'assistance freely given [between socially very close people] ... regarding which an open stipulation of return would be unthinkable and unsociable' (Sahlins 1972: 191). The mid-point is 'balanced reciprocity' in which the 'reciprocation is the customary equivalent of the thing received and is without delay' (ibid.: 194), and which occurs between people with more social distance as in the previous exchange, people who 'confront each other as distinct economic and social interests' (ibid.: 195). The other extreme is 'negative reciprocity': 'transactions opened and conducted toward net utilitarian advantage' (ibid.: 195) and in which 'the participants confront each other as opposing interests, each looking to maximize utility at the other's expense' (ibid.: 195). This continuum can be used to analyse the variation in the exchanges between the long-term partners mitigating the opposition between their interests as parties in an exchange, and the occasional partners sharpening this opposition.

Potters and muleteers who were long-term marketing partners were located on this continuum somewhere towards the extreme of generalized reciprocity. These partners adhered to the commercial context of their trade, but simultaneously reduced as much as possible the contestation of price and quality of pottery, and mitigated the commercial context as friends/kin. Muleteers accepted without bargaining 'the price of Miravet'. They bought all the pottery as they had ordered it, irrespective of its quality. They were not picky when it came to assessing the quality of the pottery as good or flawed, valuing as full or half price, and they did so in dialogue with potters. A muleteer like Llop even took slightly flawed pottery from Josep Benaiges and paid him the full price for it. These muleteers ordered their merchandise when 'their' potters started to throw new kiln content and they ordered regularly so that potters never had to worry about selling their products at lower prices. On the other hand, these potters accepted with regularity the orders of 'their' muleteers so that these muleteers would never be short of merchandise and forced to buy from other potters and pay higher prices. If these potters and muleteers were very good friends, very de confiança, and of course if these muleteers were in the position to do so, muleteers loaned potters money and thus blurred the strict rhythm of delivery and payment. Though Josep Benaiges considered all the muleteers for whom he worked as friends, he only felt comfortable borrowing money from Llop (and he was the only muleteer who loaned Benaiges money).

From the moment that potters transferred the pottery to muleteers, potters and muleteers intensified their friendship. It was 'tradition', as the interviewees put it, that on the morning a potter and his assistants had to unpack the kiln, the muleteer came over to lend a hand. When they had emptied the lower chamber of the kiln, they would have a break and the potter would offer chocolate cookies and *vi ranci*, aged wine. After having emptied that chamber, the potter put potatoes, onions and sardines in the still hot ashes to broil and when they had completely unpacked, they would have a meal together in the workshop. They had such a meal *en plan de amics*, 'as friends among each other'.[36] Then they settled their accounts, often the muleteer ordered for the next firing, and then he took the pottery.[37] (Vicen Sales and Santiago Benaiges, who told me about the muleteers supplied by Josep Benaiges, remembered these muleteers mainly because they were the ones they had seen as boys who had come to help to unpack the kiln.)

Potters and muleteers might possibly exchange other goods than pottery and such an exchange would be between parties who respected each other in their interests and who treated each other correctly and thus constituted the mid-point, or 'balanced reciprocity', of Sahlins' continuum. Muleteers could bring potters firewood cut in the mountains surrounding Miravet, or food that they had bartered on their trips. Since muleteers often combined peddling with agriculture, they could also provide their potter with produce. The exchanges of firewood were balanced by customary and stable standards, e.g. one vessel for one bundle of firewood. The exchanges of food were balanced by expressing the values in money: a vessel with the value of 10 pesetes was exchanged for a specific quantity of grain also valued 10 pesetes.

Since Llop focused more on peddling and working the little land he had, he rarely brought firewood down for Josep Benaiges. If Benaiges requested it, he would bring him food. For this food, Llop charged Benaiges the same price as a shop would. Therefore Benaiges was not very interested in it and requested it only occasionally. I asked Santiago, Josep's son, why Llop would not charge his father a lower price: 'Those two were very good friends,' as Santiago confirmed, 'but when my father unpacked the kiln and the pottery valued two thousand pesetas, Llop had to pay two thousand pesetas. To do a friend a favour is different from business, and business is business.'

Potters and muleteers who traded with each other on single occasions because a potter was forced to, were located on Sahlins' continuum near the extreme of negative reciprocity. In that case nothing mitigated their opposing interests and the muleteer contested as much as possible the price and quality of the pottery, and exploited his economically stronger position to the maximum. When a potter did not have an order, or not one sufficiently large to fill a kiln, he still continued producing. Meanwhile he looked for a buyer, and if he could not find one, he stored his products until he could. Without an income, he was in dire straits. Some muleteers waited until potters were in this condition. They knew that they could bargain over the price, that they could dispute whether a piece was flawed or not, that they could buy only the pottery they wanted, and that they could wait to pay the potter until they had sold the pottery.

Joaquím[38] *was such a muleteer. He had enough land that he only needed to make one annual trip. He was described as someone who would even sell his own mother if he could make a good price on her! He waited until a potter was forced to sell and then tried to buy as cheaply as possible. For instance, he bought from Papaseit when this potter had a hard time, but as Pepe Papaseit was quick to emphasize, he was not a regular patron. Moreover, Joaquím was very critical. As one participant put it, he scrutinized the pottery as if with a magnifying glass, and was mostly interested in buying flawed pottery. The potters of course hated this ruthless commercial attitude, and they hated to sell for less then 'the price of Miravet'. That other potters were forced to do so weakened the consensus on price. When we discussed this muleteer, one potter called Joaquím an 'exploiter'. We had this conversation about 30 years after Joaquim had last bought pottery from anyone, and in the meantime this participant had become on good terms with Joaquím. 'Exploiter' was a slip of the tongue, yet with this word, and the context in which it was uttered, he showed how he had once despised this muleteer.*

Sahlins' continuum is helpful for analysing the exchange relationships between potters and muleteers, but its applicability is limited by the fact that it is developed for 'primitive societies'. In this respect, Sahlins defines 'primitive societies' as 'the absence of a public and sovereign power: persons and (especially) groups confront each other not merely as distinct interests but with the possible inclination and certain right to physically prosecute these interests' (ibid.: 186). He looks at exchanges as both expressing and as creating social relationships, hence the interplay between social relationships and transactions. As such, the relationships and exchanges between potters and muleteers can be analysed on a personal level. The potters and muleteers did not construct these relationships within a social situation lacking a 'public or sovereign power'; they did it within the wider political-economic context of the Spanish state, in which they had specific locations as potters and muleteers, i.e. not the equal 'inclination and right to physically prosecute their interests'. Their respective locations gave them unequal positions in the negotiations over the price: all the potters were dependent on the muleteers, while conversely, the muleteers were in varying degrees dependent on the potters. Potters and muleteers situated near the extreme of the continuum of generalized reciprocity therefore had ambiguous relationships. Their relationships could simultaneously be viewed on a personal level as between friends and/or kin, as well as on the political-economic level as between parties with opposing interests and with a power imbalance.[39]

The participants were aware of this ambiguity due to the two levels on which these exchanges occurred. Generally, they described the long-term relationships between potters and muleteers as friendly ones, but some potter participants reversibly interpreted several aspects of these relationships. They viewed these friendly aspects not as mitigating the commercial context, but as falling in with it, as strengthening the strong economic position of muleteers vis-à-vis potters. When a cynical potter talked about the ignorance of potters and the shrewdness of muleteers, he gave the prices set per type of pottery and not per measurements as an example. Someone like Josep Benaiges who sold vessels larger then average for the price of average ones, was exploited by his patron. Regardless how nicely Josep Benaiges and Llop might have socialized with each other, Benaiges should have negotiated for a better, more just price for his larger vessels. Another example this cynic gave was the custom of the muleteers coming to unpack a kiln. In his view, muleteers did not come to help, but to be the first to select merchandise and hence to buy the best. This participant had no illusions: though a muleteer might be a very good friend of a potter, he would without much ado buy from another potter if it was to his advantage. Some other potters had similar opinions on the friendship between potters and 'their' muleteers. For instance, Pepe Jornet thought that a muleteer lent money to 'his' potter to do him a favour, but also to restrict him by making him feel obligated to supply him. And most potters were indignant at the muleteers' refusal to discuss the half price paid for flawed pottery.

As potters and muleteers had to balance the volume of the potters' production with the volume of the muleteers' turnover, they formed a network in which each had a different number of trade partners and accordingly marketing relationships of different natures.

Josep Benaiges sold mainly to Llop, and only the remainder of his production to other muleteers. Llop always ordered in advance, hardly skipped buying a firing, but could not sell Josep Benaiges' total production. Sometimes he had to split a firing with another buyer, but he would have priority over other buyers from Josep Benaiges. If he did not need a whole firing, he, the other buyer, and Josep Benaiges would agree on the pieces to be fired the next time and how these were going to be split among the buyers. Llop and the other buyer would both take as many good as flawed wares. In addition to Llop, Josep Benaiges was remembered as selling to four other muleteers: el Pollet, Manyenga, Barborach and Nyec. Nyec, a very important long-term partner, also had considerable turnover: he and his son each peddled with their own cart. He had 'his' potter, Avante, but often needed additional merchandise. He did not go year round, rather he would pick up the merchandise when the kiln was unpacked and store it until leaving. Also, he never gave problems with payment. As Josep Benaiges' son recalled, he once could not make a planned trip while he had a lot of pottery stored for it. Instead of making potters wait for their money, he took out a bank loan and paid his producers promptly. For many years, he and Llop were Josep Benaiges' only buyers. After Nyec's son died in an accident, he bought less from Josep Benaiges and Benaiges took another buyer. Nyec was also much older than Benaiges, and retired sooner.

Josep Benaiges balanced the volume of his production with Llop's turnover by giving Llop year-round priority, and selling only the remainder to other muleteers. Other potters had annual cycles of muleteers who ordered their products.

As Pepe Papaseit told me, at the end of the 1940s, when he was 12 years old, he started to work in his father's workshop. His father had no problems selling his products. His father was a good, if slow, potter. He had two regular buyers, Tapit and Fatera. Both were peasants with a moderate amount of land and both peddled just once a year. Tapit peddled in July and Fatera in August or September. They only bought pottery from Papaseit. Fatera and Pepe's father were brothers-in-law: their wives were sisters. Fatera was mentioned as someone de confiança. Next to these two muleteers, his father worked with some others, like Nyec, though never as regularly and for such a long period. Even having these long-term marketing relationships, Papaseit was forced, on occasion, to sell his products to Joaquín, the ruthless negotiator.[40]

* * *

By their marketing relationships potters and muleteers formed a complex network determined by the potters' different volumes of production and the muleteers' different volumes of turnover. These quantitative differences interrelated with a qualitative difference, as indicated by the lengths of these marketing relationships and the differences in how a particular potter and a particular muleteer intertwined in their relationship to trade with kin- and friendship. On one level, potters and muleteers related to each other as two political-economic categories opposed to each other by different interests and unequal power. On the personal level, potters and muleteers formed a network of dyadic relationships in which they intertwined economic relationships with kin and friendship. They varied these personal relationships along a continuum of mitigating and sharpening political-economic tension. They only mitigated this tension ambiguously: one could view their relationships as between friends and kin pulling together, but at the same time also as between two opposing trade parties. This ambiguity gave their network its complexity. Moreover, a lot of these factors, like friendship and kin ties, and the volumes of production and turnover, were changing in time and resulted in continuous change in this network.

This ambiguity in the relationships between potters and muleteers was dialectically related to the meanings pottery acquired in these exchanges. As pottery mediated between two political-economic categories, it fetched a monetary price. But as these categories were differentiated due to the ambiguity in the relationships between potters and muleteers, the category of pottery was also differentiated. The price of pottery was calculated taking distinctive physical characteristics into account: potters and muleteers agreed on and contested the price calculation. These constituted the difference in the nature of their marketing relationships and its ambiguity.

---

1 However, Spain was not so isolated that it could not experience a brief boom during the First World War because of its neutrality (Carr 1982: 497; Shubert 1990: 16; Harrison 1978: 89–103). Later, Spanish isolation ironically buffered the country from the effects of the Great Depression (Carr 1982: 400).

2 Harrison offers a map (1978: 135) with the land distribution based on the cadastral survey of 1932 that unfortunately excludes the northern provinces. Tamames offers one (1990: 26–7) for 1969–70 which includes the whole of Spain. Though this last map is not based on the period under discussion, changes in land distribution between that period and 1969–70, the date of this map, have not been that significant and hence the variety of land distribution this map shows is applicable.

3 In 1969–70 only 2 % of the surface area of Tarragona was held in 13 plots larger than 500 hectares. This contrasts sharply with the region with the highest percentage, Carceres, where 36.8 % of the surface area was held in 725 plots (Tamames 1986: 27).

4 There are no statistical numbers on emigration or migration available. This statement is based on oral information given by the inhabitants of Miravet. In a socio-economic study of the county it is mentioned that statistics on migration are lacking (Anon., *La Ribera d'Ebre* 1989: 73).

5 Note that this is not to say that each and everyone in Miravet had the same values. After all, this period includes the Civil War in which people, including inhabitants of Miravet, killed each other over differences in political values. Though different people might have different values, it was the social experience of being part of this community which was the context in

which potters and muleteers acquired their values with respect to economic activities, values to which all potters and muleteers seemed to adhere.

6 At this time, only three potters' workshops owned kilns. Four other kilns were constructed as investments by non-potters; three of these kilns next to houses with workshops rented to potters.

7 Jociles Rubio (1989: 156) reports from Gandesa, a village only 20 kilometres from Miravet and probably socio-culturally very similar, the following saying: *'Lo que té un pà buscara una esclafada'*. This can be translated as 'someone who owns a loaf of bread [of one kilo], looks for a loaf of three quarters of a kilo'. Her informants explain this saying as referring to the marriage strategy of inheritors of land looking for partners who could also inherit land. In this context, it is interesting that 'bread' is a metaphor for land. However, I did not come across metaphors linking food with property in my fieldwork.

8 In Catalan *'casa'* changes according to grammatical rules in 'ca' or 'cal' in front of the name of a casa.

9 The participants remembered hearing from their fathers that muleteers used to transport the merchandise on mules. None of the people I met had transported their merchandise in this way, unless it was for only a part of the route to reach a village high in the mountains.

10 With the construction of the flood-control dam at Flix in 1948, the traffic on the river, which was already in decline, finally came to a stop (Anon., *La Ribera d'Ebre* 1989: 232 and 368).

11 There were bargemen who lived in Miravet who were only engaged in transportation, not in trade. This transportation was organized by captain-owners of barges who employed a few hands, not to be confused with large companies who owned barges and employed captains and crew as described by Moncada (1994) for Mequinença.

12 Other merchandise muleteers from Miravet might carry were *sárríes*, saddles for mules to transport things, which were also made from palm leaves and were made in the same places as the baskets; *canizos*, braided reed plates used in house-building or for drying fruit, which were made in Miravet from a reed that grows in abundance at the riverside; and brooms, also made in Miravet.

13 The muleteers traded in several kinds of merchandise: the sheep and goat skins stood out from the other merchandise, produce, in that they were not consumed in the village. The muleteers trading in skins had created networks by buying these skins from butcheries and small local traders, transporting these to Miravet, and there sold to traders and tanners from Barcelona and nearby towns. These skins had no use in Miravet and its vicinity, and had to be sold to urban people. Muleteers trading in these skins could only do this in order to make a profit on them and to increase the money they had invested in their merchandise. Hence, muleteers doing this engaged in peddling with a capitalistic outlook. These muleteers peddled as their main source of income. Although muleteers fell into the 'poor people' category with whom the participants identified, a few muleteers contrasted with the majority in that they had 'a nickel more': it was the muleteers who traded in sheep and goat skins, like the brothers Morrete, who invested in, among other things, a kiln. Note the parallel between the farmers, people who had substantial land and worked it in a capitalist mode, and the better off muleteers who engaged in their trade in a capitalist mode.

14 The importance of peddling for the whole village can be assessed by the number of muleteers. Carts were registered by the municipality, though not if they were used in agriculture or for peddling. Furthermore, these records are very incomplete. The earliest number of mules found was 125 in 1948: these animals were not necessarily used or suited for pulling carts. However, in these archives I did find forms stating the people subjected to taxation because they engaged in *industria en ambulancia*, 'ambulatory industry', specified as the *venta de cacharrería*, the 'sale of pottery'. In 1944 seven of these peddlers were mentioned, in 1954 four, and in 1959 again four. It seems that the people who were mentioned were only the muleteers who engaged in this peddling very intensively: participants invariably mentioned much higher numbers for all muleteers. Through interviewing, I got a better impression of the number of muleteers. I asked a former muleteer about peddling.

'Before the war there were 64 muleteers here. I was once in Huesca, it was the day of Saint James, July the 21st. Well, we were there with perhaps eight or ten muleteers. Since it was a holiday, all the muleteers from Miravet went to a restaurant, and there we counted – I remember it very well, I must then have been 18 years old – we counted all the carts from Miravet that were on the road, 64 carts.'

15 The clay used in Miravet consists of *llac*, a deposit of the river and free for everyone to collect, and *terra refractaria*, a material that potters mined nearby. Several potter families, but not all, owned these mines. Potters who did not own them could extract this material for free or for a small share of the extracted product. During the fieldwork period, not only were these mines abandoned, but also the land they were on. This land clearly did not have an agricultural or other economic value. Other raw materials used for engobe and glaze were very cheap to acquire.

16 Exceptions to potter households owning a mule and cart were the Pedrola household, who had sufficient land to work in order to own a mule and cart; and two households, one of them the Jornet household, of which a son decided to peddle the products of his father and brothers and so could use the mule and cart intensively.

17 Jociles Rubio (1989: 184) presents a short poem written by a land-owning suitor of Vilalba, a village near Miravet, in which he warns the woman he was courting not to marry his competitor, a blacksmith:

No't casos amb cap ferrer,

que no t'enredo amb ferradures.

Que val més un any d'oliada

que tots los martells i encruces.

This poem can be translated as:

Don't marry any blacksmith,

don't get messed up with horseshoes.

One year of an olive harvest,

has more value than all the hammers and tools.

The preference of land above handicrafts is clearly articulated in this poem.

18 The negative connotation of travelling is elaborated by Freeman. She writes that 'the more itinerant the outsider is, the greater is the contempt in which he is held. Thus ... the most despised groups are ... itinerant tradesmen ... for they are assumed to live outside the context of sedentary – and thus "civilized" – society' (Freeman 1970: 179–80). In a footnote she adds: 'Similar feelings are widespread in all areas of Spain with which I am familiar.' (ibid.: 180). None of the participants told me about having experienced 'contempt' and being 'despised'; these expressions are perhaps too strong, but her observation corroborates the negative feelings some of the participants expressed.

19 For an elaboration, see appendix: note on values of economic activity.

20 Sons of potters had to *sacrificarse*, 'exert themselves', that is they had to give up free time to practice, but none was forced or persuaded to become a potter. The only case I came across in which some explicit force or persuasion was extended was of Pepe Jornet's father. This man had an aptitude for music and while he was fulfilling his military service, he had an opportunity to become a professional musician. When his father learned about this, he was in tears and persuaded his son to return to Miravet and perpetuate the craft of potter.

21 'The price of Miravet' contrasted with the prices that the muleteers asked consumers, *el preu de fora*, 'the outsiders' price'. These prices varied according to the region of sale, the year and the bargaining skills of the peddler. Potters seemed to have had some idea of these prices, especially in the few cases that they themselves peddled, but in general the muleteers were not very open about these. My potter participants tended to estimate these prices as two, three or even more times the price of Miravet. (My former peddler participants mentioned lower prices, for instance one and a half times 'the price of Miravet'.) Indeed, potters were keenly aware of their weak economic position against the muleteers.

22 Over time, the potters assessed 'the price of Miravet' in ways that varied, and accordingly reached varying degrees of consent. A few of the participants remembered that the potters came together and decided to increase their prices. For instance, Santiago Benaiges recalled one such meeting that took place in his father's house. At that time he was too young to participate in it and therefore remembered it only vaguely. Pepe Papaseit, who was also very young then, recalled these meetings indirectly. He recalled having seen a printed list of prices on which the potters had agreed in such a meeting.

23 The participants even told me about idiosyncrasies of potters whom they had never observed working but of whom they had heard anecdotes. These were the more sensational idiosyncrasies, like Vicen Sales' brother, killed in the Civil War, who threw a cántir in one step, while usually a potter threw this in three steps.

24 Consumers were not interested in thinly thrown pots. As Pepe Papaseit remarked, some consumers even thought that a thick pot was stronger than a thin one. Consumers were only partially interested in a good line. If two peddlers were selling at the same time in the same market place, the pottery with a good line might have some advantage over the lesser pottery. If a muleteer was on his own, no consumer could be bothered by the line.

25 Due to the techniques then in use, a significant portion of pottery came out with cracks and holes and flaws in the glaze. Pepe Papaseit estimated that if one looked critically, perhaps half of the production had some of these imperfections.

26 Vossen reports a similar use of the Castillian word *docena*, though in the case he reports, this term is qualified by *de cuenta*, 'calculation', and the Castillian word *pieza*, the latter even used in Catalonia during the Franco years (Vossen 1984: 355 and 360 and Vossen 1974: 151–5 and 158–9). These uses of these terms also seem to refer as much to units of production, probably by potters working for a wage, as to units of marketing pottery to muleteers.

27 In the local spelling of several pottery types I follow Josep Maria Saez, 1992.

28 I use the measurements that Saez (1992) gives. He does not justify these measurements and from my own experience I know that such measurements were never very fixed. Therefore, I never systematically measured wares. Here I give measurements so that the reader can form an approximate idea of the sizes.

29 In their exchanges the potters and muleteers labelled and organized the bowls, vessels, cántirs and tubs by size and form variation. The meanings this pottery acquired in this organization and labelling was exclusive to these exchanges: the muleteers did not sell the pottery to their clients in these quantities, nor did they or their clients use these labels. When I interviewed Pepe Papaseit on the several types, quite accidentally he brought up that, of course, all the potters and all the muleteers knew and used these labels for the several types, but, though Pepe had not peddled and could not state it for sure, he presumed that the consumers of pottery did not. The consumers probably asked for a vessel of 100 litres and did not use the specific word for it. Others who peddled confirmed this. The people who bought the pottery from the peddlers normally referred to the several sizes by adjectives like 'big' and 'small', or by indicating the measurements. Of course, the buyers could not know those labels that referred to the custom of selling per piece and to the packing technique, and they barely knew the labels with connoted use, like mandonguer connoting mandongo.

30 Saez (1992) gives for these two types the volume of, respectively, 200 and 150 litres, yet the participants have told me that these two types did not differ substantially in their volume.

31 Saez (1992: 96) gives this volume for the *tafol normal*. He describes three sizes of tafols, yet in the period of the muleteers only one size existed, probably the normal.

32 Since Saez does not describe this cossi, these are my own approximations.

33 Note that at the time of the fieldwork potters calculated prices based on the investment of labour and raw material, hence one might interpret what they say when they pointed out the advantage of this custom for muleteers as comparing the present with the previous situation. Still, Pepe Jornet claimed that his father refused to sell to muleteers in units of 'pieces' but for prices calculated on the labour and raw material invested in the pieces. I only have data on the end of the period of marketing to muleteers in which this practice was contested perhaps

due to being the end of that period. Still this practice must have been accepted since Vossen (1974) documents many similar (but widely varying in their units of calculation) systems of valuing pottery from all over Spain, systems which were used for the payment of labourers as well as for marketing to middlemen. Unfortunately Vossen does not place them in their socio-economic contexts. As we will see in Chapter 6, when the potters shifted to other marketing practices they immediately adopted prices based on labour and raw material.

34 It was a similar mechanism to that Sahlins describes: 'If friends make gifts, gifts make friends' (Sahlins 1972: 186; though I don't want to imply that these exchanges simply could be considered 'gifts').

In Chapter 4 I describe how potters exchanged labour amongst each other. These exchange relationships were formed according to similar patterns. There was one case of two potters being brothers who violently hated each other, and did not want anything to do with each other, least of all the exchange of labour. The participants always referred to this case to point out that a kin relationships did not necessarily imply that potters exchanged labour with each other, they also needed to get along very well. In general, brothers got along well with each other, but not always. The same holds in these relationships between potters and muleteers.

35 The year before boys go into the army, boys and girls of one village or neighbourhood and of the same age form a group to have social intercourse, that is going out for drinks and dinners together, making trips, etc. Actually the word quinto comes from the army and refers to a group of men of the same age group.

36 Many such meals emphasized friendship and co-operation. For instance, after finishing a firing the potter who owned the fired pottery offered the people who had been firing a meal of rabbit or chicken. Other professions also had such meals, and in general friendship was emphasized by common meals.

37 The pottery was yet not ready for sale. At home the muleteer had to 'soak' the pottery before he could store it. At that time the potters used a clay which contained grains of chalk that made the pottery chip easily. By bathing the pottery for some hours in water, the chalk diluted and this defect was remedied and then the pottery was ready for sale.

38 A pseudonym.

39 This ambiguity is not the Maussian one which 'characteri[zes]' the 'gift': 'the so to speak voluntary character of these total services, apparently free and disinterested but nevertheless constrained and self-interested' (Mauss, Halls translation, 1990: 3), or in another translation of this text, 'prestations which are in theory voluntary, disinterested and spontaneous, but are in fact obligatory and interested' (Mauss, Cunnison translation, 1967: 1). This ambiguity, whether described as between appearance and fact, or theory and practice, is between parties transacting on one level. If I had analysed the trade relationships between potters and muleteers only on the political-economic level, i.e. ignoring their complex social network, their exchanges would be unambiguously between two opposing parties, and were definitely not an instance of the 'gift'. If I had analysed kin and friendship as articulated in these exchanges, I would have perceived the Maussian ambiguity. For instance, Josep Benaiges sold Llop vessels which were relatively large and for this 'extra' he did not get extra money, hence he 'gave' it to Llop. By this gift he would have obliged Llop to grant him such favours as loaning money, a favour which Pepe Jornet interpreted as a means to bind a potter.

40 Potters and muleteers were two distinct social categories, but in a few cases these categories overlapped on the level of a household as well as on the individual level.

As Pepita had told me, when she married Antoni Pedrola around 1940, he moved to the Barceló household. Her father continued working as a potter, but since he acquired land, he devoted more and more time to it. Since they had sufficient land, they could have a mule and cart allowing Antoni to peddle his own products. This household combined agriculture, pottery production and peddling, and hardly needed to purchase food.

Joan Avante remembered the marketing of his father in the 1950s, which he did in three simultaneous ways. He had Nyec as 'his' muleteer; his father provided Nyec with the bulk of merchandise. Joan was sure that the relationships between his father and Nyec was the usual close relationship between a potter and his regular buyer, one of de confiança. At the same time his father was selling to a shop in Mora d'Ebre, the nearest market town: he was the only potter who, at that time, sold directly to a shop. He sold almost every firing partly to Nyec and partly to this shop. His father sold only the perfect wares to Nyec and the shop and thus got the full price for them. He fixed the flawed wares and peddled them, and thus also got a good price. The combination of having that shop and peddling, in addition to a very small bit of land, made it just feasible for his father to have a mule and cart. But Joan Avante added that it was not a very strong mule and that his father only peddled in nearby villages.

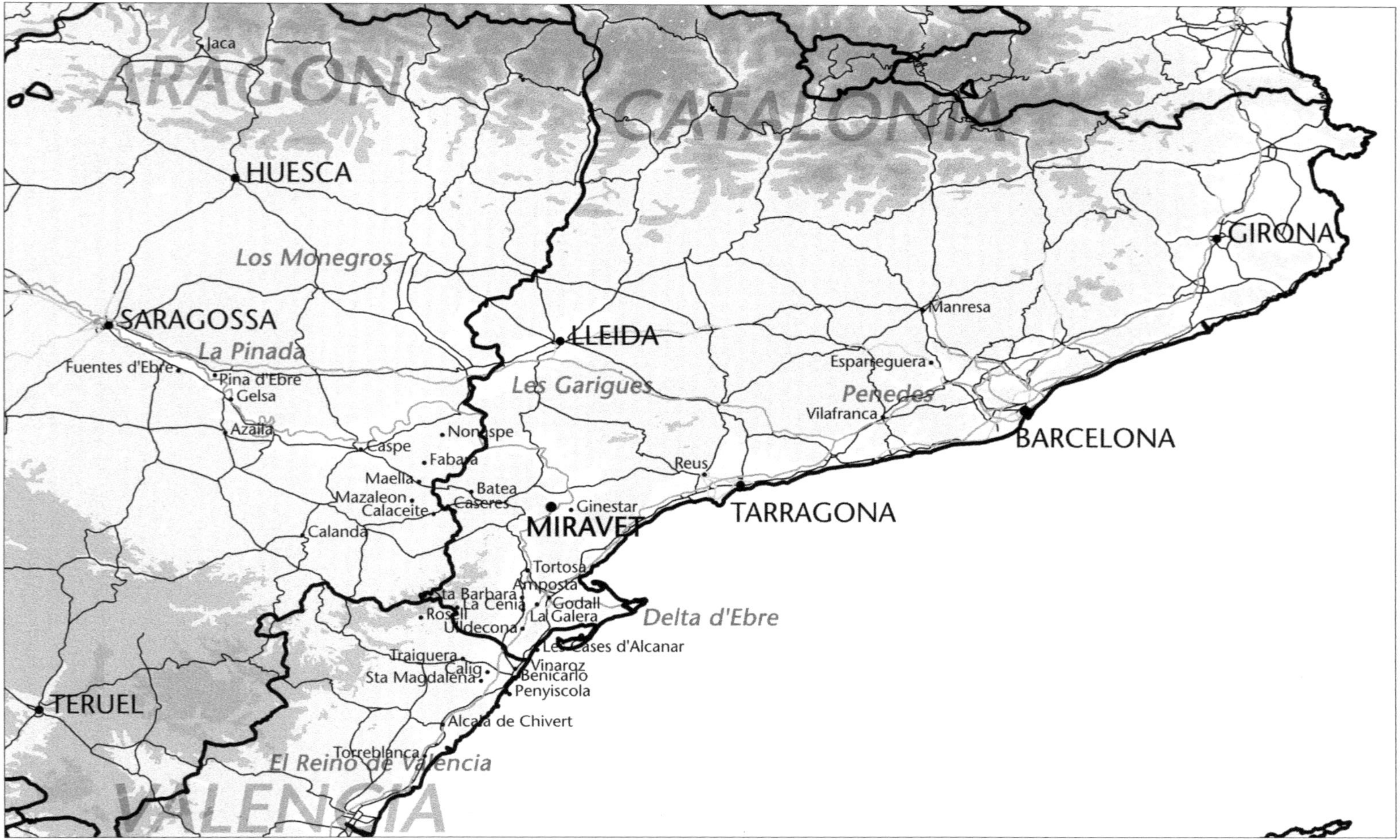

Map 3 *Map to Chapter 3.*

Chapter 3

Potters and consumers during the period of the muleteers

The national and international context of consumption in the first part of the twentieth century

Throughout the nineteenth century, one of the most important issues of Spanish political debate had been the protection of the domestic market from foreign competition with custom tariffs. In 1906, the debate on this issue was concluded with the passing of the Tariff Act, which would stay in place until 1960. In this way the Spanish State created a highly protected market. In fact the 1920s witnessed tariffs higher than any country surveyed by the League of Nations (Tamames 1986: 139–41). At the same time, Spanish exports were relatively low, especially after 1898, when Cuba was lost as a colony. Consequently, Spain constituted a separated market.

Within this national market consumers were integrated very differently. An ethnography of Ibieca, a small village in northern Aragon, describes the situation in the first half of the twentieth century as '... the material exchanges, the ecological cycles, were largely local' and 'the casa was a miniature ecosystem [and although] the degree of self-sufficiency varied greatly from large landholding to poor landholding houses, ... the casa at each extreme defined a bounded system' (Harding 1984: 81–2). Because there are several kinds of land, irrigated and dry, to a large extent people could be 'self-sufficient' with respect to their food and that of their animals. These people were also largely self-sufficient with respect to their clothing through the cultivation of hemp for linen, and the breeding of sheep and goats for wool and skins. Around the village of Ibieca they found materials, like cane for making baskets and chairs, and diverse construction materials (ibid.: 86). Harding's ethnography focuses on the self-sufficiency of these people, but by mentioning that they slaughtered pigs and produced olive oil, it implies that they consumed pottery. It is not far-fetched to deduce that pottery must have been one of the few products acquired from outsiders, particularly among the small landholders who had little produce or money for purchasing products. It was not until about 1960 that the people of Ibieca started to buy the products of others, products that they had previously produced (ibid.: 144). Hence, in the period of the muleteers, these Aragonese villagers were only integrated into the market to a very small degree, and probably only in a regional market, one that seemed not to have changed during the first half of twentieth century.

This minimal integration into the national market contrasted with large cities such as Barcelona, but also with towns within the market area of Miravet pottery. Reus, a town where muleteers marketed pottery, was also one of the Catalan towns that developed dramatically at the end of the nineteenth century due to the growth of the Catalan textile industry. In 1901 one of its inhabitants, the textile manufacturer and merchant Navas, commissioned Domenech i Montaner to design a house for him and to supervize its decoration. Domenech i Montaner was the leading Modernista architect and designer. Modernista was the Catalan variant of the international avant-garde style more generally called Art Nouveau. This house was furnished with specially designed furniture, some of which was exhibited in contempory European art and design shows. It was equipped with appliances like a stove and a water-heating system that were the best and most modern then available on the European market and were imported from Great Britain and Germany. Navas acquired the most stylish and comfortable goods available on the European market at the time, but quite probably in his kitchen and pantry used certain types of pottery made in Miravet.

The relationships between potters and consumers of their pottery

One summer afternoon, I was again talking with Santiago Benaiges about his father and his father's muleteer, Llop. I asked Santiago where, Llop peddled.

> *Llop peddled everywhere. He peddled all year round. Well now, every season had its usual region. In summer Llop told my father, 'You have to make cantirs because I'm going to Les Garigues.' This was because in Les Garigues people fetched water with mules and transported it in cantirs. At the end of the summer, when Llop went to La Pinada, in the south of Aragon, he said 'You have to make me gerrettes, those vessels thrown in three phases.' This was because people there went to fetch water with mules and stored it at home in gerretes so they had water for the whole week. That vessel was like a cistern in the house. The muleteers sold a lot of big vessels for storing olive oil, but that was in winter. Llop arranged a truck, loaded it up and went up there. He was there a month, two weeks, or 20 days, depending on how the sales went. There were muleteers who only made trips to that region and no other trips.*

Santiago explained that this trip to La Pinada was very popular and that almost all muleteers who made it did so to complement their own agricultural production by exchanging, usually bartering, pottery for produce, especially cereals.

Santiago Benaiges describes these orders intertwined with information as to when potters received them and for which specific types of pottery; where and how muleteers were going to peddle these types; and how these types of pottery were going to be consumed. This knowledge of consumption and consumers constituted his father's relationships with consumers. What was the nature of these relationships of potters with consumers?

* * *

Marx said that the relationship between producers and consumers is mediated by 'commodity' exchange. He analysed this exchange as constituting dialectically-linked meanings of objects, 'exchange-value' and 'use-value'. In these exchanges, two people equate the respective objects they own and exchange. One person has an object that he wishes to exchange with an object possessed by the second person. The first person has an interest in his object insofar as that object can be equated with the object that he wants to acquire, and he assigns to his object an 'exchange-value'. But he has another interest in the other's object: that object can 'satisf[y] human wants of some sort or another' (Marx 1984: 43), i.e. it can be consumed, hence he assigns to that object a 'use-value' (Marx 1984). A person can produce objects that directly satisfy his wants, which have use-value for him, or he can produce objects that he exchanges with others for objects that will satisfy his wants. In the latter case, such a person produces commodities, objects that have exchange-value for him, 'but use-values for others, social use-values' (Marx 1984: 48). Thus, exchange-value and use-value are dialectically related to each other.

In relationships between potters and muleteers, and those between potters and consumers, exchange-value and use-value interrelated with each other in ways characteristic of these two different kinds of relationship.

Potters exchanged pottery with muleteers in order to acquire an exchange-value; muleteers in turn exchanged pottery – again for acquiring an exchange-value – to consumers of pottery, people for which it had a use-value. Though muleteers did not consume the pottery, aspects of use-value were factors in the relationships between potters and muleteers. Potters were preferred trade partners for muleteers if they produced well-fired pottery, that is pottery without cracks and holes, and with a shining glaze, the kind consumers preferred. Potters contested several aspects of their relationships with muleteers because they knew the use-value of their products. They knew that consumers neither detected nor cared about the repairs of holes and cracks fixed by muleteers, and that consumers paid the full price for this ware while muleteers themselves paid potters only half price. They knew that consumers were interested in vessels of certain measurements and paid accordingly, while muleteers paid them according to how a vessel was labelled. They knew that consumers were usually interested in single pieces, while muleteers paid potters per peça, per 'piece' or the unit in which the pottery was organized in the exchanges between potters and muleteers.

Other aspects of use-value of pottery did not enter into the negotiations between potters and muleteers, but they did articulate the orders of the muleteers. Santiago told me about the contents of the orders his father had received and when they were received. By recounting this he located the consumption of the ordered pottery in time, i.e. that of the annual cycle, but since he was talking about the past, also in a linear, historical time, i.e. the period of the muleteers; he also explained where a muleteer was going to peddle a certain order, and thus Santiago located the consumption and consumers of certain pottery spatially. Santiago further added that a certain order was going to be bartered. For the participants, barter connoted a certain socio-economic location of the person engaging in it, and thus Santiago linked the consumption of that order to certain socio-economic characteristics and located the consumers socially. Moreover, he also talked about the consumption itself.

Potters produced their pottery to be consumed by other people. Baudrillard, in his analysis of consumption within a capitalist socio-economic context, argues that as objects are equated with each other by their exchange-value, they form a 'system' in which their physical characteristics form a 'differential logic' (1981: 67 and 75). By consuming these objects people differentiate themselves from each other in a way analogous to the physical differences between objects (i.e. men differ from women as trousers differ from dresses). Conversely, the physical characteristics of these objects acquire meanings derived from the social differences of their consumers (see also Sahlins 1976: 166–204; Miller 1987, 1995, 1995b, 1995c and 1995d; and Bourdieu 1984). By consuming an object, by realizing the object's use-value, consumers locate themselves socially. By learning how a certain type of pottery was consumed and by whom, potters related to these consumers as people located within the same social structure they knew themselves to be located in.

This process is already discernible in the above quoted conversation with Santiago Benaiges. For him, barter had certain connotations derived from his own transactions in Miravet. These connotations became linked by Llop's order with certain types of pottery, gerretes, and certain consumers, the people living in La Pinada. But this process of merging meanings of pottery derived from observation and participation in the village community with information derived from muleteers' orders for specific trips, and the simultaneous process of socially locating consumers, can be more elaborately exemplified by the following.

In a conversation in October that same year with Santiago Benaiges and Vicen Sales, Santiago mentioned as an aside, the use of the vessel as a toilet. For more than a year and a half I had been talking about this type of pottery and had never heard of this use. Both confirmed this use and I wondered where the vessel had been used as a toilet. Adroitly, Santiago answered that he never had been in my country so he did not know what the Dutchmen were doing. Vicen added that this use was common in many places per aquì, *'around here'. The large vessels, primarily used for the storage of olive oil, also had a secondary use. These vessels were placed in small closets, a bit hidden, on the ground floor of a house, in a stable, or in a patio. Santiago and Vicen had grown up using these vessels this way. I knew that these vessels were sold in a wide range and asked if consumers in Aragon also used them as toilets. 'In Aragon', Vicen explained, 'they did whatever they wanted to do. Of course they used the vessels as toilets too. If we here in Miravet seventy years ago did not have electricity, those outside Miravet did not either, nor had they toilet bowls.' Vicen continued how he had seen vessels thus used, in many houses, indeed in almost all homes. Yet he added that '... in, let's say, rich houses, they had a hole and above that they had the bathroom, but in ninety or even ninety-nine percent of the houses, we had a vessel.' The 'rich houses' were the first to have modern sanitary provisions, only later to be acquired by poorer people.*

The participants paralleled the introduction of modern sanitary provisions with the advent of electricity; first introduced in cities and then gradually into rural regions. At the time of the fieldwork, no one was using vessels as toilets, otherwise it would have been immediately apparent to me. By speaking of the use of the vessel as a toilet, participants were referring to a past practice that had receded from cities to villages, from economically powerful classes to less powerful, until its use had completely disappeared. This was a past that they vividly remembered, but they did not assume that foreigners, in this particular case Dutchmen, had similar past experiences. In this way they merged the meanings pottery acquired in consumption by them with the knowledge they had of consumption by others through muleteers, and related to these others as people located within the same social structure; all this in contrast to their knowledge of foreigners.

The relationships of potters with the consumers of their pottery existed in the mutual encompassment of production and consumption. Potters produced pottery that they exchanged as commodities: they produced objects that had exchange-value for them as they equated these objects, directly or through money, with other objects in their trade with muleteers. At the same time, they produced objects which had use-value for others, that is, they knew that these objects were to be consumed by others. To paraphrase Marx, they knew that these objects satisfied wants of those others. In this way production encompassed consumption as mutually specifying each other: potters produced objects specified by the consumption of these objects. On the one hand, physical characteristics of pottery became meaningful for the potters by their knowledge of the use-value of pottery and by their knowledge of how the physical characteristics of pottery made specific forms of consumption possible. On the other hand, by the use-value of their pottery for others (by the social use-value), potters related to these others, i.e. to the consumers of their pottery. Hence, the mutual encompassment of production and consumption is the nature and the content of the potters' knowledge of the use-value of their products – of the consumption and consumers of their products. With this knowledge, production and consumption mutually specify each other, and in this knowledge of consumption and consumers, potters related to the consumers of their pottery.

How potters knew the consumers of their pottery

Through three different channels, potters knew of consumption and consumers in three different ways. First: though potters and muleteers were in general two distinct categories of people, they overlapped with each other in some degree: a few potters peddled just like muleteers and hence knew these consumers directly. Second: like Santiago Benaiges' father, most potters, including the potters who peddled, knew of consumers through orders that the muleteers placed. And third: in addition, potters knew of these consumers because they interacted daily with muleteers and colleague potters as fellow villagers and accordingly learned about these consumers. Moreover, in the same context of daily interaction with fellow villagers, they and their fellow villagers consumed the pottery too, and as a result this pottery indexed characteristics of its consumers.

Potters' knowledge of consumers and consumption through direct contact

Some potters had a mule and cart and peddled their own production; they were potters and muleteers at the same time. These rare hybrids knew consumers of the regions where they peddled, as well as muleteers did, others peddled less extensively and therefore had a more distant relationship with consumers. During my fieldwork, two such potter/muleteers, Vicen Sales and Pepe Jornet, recounted their experiences.

Vicen Sales' father was a potter who married a muleteer's daughter and went to live with his in-laws. In order to help his ageing grandfather, Vicen accompanied him on sales trips. He began doing this during vacations while he was still in school, and continued when he started to work in his father's workshop. He accompanied his grandfather until he was conscripted into the army at the age of 18. We spoke extensively about these trips.

These trips to the southern part of Catalonia and El Reino de Valencia (as the Comunidad Valenciana was referred to in Miravet) were 40 to 50 days long, usually starting in September and lasting until the beginning of November. 'We took half a truck, and loaded fifty solseres, thirty tafols, some solserons, petxins, maybe a bit more than half a truck and we shipped it to les Cases d'Alcanar.' They sold pottery produced by Vicen's father, especially flawed pottery that they fixed up in order to fetch a regular price.

'In Les Cases we loaded our cart and went to villages where there was a weekly market. Each village had its turn, for example, on Monday we went to the market in Les Cases, on Tuesday to Amposta.' In between these trips, an inn in Les Cases was usually used as their depot, as well as a place to stay overnight. 'We returned to les Cases for loading the cart again. Perhaps there was some merchandise left from the previous market but still we had to load up the cart again. Wednesday we were in Vinaroz, Thursday in Benicarló and Friday up there in Calig and those places. Sunday we rested, and then we started the next week to go to the same villages in the same order, or perhaps to some other villages. Sometimes we loaded the cart on Saturday and we took advantage of the Sunday by going to Torreblanca, which is far way, close to Castelló. We got there at four in the afternoon, we put the cart in a place, we fed the mule and we waited until the next day. The next day we went with the cart to the market and set up a booth.' Torreblanca was not a very adequate place for them to vend their merchandise as it was located on the edge of their area. 'We vended in Torreblanca, and from Torreblanca we went to Alcalá de Chivert. In Alcalá de Chivert there was an inn where we stayed and the next morning we vended on the market, and then on Tuesday and Wednesday we did a village called Santa Magdalena which is close to Peñiscola, and then Benicarló, and the next day we went back to Les Cases. We loaded the cart and then we left again. If it was time to go to Vinaroz, we went to Vinaroz, if it was time to go to Amposta, we went to Amposta. We went there for the day that was in that town market day.'

Vicen and his grandfather found that they usually were not the only muleteers from Miravet at these markets. 'In many villages and towns we ran into three or so other muleteers. Once when I was with my grandfather in Benicarló, I had gone to bed a bit later than my grandfather. The next day he woke up a bit earlier and went to the market square to see whether other muleteers had already arrived. He returned immediately; "Come on Vicen, get up! There are three carts from Miravet!" "O my

God! three carts from Miravet!" He said, "Come on, we have to put our cart there!" And there we were together with four carts from Miravet and when we got there with our cart, another got there as well. Five carts.'

In such markets, the muleteers from Miravet were competing with each other and the greater their number, the more the competition.

At these markets there were also other vendors. 'There were altogether 15, 20 booths. One was selling vegetables, the other espadrilles, another hardware, we selling pottery, and there were booths with textiles, perhaps there were four, five of those.' In contrast to the muleteers, these other peddlers sometimes had vans. Vicen explained that 'for our kind of merchandise one would need a large van to carry it all. In contrast, textile and such things one can carry in a small van. At that time, there were few cars, and even fewer vans, let alone vans large enough to transport pottery.'

But Vicen and his grandfather did not vend all their merchandise at these markets. 'Sometimes we had been selling at a market and we were left with seven, eight pieces of pottery. In those villages there were shops we went to offer our merchandise to them.' Vicen still remembered these shops. 'In Vinaroz there was a shop which was a cart and hardware shop. They sold baskets and vessels of pottery, in short, things for the countryside. There was a second one in Vinaroz, very close to the previous one, which sold alfalfa for mules, and corn and other cereals for animals. Fodder. The third one had fine ware and our pottery.' The fine ware constituted ceramics, dining ware, cups and the like, and glassware like glasses for drinking. Because the pottery was sold to these shops in large quantities, as opposed to the usual single units at the markets, a slightly lower than market price was asked. (A former muleteer explained that sometimes these shops also peddled in neighbouring villages. In such cases they bought not just the pieces a muleteer had not sold on the market, but many more.) Thus they would return empty to Les Cases where they would again load up the cart.

Like his grandfather, Vicen was very familiar with the markets and shops in the villages and towns in this region. Vicen knew what he could sell in each of these villages and towns. In this particular region, he recalled, they mainly sold vessels. 'On one trip to this region we carried about seven, eight of the largest vessels, gerrots. In the inland villages, for instance Calig, Santa Bárbera, Ulldecona and especially the ones higher up in the mountains, villages like La Galera, Godall, La Cenia and Rosell, the people always used the largest vessel for storing olive oil.' These villages differed from the towns and villages on the coast with respect to the houses. The houses in the coastal places were usually smaller and although people living there used olive oil too, they could not store large quantities in large vessels in these smaller houses. Here they sold more solseres and tafols for the storage of olive oil. In both the mountainous and the coastal places smaller vessels like solserons and petxins were sold to be used for conserving food, especially green olives.

In this region there were two other potter villages, La Galera and Traiguera. The potters in these villages maintained a small production, but partly overlapped in their production with what was being produced in Miravet. They also made a vessel similar to the petxin made in Miravet, without glaze. According to Vicen, consumers distinguished between vessels made in Miravet and in these two villages on the basis of their use of the vessels without knowing where the vessels were made. They preferred the vessel made in Miravet because of its glaze and its strength.

Vicen and his grandfather also sold bowls in this region. 'We sold some bowls, but not many. We sold some higher up, in the mountains. For instance in Alcalá de Chivert, a village which is tucked into the mountains, we carried forty bowls and we sold them all. In one season we made three trips to this village, and on these trips we carried forty bowls and sold all forty. In other villages, we could be sure to sell one, two, four at a trip, but in a village like Alcalá de Chivert we sold all of them.'

In addition, another aspect of the difference between the mountainous and the coastal places was linked. 'The coastal places have always been nicer than the mountainous ones. People there always had a bit more money. The houses there were in better condition than high up in the mountains, in the countryside. I do not want to say that there were no nice houses in the mountains, but in the mountains it was just a single one. In Vinaroz there are now three or four buildings with seventeen or eighteen stories but a long time ago there were buildings with seven or eight stories ... and in a place like Vinaroz, all houses had a kitchen sink. That was there standard.' Though the rural houses in the mountains were larger, hence their inhabitants could store olive oil in large vessels, these lacked such facilities as kitchen sinks, facilities which indexed modernity as did a high rise. The modern facilities made apartments 'nice' and 'in better condition'. In several ways, the kitchen sinks replaced the bowls. Therefore, poor people in the countryside bought bowls, while those wealthier in the coastal places did not. Vicen associated the use of bowls with peasants and farmers: 'Although there was some agriculture, in villages at the coast, like Vinaroz and Benicarló, it was more than anything else, industry.'

The potters of Miravet produced two types of bowls, one with a narrow bottom and a half-circle design, and the 'Catalan bowl' with a wide bottom and solid red or white with green splashes. Vicen recalled that in this region they especially sold the latter, the 'Catalan' ones. 'The people in this region had a bit more money and they liked the Catalan one most ... they considered these more beautiful.'

The potters of Miravet produced a lot of cantirs, jugs used for transporting water to the home, but this type of pottery was not sold in the region around les Cases d'Alcanar. Cantirs were only used in the mountainous part of the region; the economically better off part had running water. 'With respect to the cántir, the potters of La Galera and Traiguera had this market. In the whole mountainous part, people used the cántir made in La Galera and Traiguera ... If we would have carried cantirs, since these were not from their country, they would not have bought them ... Their cántir is very different, with a design,[1] and with different handles.[2] In short, people in this region were used to the cantirs produced there.'

Vicen's grandfather had made trips to distant regions, even as far as northern Aragon, but as he aged, he limited himself mostly to the nearby region of Les Garigues. Vicen also accompanied his grandfather on trips to this region. On Monday they left Miravet for Batea, where they arrived in the evening. 'The next day we vended until half-past twelve or one o'clock, when everyone was going home for lunch, and we took the booth down. We got all the pottery together and loaded it on the cart, got the mule, put him in front of the cart and went up to Maella.' Maella was reached the same day and pottery was vended again the next morning. In a place like Maella, it was usually sufficient to set up a booth on the market square, but sometimes, if the sales were not going too well, one had to announce by walking through the village that the 'pottery seller' had arrived. In very small villages, one had to go from door to door offering merchandise. 'Here in Maella we took our stuff down at three or four in the afternoon because the people in Maella were different, they did everything a bit later ... and at five we left for Fabara, which is only eight kilometres, and in two hours we got to Fabara.' Fabara was another village where they could stay overnight and vend merchandise the following morning. The next village along the route was Nonaspe. Thursday morning they vended in Nonaspe and then went back home over Batea. This time travelling back was fast because everything was sold and the cart was empty. This trip could be varied by swapping these villages for villages like Caseras, Calaceite, Mazaleon or Caspe.

Vicen's grandfather made this trip about twice every three weeks in summer and often also in autumn and winter. If there was a good olive harvest in the autumn, and hence a demand for vessels to store olive oil, he would go there. If there was a better harvest in the southern part of Catalonia and El Reino, he would make the trip described as above, and when there was a better harvest in Baix Aragon, he would go there as well.

His grandfather was not the only muleteer from Miravet who peddled in this region. There were two or three others, one of them Llop. Because there were no market days in this region, muleteers would encounter each other accidentally. 'Perhaps you met someone else also peddling. But if in such a small village, two muleteers from Miravet ran into each, the next morning one of them would say, "well I go to that and that place", and the other "well, I will stay here". One stayed and the other left. This they did in order not to bother each other.' On these occasions, muleteers took care not to estropear, *'spoil', the market by offering merchandise for different prices. Usually these muleteers would also decide to vend their merchandise* a medias, *'fifty-fifty'. They would then conjoin their merchandise, sell it co-operatively and split the money made.*

There were also other muleteers of Miravet peddling in this region, but Vicen's grandfather remained for many villagers the preferred supplier. 'Because there was another muleteer who perhaps went, well, and because my grandfather would come the next week, some villager would say "I wait for the grandfather". And she did not buy and waited for the grandfather, but fixed customers he did not have there.' These customers were not as 'fixed' as in the distant regions, like northern Aragon, where muleteers tended to control territories exclusively, or perhaps share a territory with only one other muleteer. Still Vicen claimed that his grandfather knew every person in these villages, and was known by every person. This contrasts to the southern part of Catalonia and El Reino de Valencia, where his grandfather had no such familiarity with the inhabitants.

In Les Garigues there were also peddlers with other merchandise, the same kind of merchandise as encountered in the southern region. Because there were no regular market days in this region, one hardly ran into peddlers with other merchandise. People in this region had less to spend than their counterparts in other, wealthier regions, and these peddlers, Vicen added, rarely used vans.

In Les Garigues there were no shops. Still Vicen's grandfather had devised a way to market his merchandise in addition to vending it at the market square. 'In Batea, near the market square a lady used to live, who helped my grandfather with selling ... She knew everyone in the village. Sometimes this lady said to my grandfather, "you know what, because the sale has not gone well, you can leave me a few dozens cantirs, I think I can sell them all this week". He left them there, went to home to reload his cart and returned again. In the meantime my grandfather was peddling in other villages and some days passed. When he got to Batea another time, this lady had sold everything. In Fabara he did the same.'

In Les Garigues Vicen and his grandfather often bartered their merchandise for local products. These were produced en la casa, 'at home', and primarily for consumption at home. They were products such as barley, figs, lima beans, chickpeas and kidney beans; all the typical agricultural products of a dry mountainous area. Those consumers preferred barter above paying with cash not because they lacked cash. 'They did have money but sometimes in order not to take out money, which they did not like, they bartered with products. Even if they might pay double the value, even if they would pay a higher money value in barley than in cash, they would prefer to barter.' Barter was preferred by these consumers because their economical relationships were not primarily articulated by money and money was scarce, while agricultural products were not necessarily scarce. (When talking about these trips, Vicen explained barter this way. As we shall see shortly, in other conversations he demonstrated that his thoughts on this issue were more complex.) Muleteers had such trustworthy relationships with their customers that they did not necessarily need to pick up these barter goods upon delivering the pottery, but did so on a later trip.

In Les Garigues Vicen and his grandfather sold mainly cantirs, jugs used to transport water, the type with the wide bottom. Consumers needed these in large quantity because: 'In this region there was no water in the village. People had to fetch water with mules in the fields. In those fields they had big ponds and thus they had water. As they had to transport the cantirs with mules, well, sometimes a cántir was broken, sometimes two got broken. Many cantirs got broken.'

They also sold vessels. Gerrots were sold for storing olive oil as well as water. 'People took water from ponds and irrigation canals and stored it at home. They bought vessels to have a water deposit at home. We sold some vessels for this use, we also sold vessels for storing olive oil.' Smaller vessels were sold for conserving all kinds of food like olives and bell peppers, and especially pickled meat.

They sold bowls in all sizes except the smallest: the largest were mainly used for the frequent pig slaughtering in the region. They also sold the type of bowl with a narrow bottom and half-circles design. Occasionally they would sell eating and drinking troughs for rabbits, chicken and pigeons, and when ordered by consumers, chamber pots and simple planters.

When travelling through this region they sometimes lodged overnight in inns, or sometimes with friends. In inns they were required to pay, but not in the homes of friends. To show their appreciation, they offered a gift to their host. 'In Fabara we slept at friends. We arrived in Fabara, put the mule in the stable, which these friends had also given us access to, fed the mule, and chatted. My grandfather would ask, "What do we have for dinner? Well wait", my grandfather added, "I will go to the butchery"... and we bought this and we bought that, we bought meat. Our friends prepared dinner and there we had dinner all together.'[3] *(Usually these friends were passed from father to son. As Vicen Sales had only peddled a few years together with his grandfather he did not know how his grandfather got to know these people, but other participants have told me that they, as boys, accompanied their father and thus were introduced to these friends.)*[4]

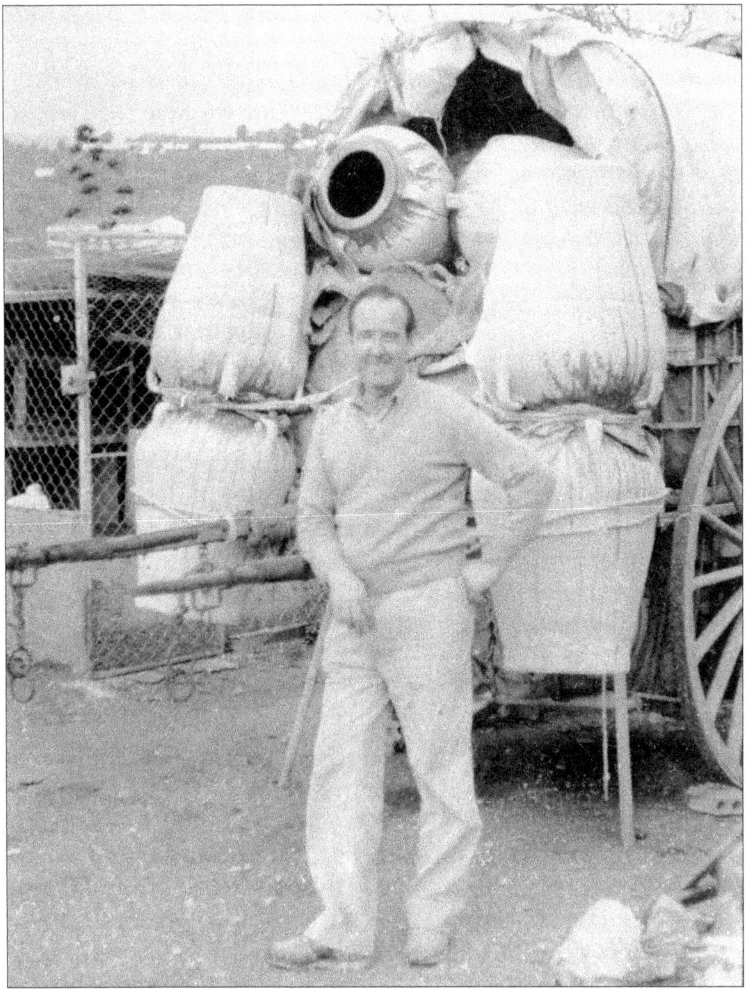

Plate 3.1 ***Pepe Jornet with a packed cart.*** To show his children how he had traveled as a muleteer, Jornet had packed the cart as he had done when peddling. Photo by Jornet family, 1990.

In addition to Vicen Sales, Pepe Jornet, my other potter-participant with much peddling experience, furnished detailed accounts of how peddlers related to consumers. He had peddled extensively, from his early teens on, though he came from a potters' family and therefore did not peddle in the systematic way regular muleteers tended to. He had employed different marketing techniques and covered different regions. His accounts were often similar to Vicen's, yet sometimes he diverged from Vicen in interesting ways. For instance, he described the familiarity of a muleteer with his clients and the selling practice *a medias* between muleteers in the following anecdote.

As a young man, Pepe Jornet had been peddling in Les Garigues on his own – that is without being introduced by his father. In contrast, other muleteers, like Vicen's grandfather, had been peddling in Les Garigues their whole lives, as their fathers and grandfathers did before them. Once Pepe encountered Llop, a muleteer who had also been peddling in the same area his whole life, and offered him, as was usual in such situations, the opportunity to vend their merchandise a medias. Llop refused this and went on to sell very well because he was well acquainted with the people in that village and knew that they would prefer to buy from him instead of Pepe, the newcomer. Llop laughed at Pepe. When Pepe told this to his father, his father responded that were this to happen again, Pepe should sell the merchandise for 'the price of Miravet'. Though a muleteer would not ordinarily make this type of price adjustment, Pepe, who was also a potter, could. Pepe did this the next time and he sold well. Shortly thereafter he again ran into Llop. Llop immediately left the village.

This is the kind of knowledge the few potters who also had direct experience with peddling, recounted. Other potter participants had peddled very little. Some potters like Joan Avante Senr. peddled only in nearby villages. Others, like Santiago Benaiges, did so for a few years: he peddled a part of the pottery production of his father but stopped when he got married and preferred to stay home.

Potters' knowledge of consumers and consumption through their contact with muleteers

Potters knew about their consumers through the orders that muleteers placed. When a potter unpacked the kiln after his firing, his muleteer came to lend a hand. Both counted the pottery and assessed its value. The potter then needed to throw the contents for a new kiln and the muleteer would place a new order. In this case a potter and a muleteer transacted with each other on a continuous basis, but there were also potters and muleteers who did this in annual cycles. A muleteer placed an order in a certain month with his potter. If a potter was so unfortunate as to have no order when he started to work on a new kiln content, he would still know, from experience, what he was likely to sell. As most potters sold their pottery to muleteers, these orders were a very important means for potters to get to know the muleteers' consumers and to relate to consumers of their products.

Santiago Benaiges described Llop, knowing where, when and what this muleteer peddled by the orders his father received. In the summer, when he went to Les Garigues, he ordered cantirs, the type with the 'narrow bottom', and gerretes. When he went to La Pinada at summers' end and on winter trips, he ordered vessels for storing olive oil and conserving food, and bowls for the pig slaughter. He peddled these in the region of Lleida. Here, Llop would make long winter trips in contrast to the short summer trips in Les Garigues and La Pinada. Llop occasionally brought back food for Santiago's father when he made the trips to Les Garigues and La Pinada – Santiago remembered in particular small loafs of bread from les Garigues, but not from his trips to Lleida where everything was sold for cash. Llop also peddled in the region of Villafranca and Manresa, yet Santiago could not remember what types of pottery he vended there, nor if, in addition to peddling, Llop also sold to shops.

Similarly, Santiago could sketch the orders of his father's other long-term trade partner, Nyec. This muleteer was much older than Santiago's father, therefore Santiago had not known him as well as Llop and could not provide as much detail. Nyec peddled a large part of the year. He mostly went to Lleida, also in summer, but of course ordered different pottery in different seasons: in summer, cantirs, and in winter, vessels and bowls. While discussing this muleteer, Santiago specified the bowls with a narrow bottom and the half-circles design. Santiago vaguely remembered that Nyec sold to a shop, but he could not recall what kind of relationship Nyec had with this shop.

This information builds on that given by Vicen Sales. Naturally Vicen Sales could talk with more detail about consumers and consumption in Les Garigues, but Santiago could talk about consumers and consumption in La Pinada. None of the potters who themselves had been peddling had done so in La Pinada, but more potters had received orders from muleteers who did. These potters complemented and added to the information on consumers and consumption in La Pinada. For instance, Pepe Papaseit's father had been working for muleteers who peddled there.

Pepe Papaseit remembered that two of his father's muleteers, Tapit and Fatera, also peddled in La Pinada. Fatera shipped his merchandise to Azaila and peddled from August until October. He mainly ordered the smaller gerres, *solserons, petxins and the medium-sized bowls, mitjaner and sangants, and a few cossis calandrins; mostly ware used for the pig slaughter. People in that region must have stored olive oil as well, and hence, also needed large vessels. Pepe guessed that they must have bought those from other muleteers peddling in that region, or that Fatera had specific clients or went to specific villages where there was no demand for these large vessels. Tapit, the other long-term muleteer of Pepe's father, peddled in another part of La Pinada. Pepe did not remember him so well, but he did remember that Tapit ordered a great quantity of the solserons and mitjaners – of the mitjaners 90–100 at a time. Tapit, in contrast to Fatera, did not carry any cossis. Both muleteers, like everyone else, did not carry cantirs to La Pinada. Tapit tended to order before Fatera, in July. According to Pepe, one went to La Pinada after the grain harvest. To peddle vessels for the storing of olive oil during the months of August and September was exceptional: the olive harvest takes place later in the year and pottery was usually sold in the season that it was used. Pepe could not explain this exception.*

In this way potters knew their consumers through orders placed by muleteers. Thus they did not know consumers personally nor as in-depth as the potters who had peddled, but still they knew details through the timing of orders, the content of the order, and where their trade partners peddled the order. Though these orders were placed by a muleteer with a potter, they were not confined to matters between trade partners only, but were topics of general conversations between fellow villagers.

Potters' knowledge of consumers and consumption through their contact with fellow villagers

In Miravet, after breakfast, lunch, work, and in the evenings, many men retired to a bar to share coffee or libation and casual conversation. I followed suit. An acquaintance would often tell me what he was presently doing, providing information about the harvesting of fruit, for example, and perhaps information about his buyers. They habitually exchanged this kind of information. Potters talked about such things as the kiln they were firing, what types of pottery was in it, by whom it was ordered, and so on. The better I knew someone, the greater the chance that I knew what he was doing. According to the villagers, the people of Miravet lived a much more leisurely lifestyle in previous times, socializing mainly within the village. If not away peddling or working distant fields, almost all men visited local bars several times a day. In the centre of the Raval, the neighbourhood where the potters lived, there had been a daily social gathering after breakfast frequented by the potters and their neighbours. Likewise the casual visits to potter workshops were more frequent, especially in winter when homes were cold and kilns were being fired. Previously, much more than during the field study, villagers exchanged local business news.

Through these daily interactions, potters knew what went on in other workshops, including who was making what for whom, and they knew each other's marketing relationships. They knew each individual muleteer with respect to the pottery trade, that is, his turnover, its volume and content, and when and where he peddled. Also, they heard from muleteers how their products were marketed and consumed, and by whom. This knowledge also had a historical dimension: many participants could even recount this knowledge for the previous generations though naturally this knowledge gradually dimmed with each successive generation.[5]

The geography of consumption

Though individual potters differed in their knowledge of consumers and consumption of pottery, they also overlapped to a large extent. By interviewing the individual potters, this collective knowledge could be reconstructed. In this reconstruction I discern several intersecting dimensions by which pottery was differentiated and categorized, each dimension with its own characteristics. These are spatial and temporal dimensions, often connected by participants as they 'spatialized time' (Fabian 1983: esp. 15) by perceiving two practices simultaneously occurring at different places as occurring at different times, one place being more advanced or modern, and the other backward or primitive. But these dimensions are also semiotic, for instance urban versus rural consumption. During the period that muleteers peddled the pottery, not many changes in consumption took place, and hence this knowledge was predominantly a spatial knowledge or geography.[6]

Bowls

The bowl was labelled throughout Catalonia as *ribbell,*[7] while it was labelled with the Castillian word *bareño* by the Argonese consumers.[8] (As stated in the

Type of pottery	use	who	where	when
mandonguer	pig slaughter	rural	Baix Aragon, la Pinada, northern Aragon, south-west part of Lleida, Les Garigues, Catalan region west of Miravet	late autumn / winter (La Pinada: late summer)
mitjaner	pig slaughter	rural	bowl with half-circle design: Baix Aragon, La Pinada, northern Aragon, southern part of Lleida Catalan bowl: El Reino, Tarragona,	late autumn / winter (La Pinada: late summer)
sangant	pig slaughter/ washing vegetables, doing dishes and laundry	rural	Bowl with half-circle design: Baix Aragon, La Pinada, northern Aragon, southern part of Lleida Catalan bowl: El Reino, Tarragona	late autumn/ winter (La Pinada: late summer)
quatreun	washing vegetables, doing dishes and laundry	rural/ town	Bowl with half-circle design: Baix Aragon, La Pinada, northern Aragon, southern part of Lleida Catalan bowl: El Reino, Tarragona	late autumn / winter (La Pinada: late summer)
grela	serving food / rustic circumstances	rural	Around Miravet and around Tortosa	not specified
pitanca	eating/rustic circumstances	rural	Around Miravet and around Tortosa	not specified
sisperun	selling blood in butcheries	not specified	not specified	not specified

Table 3.1 ***Ribell or bowl – consumption.***

previous chapter, consumers used different labels than potters and muleteers. I present those labels used by consumers which the participants mentioned; if no labels used by consumers are presented, the participants did not know them or assumed that consumers used descriptions, for instance referring to size.)

The three largest ribells or bowls, the mandonguer, the mitjaner and the sangant, were used for pig slaughter and the subsequent preparation of sausages, something done in late autumn and early winter. The pig was tied on a bench and its throat was cut. Its pouring blood was caught in the sangant, a bowl whose name is a derivative from *sang*, blood. The blood was then mixed with pork, fat, onions, spices and other condiments, boiled and kneaded. This preparation of the fillings for the sausages, of the *mandongo*, was done in the other bowls. The participants acknowledged that consumers might refer to these bowls as 'of the mandongo', but doubted that consumers would call the largest one mandonguer: this term was only used among potters and muleteers. The mandonguer, the largest bowl, was especially used for making the filling that contains rice. In one batch one could use 10–15 kilos of rice, which accounted for the large size of the bowl.

In their descriptions, the people of Miravet considered the raising and subsequent slaughtering of a pig by a household as something backward, to be superseded by the buying of pork from a butcher.[9] It had been a common practice in Miravet, and people could identify the former location of a pigsty and could describe how a pig was slaughtered as they had witnessed it. But as they had become 'more developed', in an economic sense, pig raising and slaughtering had become a professional pursuit. The participants placed the difference between regions where pigs were slaughtered either at home and or at butcheries, on a temporal scale. As one explained, 'In Catalonia the butchery where one could buy meat daily arrived earlier than in Aragon.' He added that 'If here in Miravet every other house raised and slaughtered a pig, in Aragon every house did.' The raising and slaughtering at home was perceived as an aspect of the general backwardness of villagers and related to a rural–urban distinction as well: villagers raised and slaughtered pigs at home while town and city dwellers went to the butchery. Thus, these bowls indexed rural backwardness.

The use of the mandonguer was strictly limited to the preparation of sausage fillings and therefore limited in volume of production and in regions of marketing. Of course butchers had these bowls, but since villagers raised and slaughtered pigs, they sometimes had them. Because the mandonguer was an expensive piece, not every household could afford one, and households that owned them typically lent them out to others. Consequently, the three bowls used for the pig slaughter differed in the proportions in which they were produced: Pepe Papaseit gave as an example 20 mandonguers, 100 mitjaners and 150 sangants. The mandonguer was sold in the Aragonese regions where the muleteers of Miravet peddled, in Baix Aragon, la Pinada and northern Aragon, and in the south-west part of the province of Lleida, in Les Garigues and in the Catalan region west of Miravet. It was sold during the season of the pig slaughter, autum and winter, except for in La Pinada where it was sold in the last summer months, August and September.

The three bowls next in size to the mandonguer, that is the sangant, the quatreun, and occasionally the mitjaner, were also used for washing vegetables, doing dishes and doing the laundry at home or at the wash place. By their use, these bowls were linked to the rural life. In towns and cities people tended to have kitchen sinks and were less inclined to use such bowls. Vicen Sales linked this difference to the housing in towns and cities, as they were more expensive and had more modern comforts than those in villages.

The mitjaner, sangant and quatreun were sold in the Aragonese regions of Baix Aragon, La Pinada and northern Aragon, in the northern Valencian region,[10] the Catalan province of Tarragona, and the southern part of the Catalan province of Lleida. There were two types of bowls, the bowls with a design of half-circles, and the 'Catalan' bowl. The first type was sold in Aragon and in Lleida, the second in the coastal regions. This second type of bowl was the one which the potters of Miravet copied after the bowl made in Esparreguera,[11] a potter centre a bit west of Barcelona, and called, in Miravet, the 'Catalan bowl'. In Miravet this bowl was probably taken into production in the early 1930s. The potters of Esparreguera seemed to have been producing for the market in those coastal regions, but had gradually lost it to the potters of Miravet in a northerly direction. In the 1940s the muleteers from Miravet were selling the 'Catalan' bowl in northern Valencia and in the province of Tarragona roughly up to the Ebre. In the 1950s, as Vicen Sales remembered, the 'Catalan' bowl was sold in large quantities in the region of Reus, a region where the regular bowl had been sold, though in small quantities.[12]

The potters of Miravet took over the market from the potters of Esparreguera with a copy of a bowl made by those potters, not with the regular bowl made in Miravet. The consumers of this region continued to buy the same type of bowl, even though it was made in another place. This was explained by the participants' reference to 'custom': those consumers were accustomed to purchase this type of bowl and indeed kept on doing so, and hence it was a 'custom of the region'. As Pepe Jornet put it, these consumers 'preferred what they had seen their whole life', they wanted the pottery 'of their country'. The potters of Miravet never made the 'Catalan' bowl in the largest size, the size of the mandonguer, since in its market region people did not slaughter pigs at home. The participants perceived this as indicating that those consumers were more advanced than those in Aragon, something corroborated by the slightly higher price the 'Catalan' bowls fetched: those consumers were also wealthier.[13]

The *sisperun*, the smallest bowl with a flat rim, was one in which butcheries sold sheep blood. These butcheries displayed a row of these bowls sitting on the counter. One bought its content and then returned the sisperun to the butchery. This bowl was apparently not produced in large quantities and its marketing not linked to any particular region.

Of the small bowls with upstanding rims, the *grela* was used for serving food on the table, and the *pitança* for eating. The pitança could be used when eating at a table or while standing. In the latter case one held the pitanca in the left hand and had in the right a fork or spoon. The pitança was only used when working in the field, on barges, or living in farmsteads outside the village. Life in such farmsteads and barges was considered rustic in comparison to life in regular houses: while eating one sat on stools around a low table on which the food and the *porró*[14] were placed. There, it was considered suitable for use in these settings, but for use in a house one preferred white stoneware plates.

The grela and the pitança were not produced in large quantities and were only sold in rural regions, a wide area surrounding Miravet and downstream in the region around Tortosa. The label pitança was typical for Miravet and the surrounding region, but in others it was called *napeta*.

Type of pottery	use	who	where	when
gerrot	olive oil storage mostly in mills	rural	northern Aragon, Tarragona, Leida, el Reino, Penedes	late autumn/ winter
gerreta	olive oil storage / water storage	rural	La Pinada	late autumn/ winter
solsera gran/petit	olive oil storage/ water storage	rural	Baix Aragon, La Pinada, northern Aragon, El Reino, Tarragona, southern part of Lleida, Penedes	late autumn/ winter
tafol	conserving food (occasionally for olive oil storage in cities)	rural/ town	Baix Aragon, La Pinada, northern Aragon, El Reino, Tarragona, southern part of Lleida	not specified
solsero	conserving food	rural/ town	Baix Aragon, La Pinada, northern Aragon, El Reino, Tarragona, southern part of Lleida	not specified
petxin	conserving food	rural/ town	Baix Aragon, La Pinada, northern Aragon, El Reino, Tarragona, southern part of Lleida	not specified
saboner	conserving food/ carrying soap	rural/ town	Baix Aragon, La Pinada, northern Aragon, El Reino, Tarragona, southern part of Lleida	not specified

Table 3.2 ***Gerra or vessel – consumption.***

Vessels

In Miravet, the vessel was called *gerra*, a common Catalan word though in Lleida the word *tenalla* referred to the largest ones. In Aragon, the largest vessels were called by the Castillian word *tinaja*.

The largest gerres or vessels, the gerrots and both sizes of solseres, were commonly used for storing olive oil. Upon harvesting their olives, people brought the olives to the mill to be pressed and returned to collect the oil, or they bought oil at the mill. They stored the oil at their homes, in a pantry or in the attic: something one did in late autumn and in winter. These vessels were covered with a piece of cloth and closed with a wooden lid. This lid was made out of wood in order not to chip the vessel or the cover itself.

These vessels were not equally practical. The gerrot was very big, which made it hard to store, move and clean. Moreover, if it broke or if some impurity fell into the oil, many litres of oil were lost. Therefore, the gerrot was mostly used in mills, which tended to have a dozen or so of them, and the solseres in private houses, which tended to have one to three.

Since both the gerrot and the solseres occupied so much space, they were associated with villages, because village houses tended to be larger than in cities, where people tended to live in apartments. Therefore, the tafol was occasionally mentioned as a vessel used for storing olive oil in cities.

The largest vessel, the gerreta, as well as both sizes of solseres, were also used for storing water at home. People would fetch water every so many days, store it in these vessels, and use it as needed, a custom typical for dry regions and for summer. It was not necessary that a vessel was glazed for storing water: though harder to clean, an unglazed vessel was cheaper. For this reason, potters did produce gerreta without glaze, but only rarely. They never produced other sizes of vessels without glaze, as one never stored water in these other sized vessels. Participants typically mentioned the gerreta as used for storing water in contrast to the gerrot for storing oil, though some said that one would use whichever of the two were available whenever one needed water or oil to be stored. Similarly, a smaller vessel might well be used for storing water, though for water the usage was large so one would prefer a large vessel.

Both sizes of solseres were sold in the same regions as the mitjaner, sangant and quatreun, that is in the Aragonese regions of Baix Aragon, La Pinada and northern Aragon, in northern Valencia, the Catalan province of Tarragona, and the southern part of the Catalan province of Lleida, as well as the Catalan region of the Penedes. The gerrot was sold mainly in northern Aragon, in the provinces of Tarragona and Lleida, northern Valencia and the Penedes. The gerrot had always been produced in a significantly smaller quantity than the solsera. This quantity diminished even further, as Pepe Papaseit conjectured, from 1950 onwards.[15] The gerreta was mainly sold in one region, La Pinada, and the villages of Fuentes de Ebre and La Gelsa were mentioned as places where potters knew that muleteers shipped their merchandise. Such vessels were naturally sold in places where there was no running water until the 1950s, when running water was available.

The smaller vessels, the tafol, solseró, petxin and the saboner, were used for conserving food like pickled meat, loin of pork, lard, bell and chilli peppers, gherkins, mushrooms and olives, according to the product and the method of preparation, vinegar, olive oil or salt water. These vessels could withstand the vinegar and olive oil well, but the glaze chipped off after some years when used with salt water. The smallest gerra, the saboner, was often mentioned as used by women for carrying soap, *sabó,* to a wash-place or river when they went to do the laundry, and was also mentioned as being used for storing honey. Because these smaller vessels were used for conserving pig products, they were mostly sold in villages. Yet since they were also used for the conservation of other food, it was definitely not exclusively a rural product. In towns and cities people also conserved food in these vessels. The smaller gerres were sold in the same regions as the solseres, yet the emphasis on them seemed to have been in the region around Sarragosa. They were sold less in the coastal region and in northern Aragon, and not in the Penedes.

The vessels made in Miravet were sold in regions that were probably bounded by areas in which vessels made in other potter centres were sold. Since the muleteers went to specific regions, they did not know what was sold elsewhere. But, as the participants pointed out, three market regions overlapped, in the sense that vessels which were sold in there were also made in other potter centres. These centres were Calanda and the two neighbouring villages of Traiguera and la Galera, and Esparreguera.

Calanda was a potter centre in east Aragon. The participants knew that the potters in this village produced vessels similar to those made in Miravet, yet Calanda potters did not throw on a wheel but instead gave their vessels their final shape by beating them with a bat, a much more labour-intensive technique. They therefore had a hard time competing with the faster working potters of Miravet. Moreover, these potters were working with a clay that required them to make vessels with thicker walls, i.e. they were heavier and were unglazable by their texture and production process. Because these vessels were unglazed, they served well for storing water and conserving food in salt water. As such, these vessels could compete with those of Miravet. In the period under discussion, the potters of Calanda hardly had a significant production, but still they were noted. Pepe Jornet had been peddling in the region around Calanda, using Calanda itself as his base. He told me that people would come to the market square when hearing that the pottery peddler had arrived. But when these potential buyers saw his merchandise, they declined, as they needed vessels for conserving in salt water, and would wait for the peddler selling pottery from Calanda. These consumers knew of both types of vessels – those from Miravet and those from Calanda – and which was better for which use. They distinguished this pottery by its shape and material, though they did not know that this pottery was produced in different places.

La Galera is a potter village in the south of the province of Tarragona, Traiguera is a potter village in the north of Valencia. The potters in these villages used a clay and a technique that limited the size of their products. They made vessels for conserving food, but the largest had a capacity of about five litres. Moreover, they could not glaze their products. The participants knew that the vessels made in these two places were inferior to the petxins made by them, and had noticed hardly any competition.

According to the participants, the potters of Esparreguera produced vessels of a high quality. As the Miravet potters produced the same bowls as the Esparreguera potters – i.e. the 'Catalan' bowl – they knew about the market for Esparreguera bowls, but they knew little about the Esparreguera vessels. Yet the participants did know that they, the potters of Miravet, sold the large vessels (mainly solseres, in the Penedes, the region closest to Esparreguera) and not the smaller types: perhaps the potters of Esparreguera did not produce large vessels.

Unglazed pottery

The next type of pottery is the cántir, a Catalan label; in Aragon it was the Castillian word *cantaro*. The cántir was used for transporting water from a well, cistern, fountain or river to the home. The cántir could be carried on the head, underneath an arm or by donkeys in *sárríes*, saddles made out of wood and metal, or from palm leaves. The different sizes of cántir were used

Type of pottery	use	who	where	when
cantir	transporting water to home	rural	narrow bottom: south-east Lleida, Les Garigues, Catalonia west of Miravet wide bottom: Los Monegros, Baix Aragon	summer
canteret	carrying drinking water when working field	rural	south-east Lleida, Les Garigues, Catalonia west of Miravet Los Monegros, Baix Aragon, el Reino	summer
cossi de parat	doing laundry	rural	around Tortosa, Delta de Ebre	not specified
cossi calandrin	doing laundry	rural / town	Aragon, south-west Lleida, Les Gargigues, Catalonia west of Miravet	not specified

Table 3.3 ***Unglazed pottery – consumption.***

according to different personal limitations: those strong enough could carry the heavier cántirs. At home, the cántir was emptied into vessels or placed in a kind of bench. Such a bench had open tops in which three, four or five cantirs fitted. A cántir was easily broken, for instance, by toppling over while it was put aside, or by a donkey making an unexpected move. Sometimes more than one cántir broke at the same time. Cantirs were usually bought by the half or whole dozen and every household had at least six at a time, but usually more. The cántir was only used in places where there was no running water, hence in villages, and especially in dry regions, and was mainly used during the dry season of summer.

The cántir was sold in a few regions. It was not hard to make, as it was neither a large nor glazed piece, and therefore it was made by potters in many potter centres. Hence, although the muleteers of Miravet peddled in northern Valencia and in the region of Saragossa, they did not carry the cántir made in Miravet. Consumers in those regions preferred the cantirs made in la Galera and Traiguera, and in Fuentes de Ebre. The cántir made in these other places was more or less of the same quality as the one made in Miravet, though of different shape and clay. The participants explained why the consumers in those regions preferred the cántir made in those places as a matter of 'custom', just as they had explained the different preferences for the different types of bowl. They further explained the different preferences for cantirs with a wide or with a narrow bottom similarly, though occasionally someone ventured a functional explanation. (For instance, Josep Pedrola attempted to relate it to the difference in stability, yet his explanation was not consistent.) The cántir with the narrow bottom was preferred in Catalonia, the one with the wide bottom in Aragon. The cántir with the narrow bottom was sold in the south-east part of the province of Lleida, in les Garigues, and the part of Catalonia west of Miravet; the cántir with the wide bottom in Los Monegros, Baix Aragon. Curiously, in Miravet itself both types of cántir were used.

The *canteret* was used for carrying drinking water to the fields when working the land. The strainer prevented dirt and insects from getting into the water. Because the clay was unglazed, it was porous and sweated, and kept the water cool. The canteret was explicitly used in the fields: at home one preferred a smaller piece.

The canteret was sold in the same regions as the cántir, though in smaller quantities. In northern Valencia the cántir made in La Galera and Traiguera was sold, but the potters of these places did not make the canteret.

The participants called the tub *cossi*, while Aragonese consumers used the Castillian *cuenco*. The large cossis were used for doing laundry. The tub was filled up with water. A wooden board was placed on top of it, only partially covering the tub. After washing, the garments and the cleaned articles were placed on the board. The smaller tubs were used for doing the dishes. This pottery was used at home or at washing places. The different sizes also related to these tubs' use in places of different sizes: the large ones in villages where there was enough storage room and room to use them, the smaller ones in cities where they had to be placed in sinks.

The tubs were never sold in large quantities. There was not that much demand and they broke easily while being transported. The *cossi de parat* was mainly used downstream in the region of Tortosa and in the Delta de Ebre, and also in Miravet itself. In around 1950 the use of this piece became obsolete: the participants hardly remembered its production. Its use probably became obsolete due to the introduction of running water and sinks in those regions. The *cossi calandrin* was sold in Aragon, in the south-west part of the province of Lleida, in Les Garigues, and in the Catalan part west of Miravet. None of the participants could link the preferences for the different sizes to regional or other differences.[16]

Pottery on order

Usually muleteers carried the merchandise they expected to sell. However, in the regions nearby, to which they made frequent and short trips, Les Garigues and Baix Aragon, they also took customer orders, which were passed on from muleteers to potters.[17]

The participants mentioned several pieces which used to be ordered in this manner. The *bací*, the chamber pot, often nicknamed Don Pedro, was used by elderly and sick people. This piece was not a typical piece for Miravet, but was made in many potter centres, varying in shape and size. The *cántir aiguader* could be glazed on the outside and have a small triangular hole in the spout. This piece served as a measurement for wine: one filled it up and if the wine poured through the hole, the cántir was filled up to a certain amount. (Similar pieces existed made of tin. These tin ones were always used for measuring olive oil, as oil was more expensive than wine and one was more careful with it. A ceramic measurement could always break.) The potters of Miravet also made jugs in several sizes, from one to four litres, which had the same function: they all had a hole near the top. These jugs came with a cylindrical ceramic box covered with a disk with holes in it. The jugs were placed on these boxes, which served to catch the spilled wine. Potters also made funnels used for drawing off and measuring wine. These pieces were only ordered by wine-growers who sold their wine directly to consumers in the same village. The *escafallit* was a straight cylindrical bottle used to warm the bed by filling it up with hot water and sticking it directly in the bed. The muleteers also sold, as per order, flowerpots, simple conical ones which were then called *clavellet* (after the most popular flower the *clavell* or carnation), as well as eating and drinking troughs for rabbits, chickens and pigeons; all pottery that was easy to make and was also made in other potter centres.

There was some disagreement among the participants as to whether or not certain pieces were marketed by muleteers. Of those, some participants said that they were only consumed in Miravet. Others told me that they were marketed by muleteers in the nearby regions and usually per order. One of them was the *cántir merder*, a cántir three or four liters larger than the cántir gran, and which came in two forms: one with a normal straight upstanding spout or one with a spout which widened towards the top. This cántir was used for transporting human waste (*merda*) from home to the fields in order to manure these fields. Another of these pieces was the *gargola*, a piece similar to the canteret but smaller and used at home for drinking water.[18] The same piece was produced in Ginestar, the potter village across the river, though from a clay which sweated better and therefore made the gargola from Ginestar superior to the one made in Miravet. Moreover, pieces which served the same purpose were made in distant places, in Verdú and Agost, and even sold in Miravet. The last of such pieces included the colander and a rather large cruet for the use of olive oil in the kitchen. Unlike the pottery in general, these two pieces varied in size and shape.

Barter

Potters knew that muleteers marketed their pottery in various ways. Muleteers sold pottery or bartered it for food. The participants assumed that one of the reasons consumers were interested in barter was poverty. For instance, Vicen Sales explained why people would barter: 'There were regions where the peseta did not circulate much. For instance, in the region of Baix Aragon, that was a region of misery. People there, since they did not have that much money, bartered.'

Similarly, he described the post-war period as one in which the peseta was also scarce since 'it was a hard time with Spain half in ruins'. Even though in this period barter in several products was strictly forbidden, he said that such forbidden barter occurred on a large scale. As did all the participants, Vicen linked barter with misery and poverty. Barter as an indication of poverty coincided with other indications. For instance, from the same regions where barter was the common practice, boys came to work in Miravet as mozos. Or, if some potters went to peddle and barter in villages nearby, they typically did so in those villages high in the mountains, the poorest in the region.

The participants assumed that the other reason consumers used bartering was because these people participated in economical relationships not primarily articulated by money. These people were not short of economical means in general, but of money. They preferred to acquire goods by barter,

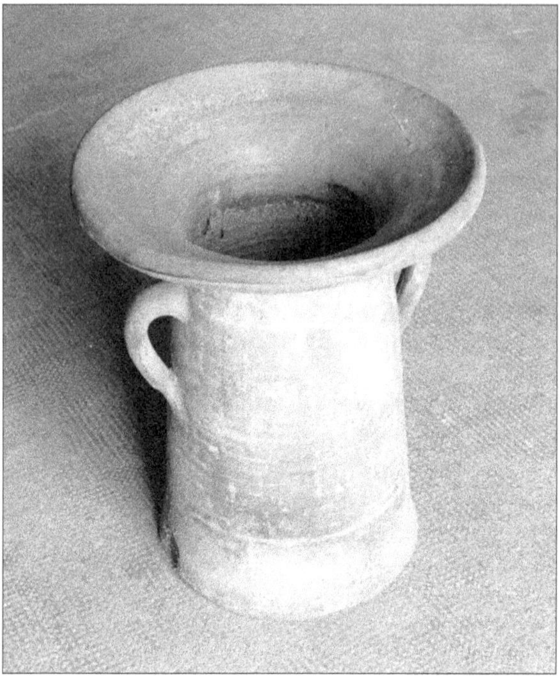

Plate 3.2 *Chamber pot.* Approximate height 40 cm. Author photo, 1992.

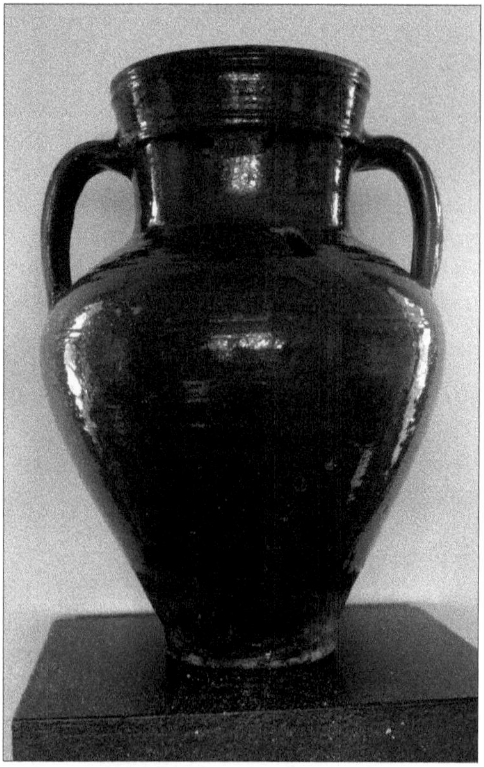

Plate 3.3 ***Cántir aiguader.*** A jug for measuring wine. Author photo, 1992.

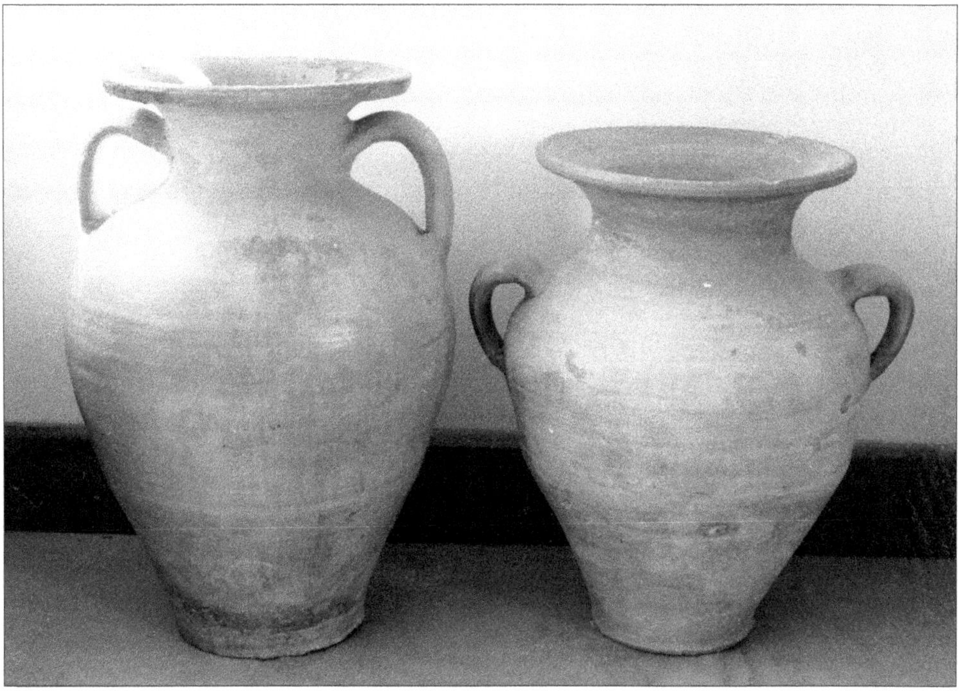

Plate 3.4 *Two cántirs de merde.* Author photo, 1992.

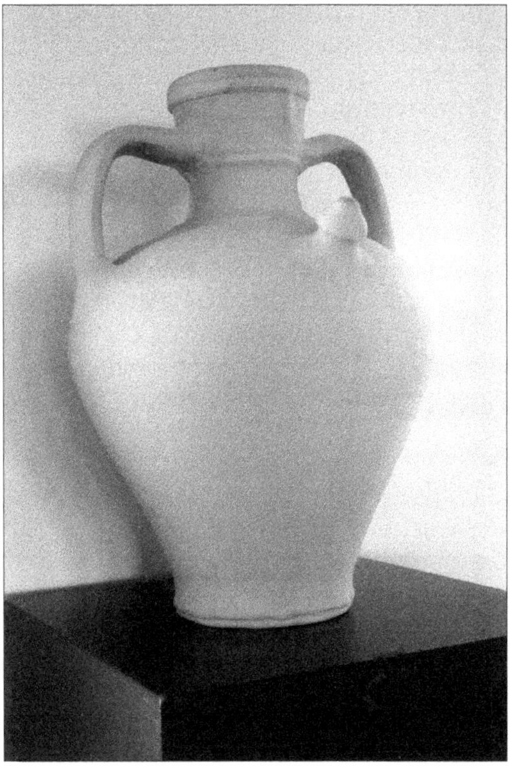

Plate 3.5 *Gargola.* Author photo, 1992.

though as Vicen Sales once emphasized, this might imply 'paying' a higher price for the pottery, in order to save their money. Potters had similar experiences as they, too, had preferred to barter firewood for their products since money was in short supply.

The participants associated Aragon with barter, especially La Pinada and Baix Aragon, the part of Catalonia west of Miravet, and some villages in the immediate neighbourhood of Miravet, the poorer villages high up in the mountains. They associated the post-war period as the most significant period in which bartering took place. The participants could not explicitly interrelate these spatial and temporal associations; for instance, none could mention a region in which up to a certain point in time people bartered. Yet they noted Aragon as the region of barter, which coincided with Aragon being the region in which pigs were raised and slaughtered at home. This was juxtaposed with Catalonia as the region of cash payment and as the region where pork was bought at the butchery.

As muleteers bartered pottery for food and skins, they returned to Miravet with these products. Details of barter were not that well remembered, yet every participant still knew that the Pinada was the region to barter mainly for cereal, and the other regions of Aragon for different products. Some potters peddled a bit for themselves. They fixed up the second-choice pottery, for which they would have fetched half the regular price from muleteers, and went on to barter this pottery on day trips. Participants still remembered well what could be acquired in which village (e.g. in Capsanes, potatoes, and in Paüls, onions). Some potters bartered their products for lead in Molá and Lloa, villages with lead mines. This knowledge of where one could barter for what proved important, since many people went to peddle not in order to trade and make a profit, but in order supply their households with products to complement their own agricultural production.

The contexts of other products

So far this geography of consumption describes how the different types of pottery of Miravet has been internally differentiated and categorized by its distribution, and its consumers and consumption. But as it was peddled and sold to consumers, this pottery was also categorized with other goods. Pottery was peddled and sold at markets together with hardware, baskets, textile, espadrilles and vegetables. Pottery was sold in shops together with hardware, baskets, ceramics, glass, flatware, fodder and articles described once to me as 'for the countryside'.[19]

Temporal dimensions

Although the knowledge of consumers and consumption was predominantly spatial – i.e. geographic – it also had temporal dimensions: a cyclical one and a linear one.

Cyclical temporal dimension

Potters talked about 'summer ware', typically cantirs, and 'winter ware', typically pottery used for pig slaughter and olive oil storage. They linked this pottery with the particular season in which it was marketed and consumed. (Save La Pinada, as noted earlier, where one peddled at the end of the grain harvest, when produce for barter was abundant.) As the pottery was peddled in different regions, the transportation of pottery and the duration of the trips varied according to the distances of these regions. By linking the pottery to particular seasons and regions, a year had a cycle with a certain rhythm.

At the end of the winter and in spring, the months from February unto May, little pottery was sold by muleteers. In the dry months of June, July and August, cantirs for transporting water were needed most. In these months cantirs formed the basic merchandise, with the addition of the smaller vessels and bowls. Muleteers could not make much profit on cantirs, therefore they sold these only in large quantities. They often loaded a cart with only cantirs and sold them on short trips of a week or less. They never shipped this merchandise for these trips by truck or train. This coincided with the relatively small region in which the cantirs made in Miravet were sold: Baix Aragon, los Monegros, the south-east part of the province of Lleida, les Garigues and the part of Catalonia west of Miravet. This was also a region where pottery was mostly bartered. When peddling was eventually in decline, the last of the muleteers were the ones who had cut firewood and bartered it for pottery, and they went to this region to barter.

August and September were good months in which to peddle in La Pinada, the plain south-east from Saragossa through which the Ebre flows. The grain harvest was bartered for pottery. This barter was lucrative for muleteers for much of the grain was sold in the same region or on the way back to Miravet. The main pottery bartered were small vessels used for food conserves, the gerretes used for water storage, and all sizes of bowls, including the ones used for pig slaughter. The marketing of the bowls used in the pig slaughter was exceptional in that the pig slaughter took place later, hence these bowls were marketed outside the season of their consumption. The participants explained that in the harvest months consumers were in the best position to barter and would keep those bowls until the months of the pig slaughter.

For the trips to La Pinada, the muleteers shipped their merchandise. Initially, they shipped the merchandise upstream to Mequinença by barge and transported it further by cart. Barges became outdated after the post-war years, especially after the construction of the dam near Flix in 1948 (Anon., *La Ribera d'Ebre* 1989: 232, 368) and the introduction of trucks. With trucks, muleteers could transport large quantities faster. Moreover, the shipment could be directed anywhere. They shipped their merchandise by train or truck to places like Azaila, Gelsa, Pina de Ebre, Fuentes de Ebre and El Burgo de Ebre, and then peddled from these places. Muleteers lost less time in getting to a site where they had a depot, and could continue peddling in the region around that site because of their larger stock. They could remain away from home longer, for a period ranging from a half month to a month and a half. Muleteers who also grew grain on their own fields had already harvested: the grain harvest occurred in Miravet earlier than in La Pinada.

The trip to La Pinada was popular among muleteers and with this trip the main season of peddling commenced: if potters had trouble finding muleteers to buy their products (and therefore had to store these products), by September all potters would have sold everything. In the autum and the first months of winter, up to Saint Anthony's feast day at the end of January, all sizes of vessels and bowls were sold. Both pig slaughter and olive harvest started in November at Saint Martin's day, but in October people could already see what they were going to need. This was the period when long trips were made. One former muleteer mentioned that he once had to travel for eight days from where he had been peddling in Navarra to get home. On average he was travelling 50 kilometres a day. Muleteers could be absent from home for several months, only returning for the *local festes* and for Christmas. They went especially to Baix Aragon and northern Aragon, to northern Valencia, to the Catalan province of Tarragona, the southern part of the Catalan province of Lleida, and the Catalan region of the Penedes. Yet not all muleteers necessarily made long trips: a few peddled in the region where they had been peddling during the summer, though now with other merchandise.

With the exception of the pottery used for the storage of olive oil, all pottery had stable markets. The muleteers knew from experience what they could sell annually in each particular place. On the other hand, the vessels used for the storage of olive oil had markets which changed from year to year dependent on the olive harvest. This harvest varied according to rainfall and frost. If there was a good harvest, people would store as much oil as possible since the following year no one knew what to expect. The muleteers needed to know which areas had or would have a good harvest. (Note that this was commercially sensitive knowledge and therefore not shared among muleteers and fellow villagers in general.) When they passed through a region, they could see if the trees were carrying many flowers or olives, and if they could not observe for themselves, they would enquire. The variation of the harvest often forced the muleteers to extend their trips in search of markets. For instance, some muleteers habitually peddled in the province of Huesca but sometimes had to make trips from this region into Navarra. In the Penedes they only sold vessels used for the storage of olive oil, which meant that they never peddled regularly in that region.

The one type of pottery that was not clearly linked by its use with a season was the cossi, the tub. The tub was sold in regions where one peddled in summer and early autum, but also, in lesser quantities, in regions where one peddled in other seasons. The tub was relatively breakable, therefore muleteers did not like to transport it over long distances by truck and train. Of the two kinds of tubs, the cossi de parat, which was marketed in the region of Tortosa and in the Delta de Ebre, had become obsolete by the middle of the twentieth century. The other and smaller tub, the calandrin, was sold in Aragon, in the south-west part of the province of Lleida, in Les Garigues, and in the Catalan part west of Miravet, regions where it was peddled in trips during autumn and winter.

Linear temporal dimension

The geographical knowledge of peddling had a commercial value, as people made their living by knowing what was sold where; hence participants could easily recall this knowledge. In contrast, knowledge of changes in peddling practices over time never had this importance. Still, participants referred to a few changes occurring in time.

One linear dimension was the practice of 'spatializing time', when participants described movements which, according to them, would take place everywhere, but which differentiated places according to stages in those developments. Examples of these are the previously described practices of using the vessel as a toilet bowl, for pig slaughter or as a kitchen sink.

Another change over time was the introduction of the 'Catalan bowl'. Around 1930, potters started to produce the 'Catalan' bowl for the market of northern Valencia, and they gradually expanded the market for this piece northwards in the coastal region.

Peddling was the oldest practice for marketing pottery the participants could recall. Based on their own experiences and stories they had heard, their collective memories went back roughly to the beginning of the twentieth century. Before that period, changes in peddling were not recalled but could be inferred due to changes in transportation. In addition to transport by mule and cart, barges had been used, downstream to Tortosa and upstream into Aragon.[20] When the nearby Mora la Nova train station was opened in 1891 (a date the participants knew because in 1991 the 100th anniversary of the opening of the station was celebrated), the volume of pottery peddled in Aragon must have greatly increased.[21] In shipping pottery by train to Aragon, muleteers added a large volume of pottery to that that they usually transported on their carts. This must also have had implications for the length of the trips they made in this area, and perhaps even expanded their distribution area. In the 1920s transportation by truck became more available, and since the 1930s there was one owner/driver of a truck who could be hired in Miravet itself.[22] This truck expanded the number of destinations that pottery could be shipped to and the volume of merchandise to be peddled from those places. It was this truck that Vicen Sales' grandfather hired to ship pottery to Les Cases d'Alcanar. Probably from then on, the cyclical time dimension of the geography of peddling as described above applied.

Another linear dimension was the shifting distribution of pottery. The market in Aragon, especially around Sarragosa, was the largest: it was one of the foci where potters started to market their products to wholesalers and shop-owners. Another focus for this shift was the southern part of Catalonia and northern Valencia, the part where weekly markets and shops were situated and where early on wholesalers took over the distribution of pottery. It was in these regions that muleteers first stopped peddling.

Finally, the history of peddling by muleteers was of course greatly influenced by the Civil War, which took place between 1936 to 1939. During the war peddling was impossible, but this exacerbated the need for pottery after the war. The post-war period was one of an extended black market in which the muleteers did very well.[23]

* * *

Potters related to the consumers of their products mostly by their knowledge of consumption and consumers, which they acquired from different channels. As this knowledge did not change significantly during the period of the muleteers, this knowledge was predominantly one of space: an actual space as well as a semiotic space in which consumers were located – a geography of consumption.

---

1 He was referring to lines made with one stroke of a brush with several – usually seven – bundles of hair, and with manganese oxide.

2 Though these handles were different in shape, these differences did not imply any other way of carrying or using these cántirs.

3 Another participant put it this way: 'I gave them something, a vessel or so, something they needed, "Here! A present!", because not everything is making profit or suffering losses. I was grateful that they let me stay at their place.'

4 These friendships at times developed over more than one generation. One participant told me that after he retired, many years after he had stopped peddling, he took a trip covering parts of the peddling routes. That trip was by car and with his wife. He visited people in whose houses he had stayed. He was received as if, as he said, he was a member of their family.

5 Saez (1992: 30–5) gives a list of 70 muleteers of Miravet and the regions in which they peddled, often with the places they shipped their merchandise. He assembled this list on the basis of what two former muleteers had told him. This underscores how, in general, people knew the muleteers and the regions they peddled.

6 Jackson and Thrift (1995) also use the term 'geography of consumption'. However, they use this term in reference to the study of spatial distribution of consumption, while I use it to refer to the potter's knowledge of consumption, which had a predominant spatial dimension.

7 Note that some of the people living in the eastern fringe of Aragon speak Catalan, though they used the Catalan labels. In the context of labels, in referring to Catalonia, it includes those Catalan speakers living in Aragon.

8 The bowl was called *lebrillo* in Andalusia.

9 This meaning of pig slaughter might be highly regionally specific (Luque Baena, personal communication), still, Harding writes on her Aragonese village 'and by the late 1970s even household pig-raising was disappearing' (Harding 1984: 136).

10 Note the participants referred to Valencia as El Reino.

11 This type of bowl was probably also made in Vendrell and Villafranca, yet the volume of the production of these places seems to have been insignificant.

12 The participants perceived their gradual taking over of the market of this bowl from the potters of Esparreguera in inverse proportion to the decline of production of those potters. Talking about Esparreguera, the potters of Miravet described this centre as more advanced than theirs. For instance, those potters had introduced the gas oil kiln long before the potters of Miravet did in 1981. But Esparreguera potters declined in number long before the potters of Miravet did, in the 1960s. Milio Treig pointed at the location of Esparreguera, close to Barcelona, and implied that since it is so near to this metropolis it must be more advanced than Miravet.

13 The previous paragraph referred to Vicen Sales stating that in the region of Reus the regular bowl had been sold, but in the 1950s the 'Catalan' one sold in large quantities. This might indicate a shift in preference by those consumers.

14 A porró is a glass wine jug with a long spout. One drinks from it by lifting the porró above one's head and letting the wine squirt into one's mouth. Since one does not touch the porró with the lips, more people can use the same porró on the same occasion.

15 He could not explain why from that time onwards, though it could have been the changing distribution of pottery practices. Moreover, gerrots were typically used in olive mills. As during the 1950s agriculture became modernized, and olive mills moved to the use of modern containers for oil storage.

16 Apart from pottery, the muleteers sold a few other products. They sold baskets, which served also as packing material, used when harvesting grain and olives. These baskets seemed to be sold mainly in Aragon and the province of Lleida. Also as packing material, they used *sárríes*, saddles for mules to transport things, and like the baskets the sárries were also made out of palm leaves. The sárríes were sold in the same regions as the baskets. They sold brooms in the regions nearby, like Baix Aragon. They also sold *canizos*, braided reed plates used in house building or for drying fruit: the latter only in neighbouring villages and only occasionally.

17 On their long trips, consumers also ordered pieces from a muleteer, who then had to determine if he had the ordered piece in his depot, and if he would pass the village another time. These orders did not link the consumer directly with the potter.

18 Vicen Sales pointed out that the gargola 'in many places', as he said, is called canteret, or 'canteret for drinking', and that the name 'gargola' is perhaps not exclusive to Miravet, but is still very typical.

19 Please note that as with most household utensils, women were the actual buyers of pottery: this is most clearly evidenced in the female pronouns and adjectives participants used in describing customers to me. However, both men and women used pottery.

20 As modern means of transportation developed, shipping by barge slowly declined and completely disappeared in 1948 when the Ebre was no longer navigable (Sorribes i Montserrat and Grau i Folch 1989: 232 and 368).

21 Ringrose remarks that due to this same train line Saragossa 'developed as an important centre for the milling, processing, and transhipment of grain and livestock destined for the coastal cities' (Ringrose 1996: 214). Hence, the opening of this station might have had an influence on the muleteers' trade greater than they themselves realized in 1990/92.

22 In the Anuario Comercial de España (anon. 1936/37: 4535) this owner/driver is already mentioned.

23 Ultimately, this knowledge with its spatial and temporal dimensions constituted one complex of people and objects, one 'spacetime' according to Nancy Munn's conception. She argues that 'sociocultural action systems (or the activities through which they become operative) do not simply go on in or through time and space, but that they form (structure) and constitute (create) the spacetime manifold in which they "go on"' (Munn 1983: 280). In another place, she elaborates that these actions can constitute a spacetime by their 'dynamics', that is by their rhythm, and by 'connecting persons and places' (Munn 1986: 9). Next to these connections, a 'subjective dimension' is also a 'factor' in the constitution of a spacetime, moreover one that is 'a condition of [these actions] continuing operation': as an example of that dimension of 'subjective "acts"' she cites 'remembering' (ibid.: 9–10). The third factor is 'derive[d] from the qualities or properties of certain entities involved in a given set of practices. These entities consist of material or nonmaterial media and products of agents' acts, and, as well, of the bodies of agents themselves' (ibid.: 10, see also Munn 1977 and 1990). The production, marketing and consumption of pottery constituted a sociocultural action system consisting of recurring and spatially diversified exchanges 'connecting' potters with others, complemented with different 'subjective acts', and occurring in a dialectic with 'qualities' of the 'medium' in which these exchanges took place, pottery. By producing and marketing pottery, potters 'structured' and 'created' a spacetime.

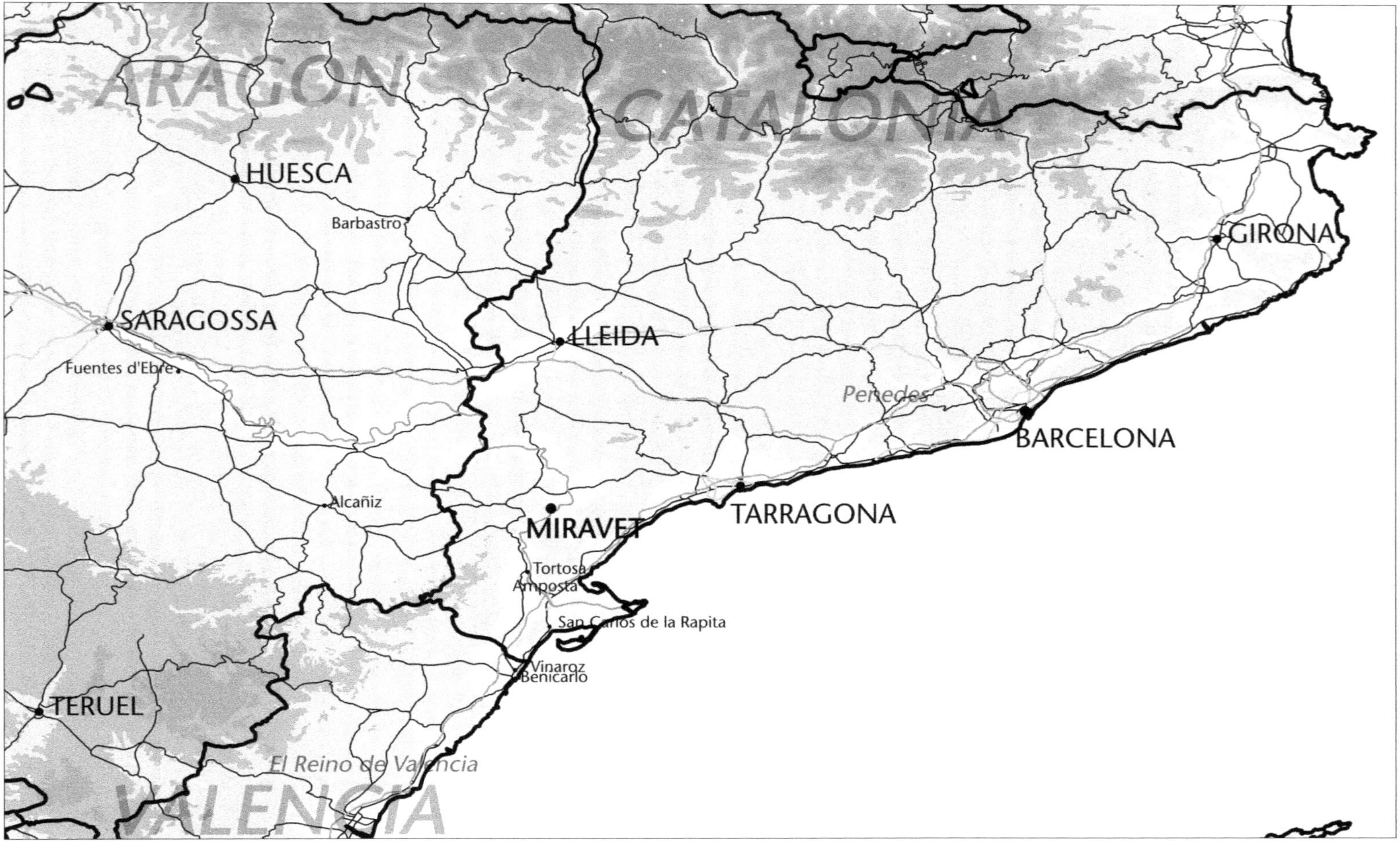

Map 4 ***Map to Chapter 4.***

Chapter 4

Selling to wholesalers and shop-owners

Participants acknowledged that not all potters marketed their products to muleteers. Some had heard, others vaguely remembered, that there had always been potters who themselves peddled or sold to wholesalers and shop-owners. They remembered the Petito workshop, the economically strongest potter workshop before the Civil War. Its main workforce consisted of three brothers and their sons, and together they produced, as well as marketed, their products. In addition to working in the pottery workshop, two of the brothers peddled and the profits were split among all three. Moreover, they had a client in Tortosa, downstream at the Ebre, and could ship merchandise by barge and thus bypass muleteers. Another case the participants remembered was that of Pepe Jornet's grandfather, who at the beginning of the twentieth century had unusual arrangements with muleteers. Then there was the Pedrola household which had sufficient land to make the keeping of a mule worthwhile and allowed Antoni Pedrola Senr. to peddle his pottery production. There had probably been more potters doing this. Potters and muleteers were not always two absolutely distinct categories. Still, these crossovers were small in number and did not change the general practice of the marketing of pottery to muleteers.

The oldest participants were in their 60s and their recollection went back to the 1940s. During that decade two potters began, independently from each other, to explore a new marketing practice. In contrast to previous occasional crossovers between potters and muleteers, these two potters set into motion a change of the general marketing practice which, beginning in the 1960s, resulted in the extinction of the muleteers, as all potters sold their products to wholesalers and shop-owners. The history of how potters came to sell their pottery to wholesalers and shop-owners is complex, as this shift relates changing social relations formed by this new type of distribution, with changing meanings of pottery thus distributed, and even changing physical shapes of the pottery.

* * *

In reconstructing the shift to selling to wholesalers and shop-owners I apply Sahlins' analysis of history in which concepts of 'structure' and 'event' are dialectically interrelated (1991), in order to overcome the opposition between a historiography based on the analysis of socio-cultural structural changes and one based on the narration of individual actions. According to Sahlins, 'structure' consists of the social structure and cultural values in their mutual integration; 'event' is the historical process coming forth from the socio-cultural structure and at the same time 'discontinu[ous]' (ibid.: 82) with that structure. An event derives its meaning from the socio-cultural structure in which it is occurring (as an 'incident' it is interpreted in terms of that structure), but it also changes that structure ['... what makes an act or incident an event is precisely its contrast to the going order of things...' (ibid.: 45)]. Therefore, the analysis of change divides into three interrelated approaches. First, one approaches an act as it is integrated in certain structures from which it derives its meaning – the continuity of an act with certain structures by which it becomes understandable. Second, one approaches that act as an event by its discontinuity with other structures – the departure of an act from previous practices. Third, one approaches an act as it occurs at the conjunction of these continuities and discontinuities (i.e. the analysis of the 'structure of conjuncture' (Sahlins 1991: 80) – the changes an act effects as it leads to new practices by its 'synergetic impulse' (ibid.: 76, see also Sahlins 1981: 67–72).[1]

The national and international context of the 1940s

In 1939 the fascist Franco won the Civil War; the same year witnessed the confrontation between democracy and fascism erupting on a global scale. Though Franco sympathized with Nazi Germany, he never supported it militarily. Therefore the Allies did not attack Spain; but when the Second World War ended Spain found itself the only dictatorship in democratic Western Europe and hence became ostracized by the Western countries: in 1946 these countries withdrew their ambassadors from Spain and excluded Spain from the United Nations (Carr 1982: 710–15).

On the one hand, forced by an international boycott, on the other hand driven by ideological goals, the Franco government embarked on a policy of creating a united Spain independent from the rest of the world. After the Civil War, Spain was in ruins. The new government intended to rebuild a socially and culturally homogenous Spain by controlling social and cultural life. Political debate was oppressed: Spaniards could not vote, express political ideas or form political parties (Carr 1982: 707–8). Labour activism and trade unions were prohibited. Instead, labourers were organized together with employers in vertical syndicates, centrally administered from Madrid (Martinez-Alier 1971: 42–3, Harrison 1985: 122, Ysàs 1994: 8).[2] Catholicism was granted a monopoly and the public practice of other religions was forbidden. All laws had to be in accordance with Catholic orthodoxy (Carr 1982: 721, Shubert 1990: 234–5). Regional cultural expressions were suppressed – like the Catalan dance, the *sardana* (Brandes 1990: 34) – or emasculated – like the Aragonese dance, the *jota*, which became a theme in slick musical revues. Castilian was considered the only Spanish language: Basque was supposed to be a language spoken only in a few hamlets, and Catalan, Valencian, Galician and Mallorcan were all considered dialects (Woollard 1989: 29). Though Catalan was in fact widely spoken, it was most severely repressed in printed form, as well as in other ways. For instance, Catalan speakers felt humiliated when speaking their native tongue, as Castilian-speakers described them as 'barking'. They were bullied by officials when they had to fill out bureaucratic forms in a language that was not their own. And they were traumatized when they started learning to read and write in a strange compulsory language.

The economic complement of this homogeneous Spain was an autarkic Spain: Spain should be self-sufficient – everything consumed in Spain had to be produced in Spain – and self-capitalizing, as the Franco government rejected foreign investments. In order to achieve this, Franco's government attempted to completely control the economy. Instead of relying on private initiative to develop industry, in 1941 the Franco government founded the Institute of National Industry, an institute that in due course turned out to be a state holding company for several industries. However, due to shortages of raw materials and energy, industrial growth was sluggish, and the proportion of the population employed in industry – as opposed to agriculture – even decreased in some years (Tamames 1986: 90–3).

Although the Franco government was above all interested in industrial development, it organized handicraftsmen in *gremios*, 'guilds': in 1946 the *gremio de alfareria de Miravet* was founded.[3] One potter showed me his identity paper as a member of this gremio, and the certificate of the test he had passed to become a member – both papers were decorated with the symbols of the Falange. He remembered that he had received a 'magazine' from Madrid, but had not kept it or remembered its content. No one could provide me with information about the function and impact of the gremio, so it must have been minimal. To the Franco government, handicrafts such as pottery were not of national or future importance, and their production was only superficially organized as an aspect of the general economic control.

During the Civil War, Franco created the National Wheat Service to develop production and to control the wheat market, especially by guaranteeing prices. Notwithstanding this service, and due to shortages of farm animals and fertilizers, the wheat production decreased, and only in 1948 was it on the level of the early 1930s (Tamames 1986: 55). Another government institution, the General Commission of Supplies and Transportation, controlled the trade in other food such as olive oil, rice and beans. Consumers had to buy bread as well as other basic food on rations (Harding 1984: 115, Herr 1971: 220).[4]

These policies of strict autarky and economic control did not improve the economic situation of Spain. The post-war period was one of hardship and, especially, of hunger. Fraser quotes a description by an participant: 'At the end of the war, in 1940 to be exact, there was a year of terrible hunger everywhere. Spain was destroyed. There were no crops, no bread, no oil, nothing. The people ate anything they could find: thistles and weeds of all different sorts. And that's how it went on, though not quite so bad, for the next seven or eight years.' (Fraser 1973: 77) Carr gives a similar description: 'Inadequate diet made tuberculosis a scourge; poverty turned prostitution into a respectable profession; in an attempt to discourage the horde of beggars the police fined those who gave alms' (Carr 1982: 742). Yet the destruction of Spain was not equally bad for everyone: for instance, since much pottery was broken (and hardly replaced since muleteers had not peddled) during the war, the demand for pottery grew. Potters and muleteers found that their wares were in high demand.

The Franco government, in attempting to control the production and commerce of essential food products, ironically caused a flourishing black market – the market of *estraperlo* as the illegal merchandise was then called. This black market encompassed the whole Spanish population and was crucial for its survival: Pitt-Rivers, in his study of an Andalucian town, writes that 'one can see ... that the agricultural produce flows in two parallel channels: the official and the illicit. But the latter is so essential, so much more extensive and so generally accepted, that it dwarfs the former ...' (Pitt-Rivers 1971: 20). Brenan, who travelled through the south of Spain during that period, called the black market Spain's 'sole thriving industry': 'Legitimate business is starved, throttled by forms and regulations and frowned on by the Falange and the authorities, whereas the black market moves on oiled wheels with the secret help and connivance of everyone' (Brenan 1950: 107).

The market of estraperlo had advantages for some, as it had disadvantages for others. People with no direct access to food, like wage earners and city dwellers, suffered from the black market; those who did have such access thrived on it. Peasants and farmers producing the controlled products did well because of the better, guaranteed and stable prices (Tamames 1986: 57). In addition, they could also take advantage of the black market, as could people trading and processing food. Harding describes how in the small Aragonese village which she studied everyone had a stake in the black market. Poor households profited by the women smuggling food, hidden in their clothes, from the village into town. Officially assigned transporters and millers of wheat and olives put in illegal extra hours. Every household that produced wheat kept a part of its production for the black market. The rich households profited by using their influence on officials to manipulate quotas set by the National Wheat Service to their advantage. Therefore, Harding argues that '... black market production, supplemented by the stabilizing effects of the National Wheat Service, resulted in the recovery of both peasants and [large landowners from the Civil War] in Ibieca by 1950' (Harding 1984: 117).[5]

The participants told me similar stories about the black market years in Miravet. One category of people stood out especially for profiting from these years: the muleteers. For generations muleteers had been bartering and trading in food: they knew where each kind of foods was available, how to negotiate for the best exchange rate or price, and where they could sell it. Of course, when engaging in illegal trade, they ran the risk of being caught by the Civil Guard. They could lose their merchandise and be fined, so not all muleteers dared to engage in it. But those who did, prospered, especially as there was also a large demand for pottery. Many muleteers made so much money in this period that they could buy land.[6] Potters had a good demand for their products, but muleteers did even better. As a result, during this decade the tension between potters and muleteers, as opposed market parties, intensified.

Josep Jornet and Josep Fabregat – exploring a new marketing practice

During the 1940s two potters began to sell their products directly to wholesalers and shop-owners.

During visits to his workshop, Pepe Jornet told me the marketing history of his household. His paternal great-grandfather came to Miravet and learned the craft of the potter. He must have sold, like every other potter, to muleteers, the same as his grandfather did, though he also deviated from this. Next to selling his products as usual to muleteers, his great-grandfather also sold them a medias, 'per halfs'. In agriculture, this was a common sharecrop arrangement: basically the owner of the land entrava, *'contributed', his land, the sharecropper his labour, and both shared the produce equally, hence the expression of 'per halfs'. Similarly, Jornet contributed the merchandise; the muleteer contributed the mule and cart. They calculated the value of the pottery according to 'the price of Miravet' and the muleteer paid Pepe Jornet's grandfather half of it. They assessed a price for the mule and cart (a price per day) calculated the value of the mule and cart over the period of the trip and Jornet paid his partner the half of that calculation. They shipped their merchandise by barge upstream and if Jornet did not have to work in his workshop, he would join the muleteer in peddling. Early in the morning the muleteer left Jornet with merchandise in one village while he went to a neighbouring village. During the day each sold his merchandise and at the end of the afternoon the muleteer returned to the village where he had left Jornet, picked him up and together they travelled to the next village. They split the money and the food they bartered between them. Pepe did not know if it made a difference in the split if his grandfather actually had peddled or not, he only knew that they split in equal parts. Jornet made this arrangement for one annual trip, a trip particularly suited for bartering. By making this annual trip, he had enough potatoes, grain, onions and so on for the whole year. Pepe added that this muleteer was the father of one of Pepe's grandfather's sons or daughters-in-law and a close friend, very de confiança.*

Pepe Jornet's grandfather maintained this arrangement at the beginning of the twentieth century. Pepe did not know, nor were there any other mentions of, other potters having such an arrangement earlier than the 1950s. It was an exceptional situation: the potter and muleteer held on to the 'price of Miravet' – i.e. the meaning of pottery as constituted between potters and muleteers, and simultaneously the cash value assessed by potters collectively as they marketed their products to muleteers, but the muleteer allowed the potter to share in the profits made by peddling. In general, potters were in the weaker position against muleteers, and therefore muleteers would not concede a share in their profits. But the Jornet household was exceptional in that, according to Pepe Jornet, his household could engage in such an arrangement since it was relatively prosperous for a potter's household. It owned houses and a bit of land, and therefore could store products and negotiate with muleteers.[7] Pepe Jornet knew very little more of this particular relationship, only that it had been with a close kin, but he did know more of similar relationships that his father, Josep, had.

Pepe's grandfather had three sons, one, Pepe's father Josep, followed him in the potter craft: another who became a muleteer, and a third who left early for Barcelona. His uncle peddled the pottery products of Pepe's grandfather and father, but when his father and uncle both had their own households and Pepe's father wanted to sell to Pepe's uncle under the same a medias arrangement, Pepe's uncle declined after two or three years: as a muleteer he preferred the usual arrangement. Pepe's father found other muleteers willing to engage with him in a medias arrangement, though none of these muleteers ever did this for long periods, only for a few years. The muleteers who entered into this arrangement with Jornet were also his friends. Jornet had probably looked among befriended muleteers, just like every other potter, if he needed a trade partner. However these muleteers also appeared to be the relatively economically weaker ones.[8]

From Pepe Jornet's descriptions of his father's a medias marketing relationships, it appears that the relatively economically stronger position of this potter household was complemented in these relationships by the relatively economically weaker positions of the muleteer trade partners. In a medias arrangements a muleteer had to share with a potter the profit he made on the peddling of the pottery. The only advantage this arrangement carried for a muleteer was that he did not need to pay upon picking up the merchandise. An economically weaker muleteer – a trade partner not preferred by potters – could, by this arrangement, still have a good supply of merchandise. Sometimes a muleteer en route could request that one of his potters send merchandise to wherever he was peddling. This allowed the muleteer to adapt himself swiftly to a current market. Pepe's uncle did have sufficient money and therefore preferred to trade with potters in the usual way; other muleteers who engaged in such arrangements did so only for short periods.

Pepe Jornet's grandfather and father were in a position to change the usual marketing relationships between potters and muleteers by engaging in a medias arrangements and consequently to also change the meanings of pottery constituted in these relationships. An aspect of the usual marketing relationships was the estimation of pottery as good versus flawed and its subsequent pricing as fetching the full versus the half price. Pepe Jornet was not quite sure – he had been very young when his father engaged in these arrangements – but he assumed that his father would mend the flawed pottery (unless it was of the poorest quality and beyond salvation) and charge the full price. As he reasoned, if his father could engage muleteers in a medias arrangements, why not also force them to accept this uniform price? Hence, pottery was no longer differentiated with regards to small cracks and holes.

Similarly, Pepe Jornet assumed that his father also changed another aspect of the usual marketing relationships between potters and muleteers, the arithmetic organization of pottery in 'pieces' or 'dozens'. For instance, Pepe Jornet assumed that his father would charge a muleteer more for 12 solseróns than for one tafol, though 12 solseróns constituted one 'piece' as one tafol did.

Although this potter changed important aspects in his trade relationships with muleteers, he still marketed his products to a member of his close knit village community, or if he also peddled, he did that within the network established by his muleteer partner. The next step his father took in the development of a new marketing practice was significant: the step outside these familiar social networks.

From his exceptional position of advantage with respect to muleteers, Pepe's father took the step to market his products to wholesalers and shop-owners. In addition to peddling, muleteers also sold their merchandise to shops, which often combined retail with wholesale. Pepe's father seemed to have joined in peddling with muleteers more often than Pepe's grandfather (though he never peddled on his own, at least not in the period of the 1930s and 1940s). On his trips with muleteers, he recognized the possibility of marketing his products to these shops on his own, that is, not in co-operation with a muleteer. He started to travel on public transportation to contact wholesalers and shop-owners and to sell his products directly to them, products that he shipped by hiring the local truck.

Next to these trade relationships within different social contexts, the direct marketing to wholesalers and shop-owners implied a new price calculation. When marketing to wholesalers and shop-owners, Pepe Jornet's father, as well as all other potters who eventually followed him in this marketing practice, did not distinguish between good or flawed pottery and between the resultant full and half price. He fixed up the flawed pieces and charged the regular price. He delivered the ordered pieces and hence the client could not, as a muleteer often did, select pieces or debate the quality of pieces. His new clients were interested in volume and not in the variation among pieces forming a lot. If one particular piece of pottery turned out to be of poor quality, usually returned to the store by the consumer, the new clients had the right to have that piece replaced by the potter. It now became the responsibility of the potter to guarantee a certain quality level of his products. Furthermore, in contrast to muleteers, these new clients lacked the knowledge of pottery to recognize fixed up pottery. Nor were these new clients even concerned with it: several potter participants told me how they fixed up pottery in front of these clients and even taught these clients how to do it for themselves.

When marketing to wholesalers and shop-owners, Pepe Jornet's father, and all who eventually followed him in this marketing practice, charged different prices for each different type of pottery, and not one price for a 'piece' or 'dozen', irrespective of the types of pottery which formed these units. Overall, this marketing directly to wholesalers and shop-owners was economically advantageous.

* * *

Josep Jornet, Pepe's father, had an economically exceptional position against muleteers: he could require an a medias arrangement which eventually brought him in direct contact with wholesalers and shop-owners. The other potter who took an exceptional position within the network of potters and muleteers was Josep Fabregat Senr.

Vicen Sales told of his uncle, Josep Fabregat Senr. and Fabregat's location within the kinship and economic network. This uncle was the brother-in-law of Vicen's father. Vicen's father, a potter, married the daughter of a muleteer. Upon marrying, Vicen's father moved in with his in-laws, the Fabregats. From then on, Vicen's father's products were peddled by Vicen's maternal grandfather and, when he became old enough, by his uncle Josep Fabregat Senr. But Josep Fabregat Senr. soon refused to become a muleteer since he had suffered humiliation when forced to ask for food, as previously described. As members of the same household, Vicen's father taught Josep Fabregat Senr. the craft of potter and they worked together in one workshop. They supplied the grandfather with pottery, even after both left the grandfather's household. As did many muleteers, the grandfather had a wholesaler/ shop-owner as client. Later, as did many muleteers, when the grandfather was old, he passed this client on to his son, Josep Fabregat Senr. Hence, Josep Fabregat Senr. acquired an exceptional position in the network of potters and muleteers as he, a potter, was introduced to a client by his father, a muleteer. Josep Fabregat Senr. learned how profitable it was to supply wholesalers and shop-owners directly and started to contact many more such clients by travelling on public transportation and shipping merchandise by truck.[9]

* * *

These two potters developed a new and economically advantageous marketing practice by selling to wholesalers and shop-owners as only muleteers had previously done, and by taking advantage of the new means of transportation. Since they bypassed muleteers with direct marketing, they entered into a new kind of personal relationship with their new clients.

Pepe Jornet remembered his father Josep's first wholesalers, one in Saragossa and one in Lleida. Both were wholesalers in chinaware and glassware who also carried vessels and bowls. (He remembered that the wholesaler in Lleida owned a machine to engrave glassware, hence combining wholesaling with production.) His father visited these wholesalers to acquire orders that were subsequently produced and delivered. Yet when his father had a lot of pottery in stock, he also visited shops in the larger villages and towns to offer them his products, places like Olesa de Montserrat, Esparreguera, Hospital de Llobregat, Fraga and villages in the Cinca valley. These shops mainly sold food but had in addition such articles as flatware and glass. The wholesalers were supplied with ware on a regular basis. These other shops were only supplied when his father had excess pottery in stock. The wholesalers bought solely from his father, while these shops would also be supplied by regular muleteers. After having received orders, his father rented a truck and delivered the products to the various clients. This direct delivery from Miravet was the only difference with the marketing by muleteers: muleteers delivered to these clients as they passed by with their carts.

Josep Fabregat's father had passed on to him a wholesale client, a man named Gasco located in Tortosa. Since Vicen Sales – my main source of information on Josep Fabregat – was very young at the time his grandfather contacted Gasco Senr., he did not know the details of how this had happened. He did know that his grandfather was peddling in the region of Tortosa and the delta of the Ebre, a region that had weekly markets. In that region Gasco Senr. was also peddling, selling articles made of esparto and articles like flatware, baskets and iron grills. Somehow Vicen Sales' grandfather and Gasco Senr. must have met and Gasco Senr. must have asked the grandfather if he could deliver a cartload of pottery. Vicen Sales's grandfather accepted this order and thus they entered into a commercial relationship that was carried over into the following generation, to Gasco Jr. and to Vicen Sales's uncle Josep Fabregat. Vicen Sales characterized this forging of a new relationship just like that of potters and muleteers who were both long-term trade partners and friends. From then on, Gasco Senr. placed orders with Vicen Sales's father just like muleteers of Miravet did. Though he did not come to lend a hand with unpacking the kiln, he did come to fetch the products, and Vicen Sales remembered very well that he did this with a large cart pulled by two horses, something not owned by the muleteers of Miravet. Often Vicen Sales's grandfather would accompany him with his cart and mule on his way back to Tortosa transporting a part of the order. At that time muleteers sold to shop-owners, but only merchandise muleteers happened to carry. In contrast, Gasco Senr. was the first outsider who placed orders with a potter in Miravet.

Gasco Senr. must have had a shop in Tortosa which developed into a wholesale house. He acquired shops in Tortosa and in surrounding villages as clients, and in so doing the muleteers of Miravet were forced out of that territory. He employed drivers with vans to deliver the merchandise. Vicen Sales saw the vans as an indicator that this region was generally economically more advanced. Gasco Jr. took over this business from his father and opened an annex in Alcañíz, in Baix Aragon, a region where apart from the vessels and bowls, many cántirs were sold.[10] *With a hired thrower, Fabregat could supply Gasco's large demand for cántirs for Alcañiz really well. During the early phase of the outlet, Gasco ordered by the truckload, half consisting of cántirs, half of vessels and bowls.*

Josep Fabregat's production greatly increased. He employed another full-time wage labourer, even though the demand in Alcañiz seemed to have declined somewhat: Josep Fabregat needed to find new outlets. Vicen Sales guessed that he might have obtained addresses of shops in Saragossa from his father, or he might have noticed through Gasco in Alcañiz that shops in Saragossa showed interest. Anyway, Josep Fabregat went to Saragossa, contacted shops and acquired orders from them. These shops were similar to Gasco and also developed into wholesale houses. In contrast to muleteers who marketed their products mainly in villages, he focused on cities and the capital towns of provinces.

These two potters explored a new marketing practice that was economically advantageous as prices were assessed differently than when trading with muleteers. It also entailed a new kind of personal relationship – a new social structure with new motivations. Previously, potters had been marketing to fellow villagers and their marketing relationships were interwoven with kin- and friendship. Now they began to market to outsiders with whom they did not have any bond other than their trade. These relationships were essentially different from the ones with the muleteers.

Every order and delivery constituted a separate interaction with the risk of no possible future interaction. A potter contacted a potential client to acquire orders, usually for the whole season. The next season he had to contact this client again to enquire if another order would be placed. Usually after a few orders, a client would send an order by mail. Still, a potter could never be sure of another order from the same client. For instance, when Pepe Jornet described how his father went to see if clients were interested in a new order, he recounted a recurring detail: that his father entered their shops and offices holding a newspaper underneath his arm. He explained this behaviour as his father's timidity. By having a newspaper underneath his arm, his father could pretend that he just happened to be in the neighbourhood and wanted to visit his client as an acquaintance. When talking about how things were going, he would interject casually if his client might need any merchandise. He thus hoped that though the client had not intended to place an order with him, the client would still do so since he happened to visit him.[11] Pepe's father – as well as other potters who followed him in this marketing practice – was uncomfortable: though he knew this client, he could never be confident in the continuation of their marketing relationship.

Delivery and payment constituted a single and separate transaction. A client was supposed to pay upon delivery. Stories of clients who did not do so were very rare: during my fieldwork I heard of only two. The participants who told these, remembered them vividly, so indignant had they been. In both cases, potters quickly stopped serving these clients. Within de confiança relationships, potters and muleteers had blurred the strict rhythm of delivery and payment as potters could rely on future orders and muleteers would pay in advance or loan money. This was unthinkable in these new marketing relationships because of their different social nature.

With these new clients, a potter had but one single contact when receiving and delivering an order, and often even this limited contact was not face-to-face. When a potter had supplied a client several times, this client might send a subsequent order by mail. When delivering, a potter had to take care of the unloading of the truck, but did not need to do it himself: he could send a worker.[12]

While these two potters developed a new kind of marketing relationship, other potters resisted forming those relationships. Potters had established relationships with muleteers, with fellow villagers, relationships strengthened by ties of kin- and friendship, by other exchanges besides pottery, and by customs such as jointly unpacking the kiln and the accompanying meal. The relationships which were strengthened by these latter ties, were de confiança, and had an internal dynamic by which potters and muleteers took care to be good trade partners for each other and by which potters became reliable producers and muleteers reliable suppliers, and in which muleteers could make advance payments and blur the rhythm of order and delivery. These relationships were preferred as aspects of the sociability with fellow villagers: potters at that time did not like to *sortír*, 'leave' the village.

Pepe Papaseit told me how he, while still a teenager, in 1948 or 1949 made a trip to the Penedés to sell pottery to shops. The Papaseits had distant kin living in Sant Sadurní d'Anoia and Pepe stayed at their house. On a bicycle borrowed from a cousin, he visited villages and towns in the region around Sant Sadurni d'Anoia, places like Piera, Gélida, Sant Quintín de Mediona, Igualada, Vallbona and Martorell. In this region muleteers peddled vessels for the storage of olive oil, the gerrots, solseres and tafols, and when Pepe Papaseit enquired at shops in these places, these vessels of Miravet were already known. At three or four shops Pepe Papaseit acquired enough orders to fill up a truck. By truck, he brought the products to Villafranca de Penedés and there with his uncle, who had a mule and cart, delivered it to the shops.

Though young Pepe Papaseit's trip to the Penedés had been successful, it was never repeated and remained a single occasion. The following year he had to go into the army and his father did not wish to make such a trip. As Pepe explained, potters of his father's generation did not have the *mentalitat*. 'mentality', to *sortir* the village, even if it could be, at least in the short-term, economically profitable.

This unwillingness to engage in marketing relationships with outsiders is highlighted in the following anecdote.

A wholesaler/shop-owner in Saragossa had been supplied by a muleteer who retired from the profession. Therefore, the wholesaler had sent this former muleteer a letter asking him if there would be potters interested in serving him. Santiago Benaiges remembered that his father Josep was asked to do so. His father refused since it would have meant stopping work for his muleteer Llop. Though he perhaps could make a bit more money serving this wholesaler, he regarded his relationship to Llop (Santiago Benaiges emphasized the closeness of this relationship by calling Llop at this point 'my uncle') as more important. Santiago Benaiges concluded by saying that such was the mentality of the potters at that time: more potters refused this client. He did eventually find a potter willing to supply him; one, as Santiago Benaiges put it, more interested in making more money.

Not only did these new relationships between potters and their clients differ from the old, potters also formed other relationships when travelling than muleteers had. Muleteers had formed networks of friends at whose houses they could stay: potters never did this when acquiring or delivering orders. When they delivered the orders, they tried to make efficient use of the truck by serving a few clients on a single trip and to keep the trip as short as possible. For instance, Vicen Sales Jr. told me that they would leave Miravet in the evening in order to arrive early next morning in Saragossa, unload the truck at three warehouses, and be back again in Miravet late that night. If they had to be away overnight, they would stay in the truck and not at someone's house. The new marketing relationships were not only discontinuous with the previous ones formed between potters and muleteers, but also with the ones muleteers had formed with people outside Miravet.

Two potters, Josep Jornet and Josep Fabregat, explored a more advantageous marketing practice by directly selling to wholesalers and shop-owners. Applying Sahlins' analysis of history, their explorations were continuous with the marketing of muleteers (i.e. the practice of muleteers selling to wholesalers and shop-owners), but discontinuous with the usual marketing of potters to muleteers. The result was a marketing relationship that contrasted with other aspects of the social-cultural structure formed by potters and muleteers, i.e. their personal relationships as intertwined with friendship and kinship in which trade occurred with a minimum of contestation between market parties. Therefore, the other potters of Miravet did not simply follow these two potters in their new marketing practice, but attempted to develop another way of supplying wholesalers and shop-owners by organizing themselves. Hence it is opportune to analyse how potters organized themselves for different productive and marketing tasks in various social structures, each with concomitant cultural meanings or values.

The organization of production and marketing

Like their fellow villagers (and many other rural and even urban Spaniards), potters organized their production and marketing of pottery primarily in units of *cases* or 'households'. One household owned the means of production, a workshop and a wheel for example, or rented for their own exclusive use a kiln for a firing, for instance. Members of one household pooled their labour: for example, adult males threw the pottery and, when needed, were assisted by the adult females. A household marketed its products and the proceeds constituted its collective income. In these respects, a household was autonomous, or at least that was the ideal. As Josep Pedrola often said to emphasize this autonomy '*cada caragol la seva casa*', 'every snail [has] his own house'.

The relationships between households, however, involved production and marketing practices, hence interplaying with the autonomy. Throughout the history of production and marketing – as recalled by the participants – potters had been organizing production and marketing on levels higher than that of the household, even in forms which partly limited the autonomy of the household as the primary unit. For instance, potters assessed a collective price when marketing to muleteers, the price of Miravet, which represented the lowest asking price. This limited their autonomy in marketing. When potters shifted to marketing to wholesalers and shop-owners, they again shifted this organization.

Potters assigned different meanings to the different forms of organization: the meaning of the autonomy of the household interplayed with different forms of higher level organization. I analyse these organizational structures and their concomitant cultural meanings using a distinction made by Sahlins, a distinction between two types of 'economic transactions': 'First, those "vice-versa" movements between two parties known familiarly as "reciprocity". The second, centralized movements: collection from members of a group, often under one hand, and redivision within this group' (Sahlins 1972: 188). He elaborates on this distinction by focusing on their 'social organization', 'pooling is socially a *within* relation, the collective action of a group. Reciprocity is a *between* relation, the action and reaction of two parties' (Sahlins 1972: 188, Sahlins' italics). Potters organized themselves reciprocally by forming dyadic relationships between households, and they organized themselves collectively by pooling the resources and products of their households, and thus formed a group. This distinction is used to analyse forms of co-operation between potter households and to explain the different duration of these forms as resulting from their different interplay with the autonomy of the household.

'Helping'– reciprocal organization between households

Potters organized themselves reciprocally by co-operating between households, a form of co-operation that was called *ajudar*. Generally, ajudar means 'to help', yet in the context of the pottery production, this word has several meanings. For instance, usually one considered the work in a workshop done by the wife of the potter as 'helping', though the same tasks performed by a *jornaler*, a 'hired worker', were called *treballar*, 'working'. This indicated that the woman was performing those tasks without receiving compensation for it, in contrast to the jornaler. Yet the jornaler might be *donar*, 'giving', clay to the thrower and taking the thrown pieces from the wheel. As a supportive task this might also be called ajudar. Ajudar was also used to refer to the mutual exchange of labour between households without the intervention of money – participants were explicit about this. A potter who exchanged labour with another was said to *ajudar amb* that other, to 'help with' that other, or potters were said to *ajudar* each other, to help each other.

As far back as the participants could recall, potters had been helping each other, but only with the production of pottery.[13] In due time this form of helping encompassed fewer stages of production, although by this diminution, the social nature of helping did not seem to have changed. Therefore, the mutual exchange of labour is analysed in a relatively a-historical perspective.

The potters mutually exchanged labour when they needed manpower from outside their household. It could be that they needed additional labour force for firing, or that they needed a specialist for packing of the kiln or throwing of the largest pieces. The workshop that received this labour had to return it. It was not necessary, nor possible in the case of expert labour, for them to return it in the same form. In principle, all the forms of labour were interchangeable, yet a potter would try to request someone to come to perform tasks so that person's labour was employed with the most benefit. For instance, if a potter knew that someone was a good assistant in packing a kiln, he would request that person to return labour in that form. Or, a potter whose wife always assisted him with throwing, might ask another workshop to return labour in the form of a strong son to assist him when he threw the big and heavy pieces, a job considered too strenuous for a woman.

The workshops did not diverge much from each other with respect to the skills of their members. Therefore, when seeking a partner for labour exchange, potters applied a similar strategy as when potters and muleteers sought trade partners: they sought among close family members, friends, or neighbours.[14] For instance, brothers-in-law often helped each other. They did not have to do so out of obligation, but it might have occurred that both needed help at the same time, and they may have begun helping each other because of this social tie. Potters who helped each other could be doing that because they personally, or their households in general, were good friends; or because one learned the craft in the workshop of the other; or because they lived close to each other; or that they used the same communal kiln. These reasons could overlap. For instance Ferran Segarra and the Pedrolas helped each other: Ferran Segarra had learned the craft in the Pedrola workshop (his in-laws' workshop) and they had their workshops close to each other.

The participants described the characteristic relationships they formed through helping as friendly: as Vicen Sales once said, 'One helped with friends, that's the best, because the thing went very well then'; or another potter said that the help was a 'favour granted to friends'. They described the ideal for these relationships as one of harmonious co-operation. So Vicen Sales told that, a week before a potter wanted to fire a kiln, he usually went to his 'helping' friend and said, 'Listen, for that day do not plan anything because we are going to fire.' The latter could then plan to be available that day. If however, it happened that he also had planned to fire around that time, he could say so: '"It's important for me to fire this next week because I have a client waiting." "Well, then you should fire, and afterwards we fire my wares."' Like long-term relationships between potters and muleteers, these relationships were based on friendship, but also created friendships, expressed and strengthened by a festive meal offered by the potter whose products were fired.

Potters who helped each other trusted each other completely. While assisting with a firing, a potter could easily damage the pieces in the kiln. Afterwards, when the kiln was opened and the damage was discovered, no one could prove that a particular person had done it. But all the participants confirmed that no one had ever damaged a kiln content on purpose. As the next anecdote will show, even the slightest calling into question of this trust was insulting.

Toni told me once how he and Tomàs stopped exchanging labour with a third potter Pasqual.[15] *Toni and Tomàs were assisting Pasqual with a firing. Tomàs had already done some turns and Toni was arriving to start his next one. He arrived earlier than expected and accidentally overheard Pasqual complaining to his wife about Tomàs. Pasqual's complaints used fairly strong expressions: Toni said, 'the sparks came off'. Tomàs who did not have much experience with firing, had apparently done something wrong and Pasqual had somehow found out about this. Although Pasqual had said nothing about this either to Tomàs or to Toni, and although Toni learned about Pasqual's anger indirectly, Toni was furious with Pasqual. Toni did not dispute Tomàs' fault in this, and told Pasqual that he felt as sorry for him as if it had happened with his own kiln. Yet Toni did not accept Pasqual cursing Tomàs, since he believed Tomàs had fired to the best of his ability.*[16]

The potters helped each other over varying lengths of time. Since workshops helped each other, that is, not individual potters, sons could get involved in their fathers' exchanges of labour and thus workshops could continue helping each other for more than a generation. Other potters had helped each other for the duration of their working lives but the most common duration seemed to be a number of years. Some potters had helped each other for shorter periods, for instance for the duration a son was fulfilling his military service, or even for single occasions, as in cases of illness or of times under pressure to get wares ready.

The potters exchanged labour between workshops in units of *jornales*, 'workdays'. For instance, one workday of throwing was equivalent to one workday of glazing. Moreover, one considered assisting firing to constitute one workday. Firing with branches took about 20 hours. One worked with cycles of three shifts and shifts were usually two hours. Not every one had an equal number of shifts. For instance, the firing could be completed after two people had worked three shifts and one person four. So a person worked about six to eight hours depending on the shifts he had. The firing with shreds of wood, a later kind of fuel, took around 30 hours. With this form of firing a single person worked about ten hours. Moreover, since both types of firing were done by shifts, one had to be more or less available during the whole period of the firing, and one could hardly do one's regular work during the breaks. So, one sometimes lost more time than one actually worked. Yet in these labour exchanges the differences due to the type of firing, different shifts, hours actually worked or those lost, made no difference. One time-assisting firing was equivalent to one workday of any other kind of work.[17] These units of exchange seemed to be non-negotiable. Although there seems to have been some attempts to negotiate the otherwise communally determined units of exchanging labour, I have come across only two concrete instances, both in which the one who attempted to negotiate was no longer dependent on outside labour. As was once explained, people trying to negotiate the units of exchange would soon find themselves ostracized, something that many could not afford.[18]

Potters often described the equivalence in their exchanges – based on units of workdays – as something that did not matter: one participant said that 'one does not look that much at the hours, nor at the times one has worked for another.' Or a potter boasted that he refused *passar el compte*, 'to pass the bill', to his 'helping' partners. This potter told his helpers to report how many workdays he owed them; he would not tell them how many they owed him. Yet every potter kept track of the times he had worked for another and the times that others had worked for him.[19] His wife in fact noted these down on a publicity calendar hanging in their workshop because her husband, as she said, 'likes to have things clear'. At the end of the year, she said that she counted the days they owed to their helpers and vice versa. At the end of the year, the potters usually told their 'helping' partners how many workdays they owed them, that is, the days the first had been working for the second, minus the days the second had been working for the first. It was the contrast between this practice and the ideal of unconditional helping that made this potter boast about not reminding the helpers of the workdays they owed him. But he, like all the other potters, varied in the degree to which he adhered to this practice or to this ideal. Potters who were not kin or good friends would definitely calculate how many workdays they owed each other. Potters who were kin and good friends would never keep track of the days they had worked for each other.[20]

Workshops were differently involved in the network of helping due to their different needs for and availability of labour force. For instance, Quim Canals and Bernat Grinó[21] had been helping each other for about 12 years. Their relationship ended with an argument over issues in labour. Quim Canals had two sons and a daughter working with him and was sometimes assisted by his wife. Bernat Grinó worked alone, only occasionally assisted by his wife. Bernat Grinó was very dependent on labour from outside. The Canals were less dependent, yet sometimes needed the help of Bernat Grinó, as he knew how to pack a kiln, which they did not. Every time Bernat Grinó asked the Canals to help in his workshop, someone was always available, yet not the other way around. Again, the participants wanted to describe these relationships as harmonious, but in the description of this case by the Canals brothers, they showed some annoyance at this unequal availability.[22]

By this practice of helping each other, potters formed a network of mutually exchanging workshops. Though potters exchanged mutually, they formed clusters within this network around the packing and firing of a kiln. To pack and fire a wood kiln one needed at least three people, but often one used more. Usually a few workshops exchanged labour among themselves for doing these tasks, and thus they formed small clusters of workshops that frequently worked together over long periods of time. These clusters could even develop their own versions of this exchange practice. In some clusters, wives of potters who offered the traditional festive meal at the end of the firing tried to outdo each other with their mastery of cooking. Other clusters did not value these meals so much, they found them a senseless waste of money, and kept them deliberately modest.[23]

Like potters, people working the land had also been exchanging labour, though they did this differently. Potters mainly exchanged labour on the occasion of packing and firing a kiln, a job which took at least two days, but sometimes also on occasions when the largest pieces were thrown, and on the loading of trucks. People working the land did it mainly for threshing, a job that took at most one day. Potters worked in cycles of two to three months, the time it took to throw pottery for one firing. People who worked the land did so in annual cycles. Therefore a potter could easily have to work for another potter one day every two weeks, while a peasant or farmer would work two or three days a year. Hence, potters were much more intensively involved in these labour exchanges than their fellow villagers (people with other economic activities hardly exchanged labour mutually).[24]

This form of co-operation had been successful because it was a relationship between households and though potters limited the autonomy of their household when exchanging labour, they did this in relationships in which they felt comfortable doing so: kin, friends and/or neighbours whom they could trust in these exchanges.[25] This form of co-operation mainly disappeared, not because of the meaning potters assigned to this co-operation, but – as we will come to see in following chapters – because of changes in production and marketing from the 1960s onwards.

The potters' co-operative – organization of households by pooling

In 1948 the potters of Miravet founded the *Cooperativa Alfarera*, the 'Potters' Co-operative', with two goals. The first goal was to strengthen their position in the old marketing practice of selling pottery to muleteers. Potters had contested certain aspects of their trade with muleteers and had strengthened their position against muleteers by collectively assessing prices, but they had lacked means to enforce these prices. By selling all pottery through one channel, the Co-operative, they hoped to force their price assessments upon muleteers.

The second goal was to strengthen marketing to wholesalers and shop-owners through collective marketing. Although all potters experienced a tension in their trade with muleteers, they resisted the shift to another marketing practice due to the individual relationships they had constructed with muleteers. As Pepe Papaseit explained, they did not have the *mentalitat*, 'mentality', for forging other marketing relationships. Though in their trade with muleteers potters had tried to strengthen their positions by collectively assessing prices, ultimately they had mitigated the opposition towards muleteers by individually intertwining their trade with other social ties. In the trade with outsiders, they were individuals, yet there was no possibility of intertwining their trade with other forms of relationship. Josep Jornet was one of the potters who had explored the new marketing practice and had learned how profitable it was. According to his son Pepe, he told other potters: 'Don't be asses. By organizing ourselves in a co-operative, we will sell to clients. We will go out and look for clients in other places and what they [i.e. the muleteers] earn, we will earn.' Collectively, potters could overcome their reservations about selling to outsiders. They could contact new clients and transport orders more efficiently. Furthermore, as a co-operative, potters would be considered for governmental financial support, though small.

The potters founded this co-operative against a background of other co-operatives and co-operative enterprises in Spain.[26] The co-operative movement in Spain started at the end of the nineteenth century in the context of the industrial revolution and the rise of the proletariat: the first were labouring class consumer co-operatives. There is no agreement among historians on the origins of this movement. Pérez Baro, the left wing historian of the 'Catalan co-operative movement', traces its origin back to Rochdale and emphasizes this movement's critique of capitalism. According to him, many of the first co-operatives were cover-ups for political activism, and in fact were linked to the international co-operative movement (Pérez Baro 1989). The principles of these co-operatives, as derived from Rochdale, were free membership, a democratic administration, the distribution of the costs and profits according to the use each member made of the co-operative services, a limited interest on credits, political and religious neutrality, and the education of the members (Pérez Baro 1989: 13). Del Arco, the historian of the 'Spanish agricultural co-operative movement' and an official in the Franco institution administrating these co-operatives, argues that *Acción Social Católica* was one of the main motors behind this movement, and that it had as its objective to 'morally and materially elevate' those working the land, attempting to achieve this in part by the founding of co-operatives.[27]

The first Spanish rural co-operatives were incidentally founded in the province of Tarragona and were principally concerned with the distribution of articles needed in agriculture, especially viticulture. The basic organizational units of these co-operatives were villages or towns, loosely organized in an overarching union. In the beginning, these co-operatives, often named or renamed *sindicat,* strove to maintain profitable prices for grapes, and in a later stage, to improve the processing of grapes into wine. Members were enthusiastic, and by taking mortgages on their land they could make large investments in equipment: the 'cathedrals of wine', as the edifices were called, were built. These were large buildings containing the equipment for producing wine, often built in architectural styles showing local pride in the co-operatives. For instance, the co-operative of El Pinell de Brai near Miravet, has one 'cathedral of wine' built by a student of Gaudí and decorated with tile tableaux celebrating the harvest and production of wine and olive oil. (Because of its architectural and aesthetic features, this building has now become a national landmark.) In addition to these syndicates, co-operative rural credit banks were also founded (Pérez Baro 1989).

The legal framework of the co-operatives developed simultaneously. In 1887 the first law addressing the foundation and organization of co-operatives was passed. In 1906 another law was passed especially for agricultural co-operatives. The 1906 law defined rural co-operatives, regulated their foundation and legal registration, the fiscal implications and the assistance provided by the national government to modernize agriculture. Assisted by this law, many co-operatives were founded, especially between 1919 and 1929. During the Civil War many of them were dissolved, but the Franco government encouraged the foundation of new ones, and – wary of possible socialist inclinations – controlled co-operatives with its 1942 law on co-operatives, a law based on the previous ones, yet organizing the co-operatives centrally in the Central Nacional Sindicalista. This was part of the Franco governments attempt to solve social and economic rural problems without addressing the issue of the very uneven land distribution (del Arco 1971: 25–7, see also Harding 1984: 59–80, Frigolé 1975, Whyte and Whyte 1988: 18).

In the county of the Ribera d'Ebre a number of agricultural co-operatives were founded. These co-operatives were primarily based on the processing of grapes and olives. Usually large landowners owned the means for making wine and olive oil, and the small landholders, sharecroppers and renters were dependent on them for the processing of their grapes and olives. By founding a co-operative, the smaller producers could become independent. In addition, they could more easily and jointly buy products such as fertilizers. In 1909 the oldest co-operative was founded in Flix, and in 1946 farmers and peasants in Miravet founded a co-operative (Anon., *La Ribera d'Ebre* 1989: 117–21).[28]

In 1948 when the potters founded the Co-operative, they did so against the background of a history of previous local potter co-operative organizations that involved first production, and subsequently marketing to distributors. At that time the participants had not yet been born or were very young, so what they knew was what they had heard from the previous generation. Their knowledge was therefore patchy. They knew that the potters had organized themselves in one or two, non-concurrent forms, which they referred to as *col.lectivitat,* 'collective', *societat,* 'society' and/or *sindicat,* 'syndicate'. They only agreed on the use of these terms as opposed to '*cooperativa*', but they disagreed whether there had been one or two separate col.lectivitats – probably one before and another one during the Civil War. They disagreed on who might have been the president of this col.lectivitat. They disagreed on how they had been set up: often one participant could not even explain the set up of this col.lectivitat consistently. Similarly, they disagreed on the reasons that they had collapsed. More important than attempting to reconstruct the col.lectivitat, is to look at what several participants did remember and to see how they perceived the possibilities and impossibilities of various organizational structures. Though this organization encompassed both production and consumption, the participants mostly remembered its organization of production.[29]

Vicen Sales and Santiago Benaiges both retold what they had heard of this col. lectivitat. All the potters worked 'collectively': each one worked in his own workshop but there was one person, the president, who told each potter what he had to throw each day. This president, chosen because he was the most 'educated' potter, organized the production according to demand. The potters were paid according to the number and the kind of pieces they threw. Since a cántir had a lesser value than, for instance, a vessel, the president took care to change his assignments so that each potter had to throw pieces of small and large value in equal proportions. Three or four potters packed kilns: each potter had to do this in shifts. The same happened with firing which was also done in shifts. Yet in this col.lectivitat, the different capacities of the different potters were considered. Santiago Benaiges gave the example that one potter who threw large vessels well, did this in the workshop of a potter who was less skilled at this. One who knew how to pack a kiln, a highly specialized task, would do this more frequently while another potter would throw for that potter, a division of labour similar to ajudar. The potters brought their products to a general storehouse where it was sold to muleteers. Some people took care of sales and administrative work: they were paid by the col.lectivitat according to the number of days worked.

Pepe Jornet also recalled what his father had told him about this col.lectivitat. His explanation diverged from the one given by Vicen Sales and Santiago Benaiges in that according to him, the potters also co-operated in the preparation of the clay. Some potters prepared the clay to be used by all. When a potter got paid for the number of pieces he had produced, the costs of the used clay was subtracted from the value of his products. Furthermore, a potter got paid half the price for the unfired products, and the other half when his products were fired. The result of paying for the preparation of clay and for the different stages in the production was that a potter would receive money more evenly distributed in time, not only when he had finished a kiln load. The col.lectivitat had therefore a pool from which to pay the potters.

These participants, as well as others, agreed that the division of tasks, with its accompanying hierarchy, was the cause of mutual jealousy. Pepe Jornet summarized the organizational basis: 'All worked together and earned according to how much they worked, how much quantity they produced.' But those earnings were difficult to assess. As Pepe Jornet continued: 'Then it collapsed because the potters, started saying "O! Jornet [i.e. Pepe Jornet's father] does not work and earns as much as anyone else" and so on and on.' Pepe Jornet claimed that his father had had an important leading function. 'The wheel in our workshop did not turn because my father did not work at the wheel as he was performing tasks for the col.lectivitat, and therefore the others had to pay him. The others were jealous because they said that my father did not work, but he did, he administered it.' With the foundation of the Co-operative in 1948, four potters refused to join, among them Pepe Papaseit's father. I asked why and he told me that he had asked his father the same question. His father had answered that at the time of that col.lectivitat, when Pepe Jr. was a young child, he could not buy shoes for him, while the sons of people who administered that col.lectivitat could. His father wanted to suggest that the latter earned more than those who worked, that is, those who produced pottery. For similar reasons, Joan Avante said his father was also one of those four potters who did not join.[30] This mutual jealousy and distrust probably eroded the control of the administrators and thus was one of the causes that brought this col.lectivitat down.

The founders of the col.lectivitat had anticipated this jealousy and therefore had regulated members to work in synchronized periods. Almost all of the participants remembered the trumpet signal that was used to indicate the working hours; often it was the only thing they remembered of it. One member was standing on top of the kiln of Sales, a very central spot in the Raval, and blew a trumpet when all the potters were supposed to start or stop work. Every participant added to this that it had been hilarious. For instance Pepe Papaseit said: 'And there were people who left a piece right in the middle of throwing it. They washed their hands and went off to have lunch.' When one throws a pot, one has to finish the throwing, otherwise the clay dries out and becomes unmanageable. When one is suddenly interrupted in the middle of throwing a pot, one may as well discard that pot. That is why the time indications were absurd.

When potters organized themselves by reciprocally exchanging labour, they did so with kin, friends and/or good neighbours. Though they limited their autonomy by entering into commitments with this practice, they did this with people they trusted. The more they trusted these other people, the less they were concerned with their commitments to return labour. In contrast, when collaborating in the form of pooling their labour, they were forced to exchange labour, or as Santiago Benaiges formulated it, they had 'passed a bill' to do so. They collaborated with people with whom they had different kinds of relationships and hence did not equally trust.[31] Therefore they saw to it that each would contribute equally to the pool of labour as measured by trumpet signals, which turned out to be counterproductive.

When in 1948 the potters founded the Co-operative, they did so less ambitiously. Potters co-operated only in the marketing of their products to advance their position against the muleteers and to engage more efficiently in the marketing to wholesalers and shop-owners. Plus they would take advantage of governmental support. Josep Jornet became the Co-operative's president. He had seen the opportunities of the new marketing practice, but he was also a man who possessed leadership abilities, serving as an assistant mayor of sorts. He also had more education than the others, for instance he studied music formally. According to his son Pepe, when the potters held a meeting to discuss the foundation, officials from Tarragona had been present. They had advised the potters to take Josep Jornet as their president. Unfortunately, he fell ill to tuberculosis a few years later and had to step down from the presidency. Antoni Pedrola Senr., Milio Treig Senr. and Jaume de la Paula succeeded him, dividing the responsibilities among themselves. They were then young potters with idealist fervour.[32]

During my fieldwork no one who had been a member could or wanted to provide information on this Co-operative. I related amicably to Antoni Pedrola Senr., who had been its president for a few years, but unfortunately he was ill at the time of the fieldwork and could not provide information. I also related well to Josep Fabregat, who had been very influential in the Co-operative, but he always avoided questions about it, as well as about his role in the history of the potters. Similarly, Milio Treig Senr., who had been the secretary for a few years, declined to talk extensively about the Co-operative. The people who did give me information were sons of potters who had been members. They began citing the short duration – only three years – and how some members ended up in debt, and all quarrelling with each other. No wonder the people who had been involved did not like to talk about it: the Co-operative had been a socially traumatic experience for the potters.

The potters tried to achieve their goals by selling all their products exclusively through the Co-operative.[33] They could completely retain their independence with regard to the production.[34] They delivered their products to a common warehouse. The warehouse clerk noted down the number and the kinds of pottery delivered by each potter and the potters were paid accordingly. Clients could come to the warehouse and buy the number and kinds of pottery they wanted. Some members visited wholesalers and shop-owners in order to acquire orders from them for the Co-operative. The Co-operative even had a permanent depot in Saragossa, the centre of their main market.[35] The wholesalers and shop-owners who had been clients of individual potters, became clients of the Co-operative. As mentioned, four potters chose to be outside the Co-operative. Still, as the son of one of them, Pepe Papaseit, pointed out, the Co-operative was advantageous for them. They could easily charge the same prices as the Co-operative, while they did not have the overhead costs the members had.

Though potters attempted to organize themselves in an egalitarian way, they had to distribute tasks and responsibilities. One of the problems the potters had with the Co-operative was that they had a hard time accepting the hierarchy implied by such a distribution, or as Santiago Benaiges put it, 'all wanted to be commanders and no one soldiers'. As has been noted, four potters refused to join the Co-operative and sons of two of them, Pepe Papaseit and Joan Avante, recalled that their fathers had done so because they distrusted the administrators. According to Pepe Jornet, people criticized his father, the first president, for making money as a president, just as they had done similarly when he had an important leading function in the col. lectivitat. Antoni Pedrola Senr.'s family explained how the backbiting by the villagers had frustrated and destroyed his idealism and his belief in the Co-operative. Several participants explained that at first the Co-operative had employed a warehouse clerk, but that after a while the members refused to pay someone for performing this task and instead they took turns doing it. The result was that no one was responsible for the bookkeeping of the warehouse and soon this was in utter disarray. Participants generally had the opinion that the Co-operative had been a good idea, yet the members lacked, in Joan Avante's words, an *esperit cooperitiva*, a 'co-operative spirit'. They wanted to be in charge of the sale of their products, did not want to give up any part of their autonomy and distrusted other people when they had to do so. As with the previous pooling of labour and resources, they had to collaborate with people with whom they had different kinds of relationships. They had to limit their autonomy because of the requirements of the organization, not as a function of the trusting relationship they had with someone.

The founders of the Co-operative took this into account when they organized it less ambitiously than its predecessor the col.lectivitat, by excluding production and limiting the co-operation to marketing. They attempted to market pottery collectively and to do it in an egalitarian way by accepting all products from all members. The Co-operative accepted pottery as specified by type and size. Variations in the actual measurements, for instance one solseró being larger than another, were ignored. The Co-operative also estimated all pottery equally. In the case where a piece contained small cracks and holes, the potter would fix it and have it accepted like pieces without flaws. These products were stored in the warehouse according to type and size, but irrespective of the potter who had made it, and irrespective of its quality. From this warehouse the pottery was distributed.

When marketed to wholesalers and shop-owners, the pottery was delivered to the clients' warehouses and shops. Clients' orders were specified to type and size: variations in the actual measurements of sizes were ignored. The Co-operative was responsible for a certain quality: if an order turned out to contain pieces of poor quality, a client returned these pieces and expected replacements for them. Moreover, clients lacked the knowledge to finely assess the quality or to recognize fixes. Hence, when selling to wholesalers and shop-owners, the Co-operative was selling pottery as standardized by type and size, and a quality for which it charged uniform prices.

When marketed to muleteers, the pottery was picked up at the warehouse by a muleteer. The attendant was supposed to supply the muleteer with those pieces of the requested type and size which first came to hand, regardless of the quality or the actual measurements of the individual pieces. However, since muleteers sold the pottery piece by piece, they were interested in the measurements and quality of individual pieces and they recognized nuances in quality, especially if pieces were fixed. Muleteers asked the warehouse attendant for pottery made by specific potters, of the best quality and of relatively large measurements. For instance, a solseró which was larger than the regular solseró could fetch a higher price when sold to a consumer. When selling pottery to muleteers, the Co-operative was selling individual pieces, though for uniform prices.

Hence the Co-operative sold standardized pieces to wholesalers and shop-owners, individual ones to muleteers, and charged everyone uniform prices. The result was that the Co-operative did not sell the relatively small and poor quality pottery to muleteers, especially the fixed-up ware, and of course could not sell this lesser pottery to wholesalers and shop-owners. Therefore, the lesser pottery was not sold, though the producers of it received the same price from the Co-operative. As almost all the participants unanimously mentioned, this steadily growing loss was the main cause of the eventual collapse of the Co-operative.[36]

Concomitantly, when selling through the Co-operative, potters lost the incentive to produce pottery of good quality. Previously when potters and muleteers traded in one-to-one marketing relationships, potters had to compete for preferred trade partners and the quality of the pottery was a factor in this competition. Now that these relationships had become collective, potters were not so concerned with the quality, and the Co-operative, because of its egalitarian structure, had to accept any kind of pottery from its members. Participants could still remember a particular potter who was more concerned with the quantity than the quality of his products. He finished these a bit carelessly, and once he delivered a batch of vessels that all had small cracks. Another potter once delivered a batch of bowls from which the glaze came off. Although the quality of their products was far below average and this pottery was never sold, these potters still received the uniform prices, which resulted in an enormous and continuously growing loss for the Co-operative.

This untenable situation eventually caused the collapse of the Co-operative. At one point, the Co-operative could no longer pay the potters for the products they had delivered. As the Co-operative went broke, many potters lost money. These losses, and the unequal distribution of these losses by the Co-operative members, resulted in a very traumatic experience for all.

The Co-operative failed because the members wanted to limit their households' autonomy as little as possible, and so did not collectively organize the production, only the marketing. This marketing straddled two different marketing practices, each entailing a certain relationship between producer and client, and hence certain meanings pottery acquired in this marketing relation. One practice was selling individual pieces to muleteers, hence pieces varying in quality and the price they would fetch when sold to consumers. The other practice was selling pieces standardized in quality to wholesalers and shop-owners.

Potters did not simply follow Josep Jornet and Josep Fabregat as these two explored the new marketing practice of selling to wholesalers and shop-owners, but tried to make this shift by organizing themselves on a level higher than the household. The potters had organized themselves on this level in two ways: firstly, by reciprocally exchanging labour; secondly, by pooling labour. Both forms implied different limitations on the autonomy of the household. The first was a co-operation between households: a household could choose another household and thus limited its autonomy but the choice was such that this co-operation was paralleled by other forms of relationship like kinship, friendship and/or neighbourliness. The second was a co-operation within an organization. Households could choose to co-operate within that organization, but if they did so, they were co-operating with households independently from other forms of relationship; a limitation of the household autonomy which was not really accepted. Taking this aversion into account, potters organized themselves in the Co-operative only to market their products to wholesalers and shop-owners. Yet this organization implied, unintentionally, ambiguous meanings of the pottery due to the combination of two very different marketing practices. Eventually this ambiguity contributed to the demise of the Co-operative.

Though short lived, the Co-operative served a function in the shift to the new marketing practice. The two potters who began exploring this practice developed the technique further and developed their network of clients because of their positions in the Co-operative: Josep Jornet had been the first president of the Co-operative and Josep Fabregat, though he had never been an official in the Co-operative, had been informally the most influential member. When Josep Jornet fell ill, he was replaced by Antonio Pedrola Senr., Milio Treig Senr. and Salvador de Paula. These three started, in the name of the Co-operative, to contact clients personally and hence became familiar with the marketing practice of developing a network of clients.

The national and international context of the 1950s

The social and economic situation of Spain worsened sharply towards the end of the 1940s. The Franco government failed to increase agricultural production sufficiently, while the black market pervaded everything, and overall industrial growth remained slow (Tamames 1986: 90). This caused the Franco regime to review its strict policy of autarky. The opportunities to renew contacts with foreign countries came with the Cold War. Franco posed as fervently anti-communist and thus increased his international standing. When the monopoly status of Catholicism was legalized in 1953, the Vatican was the first important power to renew diplomatic links with Spain (Carr 1982: 721, Shubert 1990: 234–5). In the same year, Spain signed a treaty with the United States which permitted the United Sates to build military bases in Spain in exchange for economic and military aid (Harding 1984: 118–20, Carr 1982: 715). The United Nations revised its exclusion of Spain and diplomatic ties were re-established with the international community (Carr 1982: 715). Though Franco still pursued autarky, Spain was no longer as isolated as it had been.

During the 1950s agriculture modernized significantly. In this decade, the Franco government implemented programmes to improve irrigation, to consolidate plots, to rent and sell tractors, and to introduce modern fertilizers and pesticides (Tamames 1986: 34 and 36–41, Harding 1984: 121–2, Harrison 1978: 159). The government finally got control of the wheat market after 1952, a year of good harvest, by ending the rationing of food and by substantially raising guaranteed prices (Harding 1984: 122, Herr 1971: 238).

During the 1950s, Spain began to shift from being an agricultural country to being an industrial one. In a 1948 report, the Bank of Bilbao stated that 'a year of good harvests is a year of prosperity for trade and industry' (cited in Carr 1982: 741 and Harrison 1978: 157). But between 1950 and 1957 the agricultural part of the national income declined from 40% to 25% – though it grew in absolute numbers – while industrial production expanded with a growth rate double what it had been during the Primo de Rivera government (Carr 1982: 742, Tamames 1986: 90–1).

According to Tamames, industrial production grew not so much because of governmental policy, though the government focused more on industrial development than agricultural, but because of several other factors. In the beginning of the 1950s Spain had a few really good harvests that increased the purchasing power of the population. Moreover, Spain received aid and credits from the United States and subsequently other countries. The basic industries grew, like steel and chemicals, and the first Spanish cars and trucks were produced (Herr 1971: 242). Another crucial factor was the beginning of tourism (Tamames 1986: 90).

The network of roads and railways had been severely damaged during the Civil War. In the 1950s it was restored and further developed. Between 1950 and 1962 the road network grew by 20,000 kilometres, and for the first time many small towns and villages were reachable by car and truck. During that time the fleet of vehicles grew from 9,000 buses, 90,000 cars and 60,000 trucks, often all made before the war, to 12,000 buses, 300,000 cars and 150,000 trucks (Herr 1971: 243). The trucks, cars and buses crowded the roads more and more, pushing out the mules and carts. Muleteers had to compete for the road with drivers of automobiles: an unpleasant and dangerous confrontation. 'War' one of them called it, a war which they were bound to lose. By the turn of this decade, muleteers became all but antiquated. Pepe Jornet told me that on his last trips with a mule and cart – in the early 1960s – every time a foreign car passed him, it would stop a little bit further on, and people would come out and take a picture of this remnant of the past. Muleteers often stayed overnight at lodgings and now found it progressively more difficult to find places that could accommodate a mule and cart.

Though production increased, the Spanish economy deteriorated. Industrial production was unbalanced, as consumer goods were overproduced, and needed investments were not available. By 1957, the country was almost bankrupt (Harrison 1978: 154, Carr 1982: 743, Herr 1971: 244).

Forced by economic crisis, Franco gave up his policy of autarky, and subsequently loosened control of the economy. In 1958 Spain joined several international economic organizations and opened up for foreign investments (Harrison 1978: 154, Carr 1982: 745, Herr 1971: 23). The Stabilization Plan of 1959 contained measures to cut government spending considerably: to increase taxes (regressively though) to devaluate and stabilize the peseta; to open the market for the import of certain products; and to loosen the control on industry. These measures strengthened the confidence of foreign investors, and their investments increased (Harrison 1978: 155, Herr 1971: 259, Carr 1982: 746, 749, Tamames 1986: 97).

The immediate result for the general Spanish population was economic hardship that triggered off a migration from village to city and a wave of emigration. However, instead of going to the Americas, as they had done previously, the Spanish emigrants went to northern European countries, mainly France, Belgium, the Netherlands, Switzerland and the UK, countries which experienced an economic boom and lacked labour force. Though they had previously emigrated permanently, now they went temporarily. In 1955 about 3,000 people emigrated; in 1957, 25,000; and in 1964 – the top year – 225,000 (Carr 1982: 750, Harrison 1978: 155, Nadal 1984: 207, Herr 1971: 260). As they had done in the beginning of this century, many Spaniards migrated to the city, especially Madrid and Barcelona, but also Seville, Saragossa, Valencia and Bilbao. For the years before 1962 no systematically collected statistics on this migration exist, but in 1962 out of a total population of 30.5 million about 350,000 Spaniards migrated, and in 1964 – also the top year for this migration – almost half a million Spaniards migrated to cities (Del Campo 1975: 67).

The city migrants were attracted by the opportunities offered in industry, but they were also pushed out from the countryside by the modernization of agriculture. They were mainly wage-earners and/or small landowners who lacked the means to modernize. Muleteers peddled because they did not have land, or they had little and needed to complement their agricultural production with peddling. Muleteers abandoned their profession and left for the city. The migration to the city raised rural wages, which stimulated agricultural modernization (Carr 1982: 752, Harrison 197: 160, Harding 1984: 138).

By 1961 the Spanish economy was recovering and started an unprecedented economic growth period – generally called 'the Spanish miracle' – a growth more rapid than that of any other country in Western Europe (Carr 1982: 746, Harrison 1978: 155–6, Herr 1971: 22, Shubert 1990: 208).

During the 1950s, Spain developed economically, but over the different regions this development was very uneven. The average income per capita in the richest region, Madrid, was 2.7 times that of the poorest, the inland region of Galicia (Carrion Garzaran 1976: 80–1). The centralist Franco government had never been concerned with equal distribution, but with national growth. The agricultural modernization policies were directed to regions declared to be of national interest, leaving other regions behind. The industry developed mainly in the traditional industrial centres of Barcelona and Basque country (Carr 1982: 753, Harding 1984: 121, Tamames 1986: 34–5).

The industry in the county of La Ribera d'Ebre was neglected by the government and hence underwent little development. The industry was the same as it had previously been: small and mainly based on the processing of agricultural products. The only significant growth took place in the chemical plant in Flix.[37] In La Ribera d'Ebre industrial modernization had still not begun (Anon., *La Ribera d'Ebre* 1989: 367).

The county of La Ribera d'Ebre was also left out by the projects set up to modernize agriculture: there were no large scale irrigation or land consolidation programmes. Here agriculture developed because of general market conditions, the co-operatives and agricultural loans. The kinds of cultivation did not change much (Anon., *La Ribera d'Ebre* 1989: 359–64, see also Hansen 1977: 147) and modern techniques were only introduced on a small scale. For instance, for the earliest year (1963) that such data were available, there were only nine tractors registered in Miravet.[38]

But these developments in the 1950s did affect the marketing of pottery. Modern transportation made travelling by mule and cart impossible, and the modernization of agriculture encouraged small landowners and landless peasants, typically people who had complemented their agricultural production with peddling pottery, to migrate to the city. The 1940s had encouraged potters to explore a new marketing practice, the 1950s forced potters to shift to this practice.

Marketing after the Co-operative

The Co-operative had wholesalers and shop-owners as clients. When the Co-operative was dissolved, Josep Fabregat and Josep Jornet returned to supplying the wholesalers and shop-owners they had been supplying before and some who had been acquired by the Co-operative, and Antonio Pedrola Senr.,

Milio Treig Senr. and Salvador de Paula – who were officials of the Co-operative when it was dissolved – kept the rest of the wholesalers and shop-owners acquired as clients by the Co-operative.[39]

In the period immediately following the demise of the Co-operative, Josep Fabregat focused more on marketing pottery than on producing it. He produced little, or even nothing for periods, but he bought pottery from other potters, like Josep Papaseit, Viola and Bernardo, his former worker who had started a workshop of his own. He bought their pottery even though they simultaneously sold to muleteers. He had been co-operating in one workshop with his brother-in-law Vicen Sales Senr., but had stopped doing this during the war.[40] When, in 1956 his nephew Vicen Sales returned from military service, Josep Fabregat and Vicen Sales Senr. again joined their work forces and means of production and focused on production. They thus had the largest turnover of pottery of the village.[41]

During the years of the Co-operative, Josep Jornet came down with tuberculosis, which hampered his work. For a few years his wife and son, Pepe, then in his early teens, peddled their products, alone or a medias with muleteers. When Josep Jornet recovered, he peddled, again alone or a medias with muleteers, but also sold directly to wholesalers and shop-owners in a medias co-operation with muleteers. In the latter co-operation, he used a muleteer's commercial knowledge and contacts too. Moreover, he returned to supplying the wholesalers and shop-owners he had had before the Co-operative. He also expanded his production by employing two potters from the nearby village of Benissanet, and by buying pottery from other potters.[42]

Antoni Pedrola Senr., Milio Treig Senr. and Salvador de Paula were helping (or exchanging labour with) each other as usual. Since the end of the Co-operative, they started to expand their joint efforts to also include the collective marketing of their products. They arranged their co-operation as a variant of the a medias type between potters and muleteers. From the money they made by selling their wares collectively, they first subtracted the costs of transportation, then each received the value of the pieces he had contributed, which was calculated according to the 'price of Miravet'. Finally, they split the remainder into three equal parts. This arrangement functioned well, as long as all three were producing more or less the same quantity. When after a number of years they diverged, and the one with less production was making proportionally more money on the collective marketing, they stopped marketing their products in this way. All three potters had been officials of the Co-operative, a kind of triumvirate. When the Co-operative dissolved and they started to market their products collectively, they sold mainly to wholesalers and shop-owners they had supplied through the Co-operative.

All these potters sold their products mainly in Aragon, especially the Aragonese capital Saragossa, where they competed with each other for a market share. Someone even called it a 'commercial war'. Some participants told anecdotes of this competition, always very much in favour or against one of the competing potters, making this information insufficient for reconstructing what had actually happened. Yet these anecdotes are still interesting since they show what the parameters of this competition were and how the cultural meanings of the relationships with muleteers became applied in this new marketing practice.

One of these was the price of the pottery. Previously potters assessed collective minimum prices, the 'price of Miravet'. When the Co-operative failed, they returned to assessing such prices. Some participants remembered only one meeting in which this was done after the Co-operative. This time these prices were printed and one was supposed to hang the list in one's workshop. But like the Co-operative before, the potters had no means to enforce these prices except that a potter who did sell cheaper would lose his good name.

However, most of the anecdotes were about competitors attacking each other by offering their products cheaper. Invariably, it was always the other party who was guilty of this underselling. One such anecdote was told by Pepe Jornet, and depicts his father as someone concerned for his reputation, while at the same time confronted by people trying to undersell him.

People in Miravet were gossiping that Pepe's father Josep was selling his products cheaper than the collective prices. He denied this and proved it to Miquel Borell.[43] *Before the Co-operative, his father Josep Jornet had the wholesaler Gistao in Saragossa as a client. After the Co-operative had dissolved, he returned to sell to him but for higher prices than the ones charged by the Co-operative. 'He agreed with the new prices', Pepe said, 'since Gistao did not know any other potter and did not know what the regular prices, "the price of Miravet", were.' He happened to be joined on a delivery to Gistao by Miquel Borell, who helped unload the truck. Pepe Jornet: 'When they were done, my father said "Borell, come on join us, now we are going to settle accounts and you will see that I am not selling cheaper".' Initially Miquel Borell did not want to since he considered it not his business. '... but Josep Jornet insisted. "I want you to come along because then you will see if Jornet sells too expensive or too cheap ... so that you can say, "Jornet is right, he sells for that price. I have seen the bills and it was like this." When Miquel Borell saw the prices Josep Jornet charged, and in addition that in contrast to all other potters, Jornet also charged his client the costs of transportation, he was flabbergasted.'*

Three years later, Josep Jornet again delivered to Gistao's and this time Gistao told him that 'his friend' had been there. This friend turned out to be Miquel Borell. According to Pepe Jornet, Gistao had said: 'He came to offer me merchandise. He did not tell me the prices, he said that he would be satisfied with any price, and about the costs of transportation, there would not be any costs of transportation.' Pepe Jornet explained that Miquel Borell expected Gistao to suggest a price lower than his father's, but he himself did not want to suggest a price to Gistao since the latter might tell Jornet his price, something he did not want.

The potters still considered it important to charge a price that was the same or higher than the collective price, the 'price of Miravet'. As members of the village community, the muleteers knew very well who was paying what, and potters established these collective minimum prices in order to strengthen their negotiation position against the muleteers. However, as this anecdote shows, the new clients did not know the prices others paid (they probably did not know other traders in pottery at all). When trading with these new clients, potters were negotiating individually and the collective price only made sense in case of competion for the same client, something they did in Saragossa but only in the beginning.

The potters established these collective prices as an aspect of the organization and labelling of the pottery as it occurred in their trade with muleteers. Yet in trade with the new clients, such organization and labelling did not take place; hence a potter could charge prices independently from this organization and labelling. Josep Fabregat devised an intermediate, flexible solution, one in which he acknowledged the collective price as an aspect of the customary organization and labelling of pottery, but at the same time was able to negotiate below this 'retail' price with his client independently from the other potters. He produced, especially for the Aragonese market, three new types of vessels which he called *tafolet*, *solseronet* and *petxinet*. These three were slightly smaller than the regular tafol, solseró and petxin, and their names were diminutives of these types. Participants critical of Josep Fabregat explained this introduction of new types as a marketing trick: since these types were not regular ones Fabregat did not need to sell them for the regular, collective prices. Instead he sold these for proportionally much lower prices than the 'price of Miravet'.

Next to the price of the pottery, another parameter in this competition was the respect one needed to show for another's commercial contacts. Potters and muleteers competed among each other for preferred commercial partners. The dynamic of their marketing relationships – relationships that were intertwined with kin and friendship – helped to ensure that these relationships became long-term. Everyone knew these relationships, realized that it was impossible to break them apart, and consequently respected them. However, the new marketing relationships consisted of separate interactions of order and delivery, which made them easier to break up. One now expected that fellow potters respected trade relationships by restraining themselves from interfering with others' trade partners. As in the above mentioned anecdote, Josep Jornet had never expected that Miquel Borell would visit a client of his, otherwise he would not have introduced him to that client.

I also heard anecdotes of potters contesting the respect that should be shown to each others' trade relationships. When his father Josep Jornet was convalescing from tuberculosis, he hired two workers from the last potter workshop in the neighbouring village of Benissanet. These potters had their own workshop but preferred to work for a wage at Jornet's. Jornet had expanded his production and therefore needed more outlets. These two potters offered to contact a former client of theirs in Saragossa for Jornet. At that time Josep Fabregat was selling to that client and Jornet refused this offer since he respected Fabregat's contact and did not want to interfere with his clients. These two potters replied that before this client was Fabregat's, it was theirs, hence they had more rights to sell to this client. They interpreted the case that although they were now working for Jornet, they still had rights to this client, that Fabregat should respect their contact with him. Eventually, Jornet did sell to him.

Potters and muleteers had long-term relationships because they valued each other as partners. Though the new marketing relationships lacked the kind of internal dynamic which had created the long-term relationships between potters and muleteers, potters and their new clients still valued each other. Though some clients shifted from one potter to another if the latter offered the merchandise for a better price, some were loyal to their suppliers. Of course, participants never told of clients they had captured from other potters, and they hardly spoke of clients which they had lost in this way, but they related, with pride, of clients who had refused to shift because they had appreciated trading with them. Thus, according to Pepe Jornet, his father's client in Saragossa to whom Miquel Borell went to offer wares refused this offer, saying: 'No, I'm sorry but I'm very satisfied with Jornet, he serves me well and when I need merchandise I will order it from him. It's already many years that we have been doing this with each other. I'm very satisfied and I do not want to change.' Or Pepe Jornet explained how eventually his father sold to the former client of their workers from Benissanet: 'This wholesaler in Saragossa bought from us because of the one potter from Benissanet who visited him. He would not have bought from us because he did not know us. If my father had contacted him, he would have told him that he was satisfied with Fabregat, but because he had such a friendship with those potters from Benissanet, he had been a client of their father, he accepted the offer.'

Hence, in their trade with the clients in Saragossa, potters continued to uphold their old practices by maintaining a collective price, the customary arithmetic organization and labelling of pottery, and the respect for each others' partners. But these new marketing relationships were discontinuous with the old ones by being relationships with clients who did not know each other (hence with whom potters negotiated prices individually unlike with muleteers who knew which prices other muleteers were paying) and by an internal dynamic that made each transaction singular.

* * *

As muleteers declined in numbers because of increasing modern transportation and migration to the city, potters who had not shifted to the new marketing practice encountered problems selling their products and had to peddle themselves

Pepe Papaseit recalled how his father gradually had problems selling his products as his muleteers stopped peddling: Tapit because his mule died, and Fatera because he was getting old. In the middle of the 1950s his father hit on hard times. He regularly had to produce pottery without having an order for it. He tried to come through these hard times by engaging a kin peasant in an a medias contract, someone who was not a muleteer but was willing to help his kin. Pepe himself made four or five short trips with this uncle to barter flawed and fixed up pottery for food which they later split. Then Pepe started to contact wholesalers and shop-owners.

Hence, Pepe Papaseit devised a method that was continuous with the old marketing practice in that one peddled with mule and cart, but discontinuous in that it was not the muleteer who peddled but he himself and a kin who owned a mule and cart.

In the early 1950s, because of his illness, Josep Jornet could not market his products a medias with a muleteer nor contact clients. As a makeshift solution, his son Pepe and his wife peddled by truck. By truck they shipped the merchandise to several villages where they deposited it. Then Pepe on his bike and his wife by bus would make the route and peddle in each village on a different weekday. At the end of the season, they hired a truck to pick up the left over merchandise. They did this for two or three years. Josep Jornet devised a method which was continuous with the old marketing practice in that the pottery was peddled, but discontinuous in that trucks and buses were used, and that his son and wife actually did the peddling.

Potters might combine this peddling with marketing to wholesalers and shop-owners. Then they would peddle the flawed and fixed up products that they could not sell to their clients. (In the same period the amount of flawed pottery declined since a new model kiln was introduced in 1948 and the technique of packing kilns was improved a little bit later).

Antoni Pedrola Senr. had for a long time peddled his own products but had also shifted to marketing to wholesalers and shop-owners. Still he continued peddling the flawed and fixed up wares to barter for food. His young son Josep joined him on one of these and hence told about these trips.

Eventually all peddling came to an end. One muleteer continued peddling pottery, but switched over to a van: he made only a few trips with it. A few muleteers trading in skins modernized themselves by also acquiring vans, but they focused exclusively on that trade and quit marketing pottery. The last muleteers had become relics of a past socio-economic order: they were the ones who cut firewood in the mountains surrounding Miravet and bartered it for pottery, which was again bartered for food. The last muleteer made his final trip in the mid 1960s.

* * *

As all potters shifted their marketing to wholesalers and shop-owners, a new structure of social relationships and cultural meanings emerged. Potters and their new clients now formed a network beyond the village and the new clients' orders became spatialized: different regions constituted larger or smaller markets. Consequently, potters with a large volume production had the large markets, and potters with a small production, the smaller ones. The latter still sold to muleteers or to the large ones, that is, to Josep Fabregat and Josep Jornet. Because the muleteers gradually stopped peddling during the 1950s, the smaller workshops were having difficulties with marketing their products. They too began selling to wholesalers and shop-owners, but they did this typically in regions with smaller markets, and typically regions closer to Miravet, the regions last supplied by muleteers. These potters had less economic stamina and therefore needed to deliver more frequently.

When Pepe Papaseit returned from military service, about 1957, the muleteers were disappearing. As he had done once previously, he again searched for clients outside the village. He went to the southern coastal part of Catalonia and El Reino, to Tortosa, Amposta, San Carlos de Rapitá, Vinaroz and Benicarlo, and acquired in each of these places two or three clients. After a short while, he could anticipate what these clients needed and he prepared a kiln content accordingly. When he had fired a kiln and the pottery was finished, he made a trip by bus visiting these clients and knowing what he had in stock, he could offer them his products. A few days later he made the same trip, this time by truck – a truck loaded with the ordered products – to deliver the orders to the subsequent clients. He made these trips every two to three months, whenever a kiln content was finished. If he did not sell all the products immediately after the firing, he kept them until September when he would sell them as the season of the pig slaughter and the olive harvest began.

With the exception of one client in Tortosa, all these clients were shop-owners and constituted a small market. These shops were the same kind as had been served before by Vicen Sales' grandfather when he was a muleteer who sold the leftovers of vending in markets (see Chapter 3). They were the small shops that mainly sold food and household utensils and things like baskets and espadrilles. One even combined this with a lodging house and a bodega. These clients constituted a small market in comparison to the market in Saragossa and Lleida which consisted of wholesalers and large shop-owners. Pepe Papaseit, who had a workshop in which only he and his father worked and produced comparatively little, could handle this market: for years he was the only potter who went to offer his products in this region. Except for one instance, he had no competition and could ask the price he wanted. The exception happened in one year when the olive harvest in Lleida was bad and one of the potters usually selling in that region visited Papaseit's clients and offered the pottery for a lower price.

Initially, potters perceived the new marketing relationships as they had previously perceived their relationships with muleteers: they interacted with the new clients as if they, the potters, were still a socio-economic category collectively confronting clients; and they mutually respected each other's marketing relationships as if they and the clients were still members of one community, that of the village. But at the same time, potters engaged in marketing relationships with clients who were spatially distributed, hence their clients did not constitute a socio-economic group – only a category – and therefore the marketing relationship became individualized.

Furthermore, these personal relationships also differed with respect to the kinds of ties involved. As members of a small village community, potters and muleteers had been intertwining their marketing relationships with other ties: their relationships were simultaneously on the socio-economic level as between parties with opposed interests, and on the individual level between friends and/or kin – a level on which the opposed interests were mitigated. Now these relationships were only based on trade, only between individuals with opposing interests.

Pottery sold to wholesalers and shop-owners

When marketing to muleteers, potters contested the judgment of pottery as good or flawed, and the categorization of pottery in 'pieces' or 'dozens'. These contestations, sharpened by the increased tension between potters and muleteers during the black market years, had motivated potters to explore and develop the new marketing practice employing modern means of transportation and distribution of goods. But as potters developed the marketing relationships, their products also acquired other new meanings. What were the meanings pottery acquired in the exchanges between potters and wholesalers/shop-owners?

Potters and muleteers knew and used labels for the different types of pottery. The new clients did not know these labels. Some of the new clients learned the labels, but since their interaction with pottery was less intensive, they usually never bothered to learn these labels well. These labels caused confusion for them. Potters and clients simplified their communication by numbering the different sizes, going from the largest type, number one, to the smallest, the highest number. The potters wrote or stamped the numbers in relief on the pottery: Josep Jornet was given a stamp to do so by a client who was eager to use numbers. Eventually the labels fell into disuse. For instance, when I asked Josep Pedrola – a potter who was in his early forties during my fieldwork and who had hardly known the period of marketing to muleteers – he was uncertain to which sizes a particular label referred.

A vessel was made in different sizes mainly indicated by the number of handles. When sold to muleteers, the actual measurements of vessels labelled as a certain size had varied, but when sold to wholesalers and shop-owners, the sizes and their measurements had to become standardized. The muleteers had paid the potters according to the type or size of the vessel regardless of its actual measurements, but when the muleteers sold them to consumers, they had charged prices according to the actual measurements. The muleteers usually sold the vessels one by one and had assessed their sale prices by negotiating with individual consumers. Therefore, the muleteers had been interested in the variations between vessels of one size and because they had not paid a potter more for it, a potter who regularly threw vessels larger than usual had been an attractive trade partner for them. A small shop-owner might still have been more interested in relatively larger vessels, but in general these new clients were not interested in vessels larger than their usual measurements since they sold these in lots and for standard prices. As Milio Treig Senr. told me, at one time during the Co-operative he had delivered an order of vessels which were made by different potters and therefore the measurements of several had varied in size. The client to whom he delivered this order had remarked that it was easy to place a handle on a vessel, but that he was not going to pay for it. He refused to pay a certain price for a vessel because a certain number of handles indicated its size, and instead wanted to pay for a vessel according to its actual measurements. Vessels were categorized in standard measurements, and indeed, their physical characteristics changed as they became standardized in their measurements.

Pottery was judged by potters and muleteers according to their own criteria; consumers, and now the new clients, judged it differently. Potters and muleteers valued pottery that presented a *bona linea*, a 'good line', weighed little, and had a glossy glaze. Also, pieces thrown by one potter in a series had to be of equal shape and measurements. For instance, Pepe Papaseit told me that once when he was young, he had, in absence of his father, been throwing bowls an entire afternoon. Without noticing it, his pot gauge had moved and so the bowls varied a little in measurements. Upon his father's return, his father had immediately noticed this variation and, although these bowls were well thrown, had them destroyed. For his father, as was common then for potters and muleteers, this criterion was very important. When the measurements were exact, bowls nested, but as the bowls were never sold and hardly transported as nests, consumers did not bother with this criterion. Pepe said that at present he still tried to throw a series of equal shape and measurements, but since no one cared any longer, he would never make such a fuss if pieces were of inconsistent measurements as his father had. Consumers had only been interested in the lack of cracks and holes, and the glossy glaze. The new clients' criteria differed little from those of the consumers, and they had no way of knowing if all pieces thrown in a series were of equal shape and measurements.

The 'price of Miravet' was the price muleteers paid upon picking up the ware; now the price began to include the delivery to warehouses and shops. The muleteers had come to help the potters unpack the kiln. Upon unpacking, they were obliged to pay for the ware, and were expected to take care of its transport. Gasco, the first wholesale client of Vicen Sales' grandfather, had come to Miravet to fetch merchandise: later wholesalers and shop-owners did not do this and the potters had to deliver the pottery to their warehouses and shops. These wholesalers and shop-owners were supposed to pay the potters upon delivery, but as the potters themselves had to deliver their pottery to the warehouses and shops of these new clients, they sold it upon delivery and the price included transportation. The participants indicated this with expressions as *era vendut allí*, 'it was sold there'. (The only exception was Josep Jornet who charged some of his clients separately for the transportation.)

The pottery sold to the wholesalers and shop-owners acquired different meanings than when sold to muleteers. Potters and muleteers had, as opposed parties in trade, contested the prices, and the calculation of prices as based on 'pieces', and the differentiation between good – full price – and flawed – half price – pottery. As socio-economic groups they had contested the prices and the calculation; on the individual level, they had modified, often mitigated, these contestations. When potters shifted their marketing to wholesalers and shop-owners, they no longer constituted relationships on two levels, but only as between individuals with opposing interests: they no longer mitigated the contestations. Moreover, as the potters shifted to the new marketing practice, they held on to what they considered a fair calculation of a price, i.e. the point of contestation with the muleteers on price calculation, and based their new marketing relationships on that. They did not contest the method of price calculation with their new clients. When marketing to wholesalers and shop-owners, potters only contested the price itself, and they did this individually. The 'price of Miravet' for which the pottery was sold to fellow villagers – muleteers – became the price for which pottery was delivered to one's warehouse or shop in another village, town or city.

At the same time, meanings of pottery typical and exclusive for the marketing relationships between potters and muleteers became replaced by simpler ones. The labels were now replaced by numbers. Of the four criteria, line and weight now became irrelevant. (The other two criteria, lack of holes or cracks and of glossiness, remained.) The individual variation of vessels of one type disappeared as these vessels became standardized in their measurements.

Consumers served by wholesalers and shop-owners

During the period of the muleteers, potters had related to consumers of their products as they learned about them and their forms of consumption. As they shifted to marketing to wholesalers and shop-owners, their knowledge of consumption and consumers also shifted. How and what did the potters now learn about the consumption and consumers of their pottery?

Potters continued learning through orders they received about consumers and consumption, but since they now had to deliver these orders, they saw warehouses and shops. They learned that consumers of their pottery did not change much from previously, but the way they were served did. Again, the participants 'spatialized' time: they perceived the concurrent different forms of marketing to consumers occurring at different regions, as one linear development from peddling by muleteers to selling at markets and at shops taking place in more or less advanced regions.

Vicen Sales's first wholesale client was Casco. Established in the Tortosa region, the most commercially advanced market at that time, Casco came to Miravet by cart pulled by two horses. Selling a wide variety of products at weekly markets, he added pottery to this. Eventually he opened a shop and developed into a wholesaler, and employed assistants who sold at local markets from their vans. Casco's eventually prevented Miravet's muleteers from peddling in the Tortosa region.

The first potters to sell to wholesalers and shop-owners did this in the regions which constituted the largest markets, northern Aragon, Lleida and a bit later the Penedés, a market region that now came to include Manresa. These were the ones where muleteers had made long trips at the end of summer, in autumn and in winter, and where they mostly sold pottery as opposed to barter. In these regions, similarly to Tortosa, peddlers developed into shop-owners, and shop-owners developed into wholesalers located in cities. Their shops in cities became more important for the main consumers, villagers who regularly visited cities using public transportation. Wholesalers supplied village shops, often newly founded, as these shops expanded their offerings to include pottery. These shops also pushed muleteers from the market.[44]

By delivering their merchandise, potters learned the contexts of other merchandise in which their pottery was offered:

Vicen Sales usually delivered the wares, and hence could describe clients of the Fabregat-Sales workshop well. He described the most important clients. In Saragossa they had two wholesalers and two shop-owners as clients. These four clients carried similar kinds of merchandise: hardware, kitchen utensils, like pots and pans, jugs, lamps, flatware, cutlery, glass and what Vicen Sales called the fine ceramics from Manises. These ceramics were mostly industrially made, hence finer than the less technically advanced produced pottery of Miravet but definitely not porcelain, and objects like plates, jugs, chamber pots and canisters. Moreover, all these clients carried pottery from the nearby village of Fuentes d'Ebre, pottery like troughs and drinking jugs. One particular wholesaler he remembered also carried pottery from Navarette (La Rioja), a kind of pottery that was marketed west of Saragossa. It was similar to the one produced in Miravet though not as good in quality or as large.

This client also sometimes placed orders for vessels in the same shapes and sizes as the Navarette ones, which were to be marketed especially in that region, as the customers there preferred vessels in the shapes and sizes they were used to because that was 'the pottery of their country'. The merchandise of these clients was diverse. For instance, Vicen remembered that shortly before All Souls' Day, one client carried plastic flowers to be put on graves. In the capital of Huesca the Fabregat-Sales workshop had a wholesale client who also carried similar merchandise, but instead of the pottery of Fuentes d'Ebre, it carried pottery from Naval, a village near Huesca, where pottery like casseroles and cooking pots was produced. As often happened in smaller places, Barbastro had clients who mainly sold food, such as sausages, rice, cheese, ham and potatoes, in addition to domestic utensils, including pottery. Often such a shop-owner occasionally peddled in nearby villages.

The potters with a small production volume began to explore regions with smaller markets much later, at the end of the 1950s. These were the nearby regions where muleteers had made short trips and where now potters sold to shops who did not combine their retail with wholesale. For instance, Pepe Papaseit now supplied shops in the southern coastal part of Catalonia and El Reino previously supplied by Vicen Sales' grandfather, a muleteer. These shops carried baskets, fine ceramics like flatware and cups, glass and fodder, but, especially in smaller places, their offerings might include many more necessary everyday products.

* * *

During the 1940s, individual potters began to explore marketing their wares to wholesalers and shop-owners instead of to the muleteers of Miravet. This new marketing practice was economically advantageous as they could charge better prices, but also entailed another social relationship as these clients were outsiders to the village community. This social aspect made other potters hesitant to follow this new marketing practice. In order to overcome this hesitation, as well as to strengthen their positions vis-à-vis the muleteers, potters organized themselves by pooling their distribution in the Cooperativa Alfarera. Because of this pooling, the Cooperativa collapsed within a few years. However, potters continued to develop the marketing to wholesalers and shop-owners, not only because it was more advantageous, but also because muleteers disappeared around 1960 due to the modernization of the Spanish economy. The marketing to wholesalers and shop-owners entailed changes in the pottery. For instance, the measurements became standardized. But potters also acquired other knowledge of the consumption of their wares. Most importantly, potters delivered to the warehouses and shops of their clients and learned the contexts in which their products were offered.

---

1 Applying this analysis, Sahlins synthesizes the socio-cultural structural changes with individual actions by his concept of the 'social-historical individual' (ibid.: 63). In his analysis of a Fijian war, this concept refers to individuals whose 'acts transcend self-reference' (ibid.: 63) – as opposed to acts of other individuals – due to a hierarchical social structure and the cultural meanings – i.e. 'political and cosmological ... import' (ibid.: 64) – assigned to acts of persons at the apex. The Fijian authorities effected changes because they were authorities – that is, their actions were continuous with the structure by which they were authorities – and at the same

time their actions were discontinuous with other structures – they caused a war. Furthermore, they effected changes because of the conjunction of continuities and discontinuities: as one chief wanted a pig to which he as a chief was entitled to, he upset other social relationships because he wanted that pig in a certain politically and cosmologically structured moment and place. In this case, it is by the specificity of the Fijian social structure and cultural meanings – at the time that this war began – that certain individuals at a certain moment and place could bring war to their peoples, but the general implication is that as people are socially structured, the acts of different people affect a social structure differently by their different cultural meanings. Taking place at a moment and place with critical structural characteristics – i.e. a structure of conjuncture of continuities and discontinuities – acts of people occupying critical structural positions effect changes in the social structure and cultural meanings. Hence, Sahlins synthesizes individual acts with socio-cultural changes by placing these acts and actors in the socio-cultural structure of which the actors are constituents. An event is not a chance conjuncture of all kind of structures, but the act of a 'thinking subject' (ibid.: 48), that is, the 'actuali[sation of cultural categories] in a specific context through the interested action of the historic agents and the pragmatics of their interaction' (ibid.: 81). Or as Sahlins writes a bit later on, 'the structures interact in the medium of people's projects' (ibid.: 83). There are two dialectically related and complementary forms of agency: individuals act because of their – explicit and implicit – motivations and structures 'act' by relating actors with each other, and by providing meaning – motives – to actors.

2 Anderson cites from the programme of the Falange, the fascist-nationalist organization providing Franco with its ideology: 'We conceive the economic organization of Spain in terms of a gigantic union of producers. We shall organize Spanish society on a corporative basis by means of a system of vertical syndicats, arranged in accordance with the various branches of production in the service of the national economic integrity.' (Anderson 1970: 30)

3 The only archival source on this gremio found in the Arxiú Provincial in Tarragona was with regard to its dissolution in 1971. When the provincial delegate wrote that he had decided to dissolve this gremio in April 1971 (to the national director of the *Organización Sindical*), he mentioned that it was founded in 1946.

4 On the village level the state control of agricultural production and trade meant that large landowners had additional sources of power: they had influence over quotas and other villagers were obliged to use their mills (Harding 1984: 117). The participants told me of the enormous power of the largest landowner in Miravet. Many villagers were dependent on him since they rented or sharecropped land from him, since they had to use his mills, and since he had connections with agencies controlling the agricultural production and trade. With his power, this man could also exhort outward forms of respect. For instance, on the saint's day of his family's patron saint, he invited all subordinate men to his house and treated them to cigars and wine: people hated this condescending invitation but felt forced to pay their respects to him – an extra humiliation to their inferior political economic position. Potters were usually exempt from this kind of degradation, but could be cowed by other means. For example, when Josep Benaiges was ill, he was forced to borrow money from this landowner and hence was forced to pay him respect, such as attending the saint's day celebration.

5 Arnhold claims that in the Castillian village which he had studied:

> The post-war years of national hardship were perhaps the best ever for Hyales, to judge from the fact that almost all the good houses in the village were built then. Even the poor were better off than elsewhere, for at least they were not starving. (Arnhold 1981: 117)

6 Brenan presents a similar case of a person who made a lot of money during these years by illegal trade. He met on his travels a woman…

> … who has risen from absolutely nothing to be the owner of a neat little shop in Málaga and of some landed property. I asked her how she had managed this; entirely, she said, by very hard work. Every morning over many years she has risen early to take the train or bus to some distant village to buy from the peasants. She has not got back till late that evening or early on the following day, after a night spent uncomfortably on a bench in the railway waiting room. On every journey she had to run the gauntlet of the police with her boxes and baskets and to risk a considerable

fine if she was caught. But she was never caught; according to her brother, her air of quiet respectability disarmed the Civil Guards, who were not out to enforce the law but to keep up appearances. For everyone realizes th at without the black market life would simply come to an end. (Brenan 1950: 110–11, see also Herr 1971: 234)

7 Other participants neither confirmed nor denied that the Jornet household had such an economically strong position. The economic position of a household was, and still is, linked with its social status. At the same time, participants, in general, wanted to depict themselves as equal. Therefore, for instance, Joan Avante had said that they had all been poor. There are some other indications of the relative wealth of the Jornet household: the Jornets did indeed own, as everyone confirmed, their own house while the majority of the potters did not.

8 One of these muleteers told me that he had engaged in such an arrangement with Jornet in order to help him out, to do him a 'favour'. Other participants confirmed that this muleteer was a friend of Jornet, hence that this muleteer was correct in perceiving his engagement as a favour, but they also confirmed that this muleteer was economically weak.

9 The Ventura household seemed to have had an early trade contact with wholesalers and shop-owners, but I could not gather any information about that contact. The Avante workshop also had a shop-owner as client. This client was located in Móra d'Ebre, the nearest town. Though this potter had a mule and cart and transported his products, he did so only to this shop and did not try to acquire other shops as clients because his mule was not strong enough for long distance transportation.

10 Since so many cántirs were produced, Vicen Sales could date the development of this outlet in Alcañiz as happening right after the war. After the war, Vicen Sales's parents left the household of his grandfather and for some years Vicen Sales's father and Josep Fabregat stopped working together. It was in these years that Josep Fabregat employed a wage labourer full-time, the first potter to do so. This labourer was a potter from Andalusia who for political reasons due to the Civil War had had to leave Andalusia and who specialized in throwing cántirs.

11 In the 1950s, when Pepe Papaseit visited shop-owners, he used a similar strategy.

12 Although every order and delivery constituted a separate interaction with no necessary follow-up, and although these orders and deliveries were the only contact between potters and clients, some potters developed lasting relations. Just one year before I began my fieldwork, the Fabregat-Sales workshop had for the last time supplied a client in Saragossa whom Josep Fabregat Jr. had first contacted in the 1940s. This trade relation had been exceptionally long, but potters frequently had relationships spanning more than two decades. Though these relationships could last many years, and hence were for both parties satisfactory, the partners never constituted long-term relationships of the kind potters and muleteers had.

13 There was one instance of three potters helping each other with the marketing of pottery but then only for a couple of years.

14 Participants emphasized that neither kin, friends nor neighbours were obliged to engage in these exchanges. There was a well-known case of two brothers hating each other so violently that they did not want to have anything to do with each other, least of all exchange labour. Brandes writes that 'in Becedas, two men who are friends, neighbours, or relatives are under moral pressure to assist one another in haying' (Brandes 1975: 84). This was definitely not the case in Miravet, where a household was, like in Valdemora, 'free to associate with others as it sees fit' (Freeman 1970: 199, see also Freeman 1968).

Freeman emphasizes that in Valdemora people prefer to exchange 'services and goods' with kin and friends above exchanging these with neighbours (Freeman 1970: 145–6). In Miravet, kinship, friendship and being neighbours often coincided, but there was also a very practical reason why potters would prefer neighbours. Until recently, potters used common kilns, usually the nearest to their workshop. Hence neighbouring potters were firing in the same kiln and thus knew all the idiosyncrasies of that kiln, a knowledge that came in handy when exchanging assistance in packing and firing with a neighbour.

15 Pseudonyms are used in relating this anecdote.

16 The participant related this incident only once. Later when I asked him to elaborate on it he refused to do so. In general, the participants did not like to talk badly about others, at least not with me. Therefore this is one of the very few cases of breaking off helping relationships due to an argument. Perhaps these relationships were not as harmonious as they were described to me.

17 Some reservations should be made about the units of exchange being workdays. Talking about specific cases of helping, I sometimes came across discrepancies between, on the one hand, the equivalence of labour based on quantities of time, and on the other hand, on the form of labour. The workshops of Benaiges and Petito had been helping each other: they assisted with firing, Josep Benaiges went to load the kiln for Petito, and so on. They also included in this exchange the throwing of pieces. Josep Benaiges's son Carmel recalled that he went to throw solseres at the Petito workshop, 20 per day starting from scratch, that is, he also had to throw the comienzes. A potter from the Petito workshop came to return labour by throwing petxins, 33 per day. Solseres and petxins are vessels of the same shape, yet a solsera has a content of from 75–90 litres and a petxin of five. Though these workshops had been helping each other the equivalence of labour in terms of time was accepted. However, in talking to me, Carmel was still indignant about the fact that the slow potter of the Petito workshop came to return his labour by throwing so much less, only small petxins. Although he had been, and still was, indignant about it, he had to accept it. Although the units were basically workdays, the situation might be more complex. Pepe Jornet and Pepe Papaseit helped each other firing with wood. Pepe Jornet had a small tractor. With this tractor he threw the firewood down into the pit from where it was thrown into the kiln. He did this easily with his tractor, moreover he did not lose much more time with it since he was already present to assist with the firing. Since assistance with a firing is equivalent to one workday, this use of the tractor constituted extra 'help' that had to be returned. In this case Pepe Papaseit returned it by assisting with loading trucks. Though Pepe Jornet's task was not as time consuming as Pepe Papaseit's loading of a truck, these potters equalized both by taking into account that a tractor, an investment, was used, and that the tractor consumed fuel. Thus, this particular exchange shaded over into an a medias arrangement. This is the only instance of such a transaction I came across.

18 Until modern labour legislation was put into practice, agricultural labourers had been hired for a jornal, regardless of how many hours that entailed.

19 See Pitt Rivers (1971: 138–9) for the observation of a similar 'paradox', as he calls it, in an Andalucian town.

20 Brandes also analyses a difference between two forms of obligation to return an exact amount of labour, but he perceives this as related to the 'distinct origins of both types of co-operation… Cooperative hay collection is a completely traditional practice, passed on through the generations to meet a common labour shortage. For hay collection, the nuclear family would never be sufficient.' In contrast, 'the harvesting of fruit and potatoes ... can be and frequently is carried out by the single nuclear family. To work for others or to take extra hands oneself is merely a wise management of time in a peak agricultural season' (Brandes 1975: 86). When assisting in the hay collection, one can never 'calculate' one's labour 'as if it were an economic commodity' (Brandes 1975: 84–5) because it is a favour granted to a friend, and as Brandes notices, indeed these exchanges are unbalanced [though these imbalances are 'resented' (ibid.: 100)]. But when assisting in the fruit and potato harvest one can assess reciprocity since previously workers were hired to assist and that calculation is 'carried over' into the recently developed exchange of labour in this harvest.

Potters hired workers for some tasks, but that has always been exceptional. For instance, in the period covered by the recollection of the participants, only four men had been employed as throwers. For other tasks, hiring was even more exceptional: only one man, and only during a part of his life, had been hired to pack a kiln – the main motive for mutually exchanging labour – and this person was often 'paid' with pottery which he went to peddle. For tasks that did not require expertise, like assisting with the loading of a truck – though the loading itself required skill – there were often some people hired in addition to 'help' partners. Hence the distinction between obligations to reciprocate labour that Brandes perceives did not exist in the case of the potters. I analyse the variation in the obligation to return labour as related to the social distance between the exchangers (see the discussion of Sahlins' 'interplay between reciprocity, social relations and material circumstances'), a dimension which is ignored in Brandes' analysis.

Both Brandes and Silverman relate the mutual exchanges of labour to the shortage of workers that can be hired. They argue that people are forced to exchange labour since it is not available on the market (Brandes 1975: 87, Silverman 1968: 9). This was not the case in Miravet. At the times that the pottery production was labour intensive, potters could not afford to hire workers. At the time that the production had become capital intensive, extra hands were hardly needed, and if they were, they were hired. Another factor is the skills of extra workers: often only potters had these skills and they, with those very few exceptions, did not work for a wage. The only cases in which potters provided labour to each other but refused to accept labour in return, were cases of illness. When a potter suddenly fell ill, it could happen that he had thrown comienzas. If these were not finished, they would become too dry and had to be discarded. A befriended or kin potter would come to finish these comienzas and the ill potter would have no obligation to return that labour, but he would of course be expected to do the same if the other found himself in a similar predicament.

21 Pseudonyms.

22 This case further diverged from the ideal of harmonious co-operation. It happened that the Canals needed Bernat Grinó to pack their kiln on the same day as a third potter needed him. The third potter insisted that Grinó should come to pack his kiln first. Because Grinó did not come, the sons of Canals were forced to pack their kiln on their own. In the same year this happened twice more: Quim Canals wanted Bernat Grinó to come to his workshop, but Grinó had other obligations. Finally, at the end of the year when Quim Canals went to Bernat Grinó to settle the workdays their workshops had been exchanging, Bernat Grinó turned out to owe Canals four or five days. Quim Canals said that Bernat Grinó need not any longer come to work for them, that he was fed up with the other being unreliable, and that his sons would continue to pack their own kiln.

23 As Brandes writes, in Becedas (Avila) people exchanged labour very similarly but they did not form such clusters but a 'string of work teams, each with a different but overlapping membership' (Brandes 1975: 83).

24 Note that in Miravet potters helped potters and people working the land helped people working the land; people only helped colleagues and unlike in Becedas (Avila) where people exchanged or helped each other because of friendship and kin ties regardless of their occupations (Brandes 1975: esp. 85). That people working the land were less intensively involved in these exchanges is corroborated in the ethnographic literature: see for instance Freeman 1970: 145–6, Brandes 1975: 82–7, Harding 1984: 87–91. Pérez-Díaz compares the intensity of this exchanging among farmers and peasants he studied with the one Brandes studied and he relates the differences to the forms of cultivation. In his analysis of 'factors' 'fostering' the co-operation between Castillian peasants, Pérez-Diaz conflates the difference between 'reciprocal' and 'pooling' and the success or failure of a co-operative becomes dependent on the extent peasants have been reciprocally exchanging labour (Pérez-Diaz 1991: 63–8), while my analysis explains how potters had been very much involved in reciprocally exchanging labour, but nonetheless failed to form a co-operative. Moeran similarly observes that the Japanese potters he studied '... have been more closely linked through forms of labour co-operation than is usual for rural communities' (1984: 92). It is interesting to note that the forms of labour co-operation between Japanese potters and Catalans are similar due to the nature of the potter's craft and its position in an economic context characterized by labour-intensive and capital-extensive production. However, these forms are also different due to such factors as the meanings pottery has for its Japanese producers and consumers in contrast with the meanings it has for the Catalan producers and European consumers.

25 See also Freeman's observation that the autonomy of the household was 'achieved' in 'the networks of alliances' as opposed to 'compulsory membership' in organizations based on pooling (Freeman 1970: 199).

26 Ethnographers have frequently described how Spaniards formed collectives administrating common resources like land and water in organizational forms for centuries. They described how their participants functioned within these collectives which were a social given. The

interest here is in those forms of co-operation that originated from the wish to reorganize production and marketing. Unfortunately, these forms are far less frequently described. Frigolé (1975) is a good exception.

27 Pérez-Díaz gives a short comment on how the 'Franco establishment' 'frowned on' forms of co-operation that had been the 'objectives of the socialist and anarchists organizations' (Pérez-Díaz 1991: 68).

28 These co-operatives were founded to improve the market position of their members. During the Civil War, in Republican-held territory co-operatives, or other co-operative enterprises, were founded in relation to another political-economic context and with socialist, communist or anarchist goals. During the period from Franco's revolt in 1936 until the conquest by the Nationalists troops in 1938, the people in the county of La Ribera d'Ebre organized themselves politically and economically in several ways. The CNT (Confederación Nacional del Trabajo), an anarcho-syndicalist trade union, prevailed and forced the collectivization of all the land in a number of municipalities. But subsequently, the UGT (Union General de Trabajadores), the socialist trade union federation, gained more and more power and this party espoused a politic of only collectivizing land of fascists and large land holders, and a continuation of such practices as the mutual exchange of labour and co-operatives. In many municipalities both political parties clashed, in Morá d'Ebre even violently, but Miravet was an exception for the absence of such clashes. This village was also exceptional in that it had a 'programme of "municipalization" whereby the [municipal] council was the local paymaster for both those employed on expropriated land and those providing services. A money wage was paid.' (Bailey 1990: 192). Participants showed me money printed by the municipality during that period.

29 The Petito workshop was the largest workshop before the war and was exceptional because it was formed by three brothers (each living in a separate household) and their sons. One of the sons is still alive. He was then an adolescent but still remembered how during the war a group of potters had driven up – in a car with skulls painted on it – and had required his father and uncles to join them in a collective, probably an anarchistic one. His father and uncles had replied that since they were three households working together, they were already in a collective. Upon this refusal to join, the others had locked the Petito workshop. This participant named potters who had forced them to join, but at that point in the interview he broke down in tears and stopped: the war, but more the period immediately following it, had been very hard for him due to false accusations of fellow villagers. This information about the potters seems to contradict the information presented by Bailey (1990). She seems to imply that Miravet was a village in which forced collectivizations of land did not take place, hence probably also not of potter workshops.

30 No data is available on the other two potters who refused to join.

31 Brandes described a failure to co-operate between farmers when collectively spraying insecticides: people could not trust each other with respect to the 'care' one would give in this spraying (Brandes 1975: 101).

32 Harding summarizes the debate on analysis of resistance to the formation of co-operatives as one between 'mental and material structures' or 'culture and political economy', and instead of entering this debate she offers a third point of view, that is, of 'historical experience and social relations of villagers, structured by culture and political economy' and by their embodiment in 'personalities' barring villagers from forming a co-operative. She elaborates her position by pointing at the class conflict which the people had experienced and which barred them from forming a co-operative which should have to include members of the two antagonist classes (Harding 1984: 185–94). I argue that potters failed in their co-operatives due to conflicts on personal level, but those conflicts were not, as in Harding's case, between people of different classes. Though Josep Jornet differed in some aspects from his fellow potters, he still had the same socio-economical position as his colleagues. Freeman too sees that one of the reasons why peasants could not form a co-operative was due to the unequal economic interests, as resulting from unequal landed property, that the members would have in such a co-operative (Freeman 1970: 157–67).

33 The only exception was the pottery they had to barter for firewood and for settling the rent of kilns.

34 Since the members retained their independence with regard to the production, they could continue to organize their production according to the custom of helping, yet with one reservation. On 15 February, 1950, it is written in the minutes of this Co-operative: 'It is agreed that all members will mutually help each other with all kinds of work, and that they will not do this with those potters who do not belong to the Co-operative.' During the three years that the Co-operative existed, the potters who were not members were forced to help each other. I have no more information on this other than that Pepe Papaseit recalled that the help of his father, not a co-operative member, was refused on the occasion of an accident with a firing and that his father was very indignant about it. This overlap between the Co-operative and the helping arrangements probably indicates the kind of conflicts with which the history of the Co-operative is fraught.

35 Part of the remaining books of the Co-operative exist. There are bills with the address of this Co-operative depot printed on it.

36 The Co-operative was officially registered: all the participants agreed on that, yet none could explain what the official recognition meant. Pepe Jornet knew that officials from Tarragona came and contacted his father, but about what he did not know, except that at one point his father and other potters as members of the Co-operative were looking into the possibility of receiving subsidies for building workshops. Unfortunately the Co-operative collapsed before they could actually apply for these.

37 In 1948 a hydroelectric centre was inaugurated. The only industry that was added to the county was a coffin factory located in Móra d'Ebre, founded at the end of the 1940s.

38 In Anon., 1963, *Censo y Classificación de Vehiculos de Motos y Bicicletas*, for the fourth trimester 1963, Municipal archive of Miravet.

39 At that time the Ventura workshop also had wholesalers and shop-owners as clients. Unfortunately I have little information on this workshop, though it appears that it was no different from the others who had such clients.

40 He bought their pottery finished, or he bought it unfired and fired it himself in his own kiln. If he bought unfired ware, he paid half the usual price, which was advantageous to the buyer.

41 So far the potters had actively sought clients by visiting them; Josep Fabregat added to this the stamping of products to give potential clients a possibility to contact him. After having thrown the pottery but before firing it, potters at his workshop stamped the products. This stamp indicated the name of the Fabregat-Sales workshop and Miravet in the province of Tarragona. Fabregat hoped that potential clients would thus get to know his workshop and address and contact him by mail. Vicen Sales, who at that time was working at that workshop, said that it actually worked, but also added that the old clients of Fabregat disapproved of this stamp since their customers could contact Fabregat and buy directly from him. Therefore Fabregat stamped his products only for a short period. Joan Ventura similarly stamped his products at that time: apart from knowing that he did, I have no more information on this. I have seen pottery made at the Petito workshop which was also stamped. In addition to their name and address, their stamps mentioned *barro colado*, a preparation of clay, which they were the first in Miravet to use, which prevented chipping due to small grains of chalk in the clay. They started preparing clay in this way shortly after the Civil War, yet due to personal circumstances they also stopped their production completely not long after the war. Again, I have no more information on their stamping, but note that this workshop was the first one to market their products to wholesalers and shop-owners.

42 There had been pottery production in this village, but apparently on a very small scale in terms of volume and types of pottery. When these two potters came to work for the Jornet workshop, pottery production in Benissanet stopped.

43 A pseudonym.

44 According to Breen, in the last half of the eighteenth century in colonial America the opposite happened: 'As the number of stores increased, so too did the number of peddlers' (Breen 1986: 494).

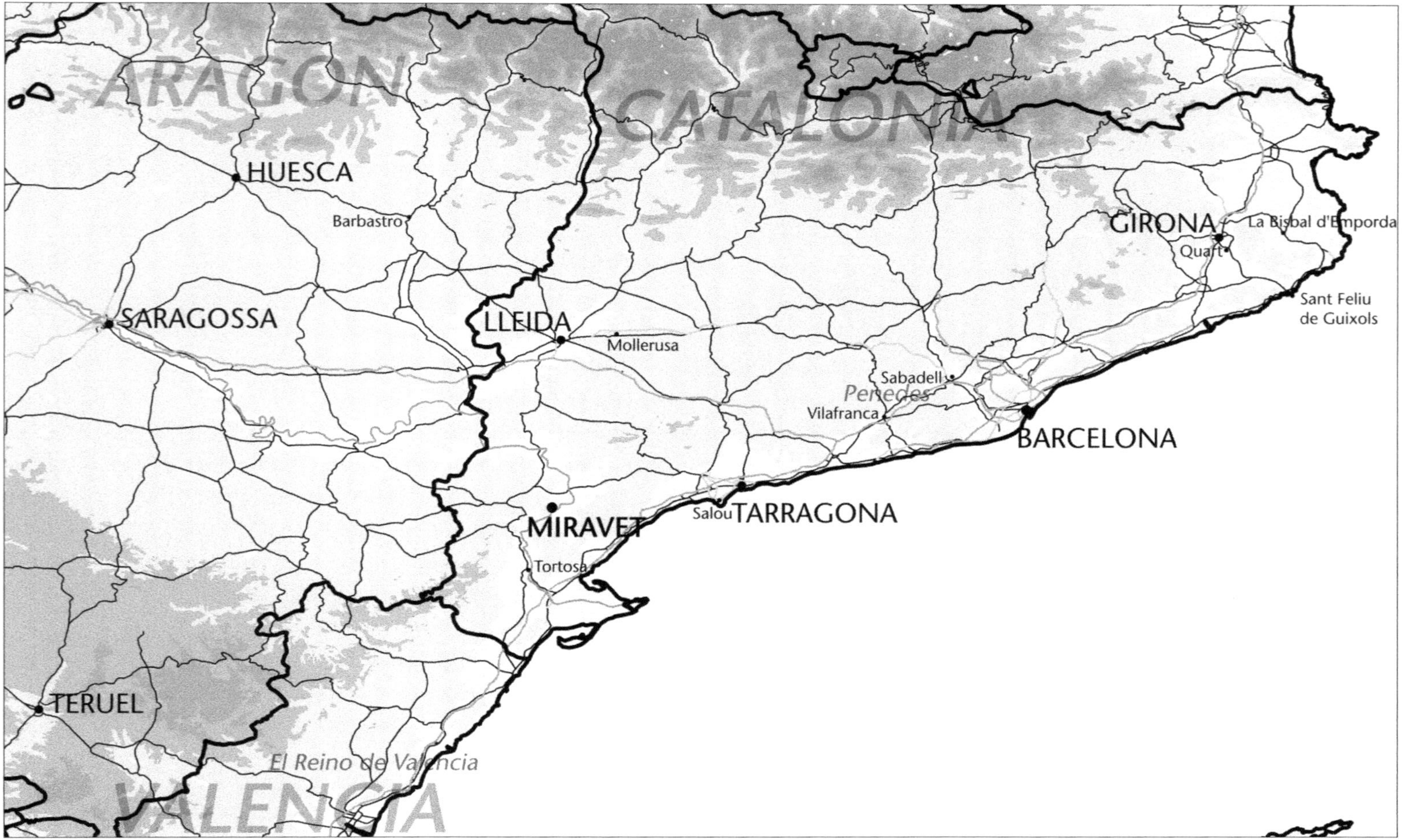

Map 5 *Map to Chapter 5.*

Chapter 5

New techniques, new products, new clients, new consumers

The national and international context of the 1960s

The economic crisis at the end of the 1950s forced Franco to open the Spanish economy for foreign investors and to ease emigration. The subsequent Stabilization Plan of 1959 was the first long-term economic plan; it was followed by several Development Plans and other government directives. During the 1960s, the Spanish government aimed to liberalize the domestic economy, to make this economy competitive on the European scale and to expand integration into the Western European economy (Anderson 1970: 190, Harrison 1978: 152).

In 1957 the EEC (European Economic Community) had been formed and Western Europe was about to experience a tremendous economic boom. In the early 1950s the Franco government had been forced to give up its strict policy of autarky, but now that this economic development began to take place in Western Europe, Spain feared it would miss out on this boom and sought to actively integrate into the international market.[1] Finally in 1970, Spain was granted a Preferential Commercial Agreement with the EEC (Harrison 1978: 162–3). The results of this increased integration into the international economy were stunning: between 1959 and 1973 imports increased by a factor of 13, exports by 14 – most of this international trade was with EEC members (Harrison 1978: 145), and foreign investments rose from $160 million in 1960 to $5,980 million in 1973 (Carr 1982: 746).

During the 1950s Spain slowly converted from an agricultural country into an industrial one: in the decade of the 1960s Spain completed this conversion rapidly. At the beginning of the 1960s the portion of industry in the total of economic production already outweighed that of agriculture, and during this decade the dominant portion of the population also flipped over from being employed in agriculture to being employed by industry (Tamames 1986: 14, Harrison 1978: 145 and 160). The modernization of agriculture accelerated due to several interrelated factors. With the development of industry in the urban centres in the 1950s, the poor rural workers and small landowners were offered a lucrative alternative in the form of urban, industrial employment. The sudden massive rural exodus at the end of the 1950s caused a steep rise in rural wages: according to the official numbers, rural wages doubled just over two and a half times in the period between 1957 and 1967; Harding estimates that in her Aragonese village wages multiplied five and half times between 1953 and 1973 (Harding 1984: 141). This rise of wages stimulated farmers to invest even more – still with government support – in mechanization (Harrison 1978: 161). Plot consolidation progressed and plots that were not profitable were abandoned (Tamames 1986: 41). Between 1962 and 1972 about half a million farms disappeared (Tamames 1986: 49, Harrison 1978: 159). Farmers cultivated market crops, the staple ones with state guaranteed prices. In this decade peasants disappeared and farmers acquired an entrepreneurial outlook on their production (Harding 1984: 114–202, Hansen 1977: 56–106).

Because of the economic boom in Western Europe and the United States, their citizens went on vacation more frequently and Spain, with its low prices and excellent weather, was a favourite destination: in 1960, 6.11 million tourists visited Spain, in 1973, 34.56 million; in 1960, the receipts from foreign tourists were $297 million, in 1973 $3,091 million, a crucial and vital contribution to the national economy. To accommodate this influx, the Mediterranean coast developed a tourist infrastructure (Tamames 1986: 177–9, Harrison 1978: 155). During the 1950s tourism had gradually grown; in the 1960s it exploded.

Dubbed the 'Spanish miracle', the industrial growth, modernization of agriculture and tourist explosion propelled per capita income to double between 1961 and 1973, while GDP increased by a factor of five (Tamames 1986: 192). This prosperity was unevenly distributed over the regions: Catalonia, Basque country and Madrid prospered more than the others (Shubert 1990: 258). The prosperity was also unevenly distributed between rural areas and cities: rural areas were poor in comparison with urban ones, with the exception of some suburbs – actually shanty towns – around big cities like Barcelona, Madrid and Bilbao (de Miguel 1972: 213–21).

Because of their economic progress, Spaniards acquired access to new consumer goods. The number of cars sold multiplied more than ten times between 1960 and 1970 (Harrison 1985: 161); the number of households that owned a refrigerator rose from 4% in 1960 to 63% in 1969; those that owned a television set rose in the same period from 1% to 62% (de Miguel 1972: 91, see also Barrett 1974: 55). As the prosperity was unevenly distributed over the regions, so was this new consumption. Furthermore, the city, as opposed to the countryside, offered more of the new consumption possibilities like modern apartments and entertainment: villagers not only migrated to cities because of employment, but also because of the consumption opportunities (Barrett 1974: 52–3, Carr 1982: 751). Unfortunately, with the global oil crisis of 1973, the Spanish economic boom halted.

These national economic developments took place, in varying degrees, in the county of la Ribera d'Ebre. The total population of the county declined due to emigration to the city. Though industry also developed here, the majority of the population was still employed in agriculture.[2] The industry was primarily family business and of the same nature as before.[3] Agriculture was also modernized. Farmers abandoned land that was unprofitable, but there were no land-consolidation programmes.[4] In the first half of this decade farmers decreased the cultivation of cereals, and increased growth of wine and olives, which were more profitable; i.e. they directed their cultivation towards the market. But the most significant shift began in the middle of this decade. Using modern irrigation techniques, farmers began to cultivate apples, pears and, especially, peaches (Anon., *La Ribera d'Ebre* 1989: 101, 364). Due to this economic development per capita income went up to 68,000 pesetes in 1973, though it was still relatively low: Catalan per capita income was 107,000 pesetes, while the national per capita income 85,000 (Anon., *La Ribera d'Ebre* 1989: 186, Tamames 1986: 192).

Miravet was no exception to its county's trends. Its population declined towards the middle of the1960s but recovered to almost the same level as at the beginning (see Appendix B). The portion of its population employed in agriculture declined as agriculture modernized: in 1963 there were nine tractors registered,[5] in 1965 there were seventeen (Anon., *La Ribera d'Ebre* 1989: 365). As in other villages of this county, farmers redirected their cultivation towards market crops. The cultivation of peaches became particularly important.

These large-scale developments affected the potters. The costs for local wage labour rose while money for investment in new labour-saving techniques became available. The demand for pottery shifted to new types of pottery, and potters marketed these new products to newly acquired clients for consumption by new consumers. Each shift in itself might have been minor, but as they were conjoined in a manner that had an 'effect of a structural synergy' (Sahlins 1991: 68), their result was pivotal.

New techniques

Potters had been embedded in an economy in which money was scarce and in which people used it only minimally in their exchanges. For their production, potters extracted raw materials using their own labour force – and only occasionally that of mules and donkeys. When this was not possible, they preferred to barter products for raw materials. Only as a last resort did they buy raw materials. Because they lacked money, they used techniques which required the least possible monetary investment, and were labour intensive. Within these constraints, production had been changing. For instance, shortly after the Civil War a refugee from the south introduced a kiln model that improved the firing. These changes were minor compared with the ones that began in the early 1950s, when potters directed their resources towards more capital-intensive production.

Until the early 1950s the potters had extracted *terra refractaria,* one of the two components of the clay they used, and silica, a component of glaze, from underground mines. This was a very unpleasant and dangerous job, and one which required collaboration between several households, although the ideal household was an autonomous unit of production. In the early 1950s, they shifted to buying these materials from companies which mined these components in the neighbouring village of El Pinell de Brai.

Muleteers, and men co-operating with them, had cut firewood in the surrounding mountains, brought it down to the potters, and bartered for pottery. At the end of the 1950s muleteers and their companion cutters declined in number and potters had to shift to firing their kilns with the waste from wood-processing factories. An additional benefit of this shift was that this new kind of firewood turned out to speed up the firing process, hence saving labour.

Initially, potters acquired lead from the mines in the nearby villages of Molar and Bellmunt. The extraction of lead was specialized work. The shafts could be as deep as 900 metres, and therefore potters, or muleteers acting as middlemen, bartered or bought it from miners. During the Co-operative the potters bought lead collectively. After its demise, Fabregat and Ventura became distributors of lead. They had the financial means to invest in a large quantity, and sold lead, for a small profit, to potters who could only buy as much as they immediately needed. When those lead mines closed around 1960, all potters switched to purchasing lead, individually or collectively in ad hoc groups, from Linares in Andalusia. Again, an additional benefit of this last shift was that the mines in Linares offered ground lead, thus freeing the potters from the very unhealthy job of grinding it.

It was common in the village that economically stronger persons invested in equipment used in production, kilns, but also in equipment, such as olive mills, that others could rent. In the 1950s Fabregat, as the economically strongest potter, invested in one grinder for lead and one for terra: other potters could pay to use these on a timed basis. The less economically strong potter workshops also began purchasing small machinery, though slightly later.

Until the end of the 1950s the potter workshops did not differ much between themselves regarding materials and equipment employed, whether they owned or rented the equipment. As their labour force consisted of household members, they only differed in this respect with regard to the number and ages of their members (especially the men, as women performed only infrequent, light tasks). At the end of the 1950s the potters continued to take steps towards more capital-intensive production, but with these steps potter workshops began to differ from each other.

Most potters only occasionally employed wageworkers for specific tasks, often boys who made a bit of pocket money, and only the few 'stronger' potters like Fabregat, Jornet and Ventura did this full-time, and always employed the unskilled.[6] In 1958 Josep Fabregat conceived another way to expand his labour force. He had refused to become a muleteer like his father and had instead learned the craft of pottery from his brother-in-law Vicen Sales Senr., who was at that time a member of his household. Both had worked together until Vicen Sales Senr. left the household. In 1958 his nephew Vicen Sales Jr. returned from fulfilling his military service and Josep Fabregat proposed that he and Vicen Sales Senr. collaborate again. They joined their workforce and equipment on a formal and permanent basis, and thus the *Societat*, 'society', Fabregat-Sales was created. This form of co-operation was a major break with the local socio-cultural notion of the household as the unit of production. Josep Fabregat took this form from urban enterprises. During the fieldwork this society still continued, but none of the other potter households had followed this arrangement.[7] In gossip, participants still showed their uneasiness with this form of co-operation. Shortly after joining together, Josep Fabregat hired a skilled potter, Carmel Benaiges; and Josep Jornet hired two potters from the neighbouring village. None of these potters wished to continue their fathers' workshops after their they had passed away, but instead preferred to work for a wage, since rural wages had become higher. Thus, in addition to employing unskilled wage-workers, for the first time in Miravet, Fabregat and Jornet expanded their workforces with skilled wageworkers.

When the Societat Fabregat-Sales was founded, they bought a jigger for mechanically shaping pots. Shortly afterwards, Josep Jornet and Pepe Ventura did the same, and much later, in the mid 1960s, Joan Avante bought one too. As a jigger 'throws' much faster than a thrower, it consumes much more clay. Employing a jigger caused problems with clay preparation, so they needed to employ machinery for that as well. Instead of mixing the components of the clay in a pit by treading it bare-footed, these potters built ponds with electric mixers in which this could be done. An additional advantage was that these ponds also made the grinding of terra superfluous. Instead of wedging[8] the clay to prepare it for the throwing, they bought electric clay mills for the task. Furthermore, the employment of this machinery made it possible to rely more on unskilled workers. The smaller workshops that did not employ jiggers lacked the incentive to invest in clay-preparing equipment; neither did they have the economic means to make such investments. They followed in the purchase of this labour-saving machinery much later, the last ones beginning to make these investments at the end of the 1960s.

Hence, during 1960s, the varying degrees of change in workforce and equipment as well as the introduction of new techniques resulted in potter workshops becoming increasingly different from each other in terms of production.

New products

Because of Spain's industrial and economic growth, new products appeared on the market, and old products became cheaper. This industrial growth also caused many people to leave the rural areas and achieve an urban lifestyle with associated consumption patterns. How did these developments affect the demand for pottery?

The pottery bowls came to be replaced by plastic and metal ones. These weighed less, broke less, were easier to store and were cheaper. As people moved to the cities and as butcher shops in villages became more common, pigs were slaughtered less frequently at home, with the result that larger bowls also declined in use. The demand for pottery bowls almost disappeared.

The large vessels used for storing olive oil came to be replaced by small containers – bottles and cans – in which olive oil was sold. Furthermore, the large vessels took up a lot of space, space which was available in the large open lofts and in the large pantries of village houses, but not in smaller city apartments. The small vessels came to be replaced by glass, and in general storage in pots came to be replaced by deep-freezing and cans. The demand for such vessels declined sharply.

The large tubs were used for doing laundry, the smaller for doing the dishes. They were now also replaced by plastic and metal tubs, and, mostly, by modern ceramic sinks and washing machines. The demand for tubs disappeared.

Though rural areas lagged behind in their participation in the economic prosperity, all villages, even the smallest, eventually had running water. Because large vessels had also been used for water storage, and cántirs had been used for fetching water at the fountain or river, both types of pottery became obsolete.

When at the end of the 1950s this sharp decline in demand for previously main products became apparent to the potters, it had an enormous emotional impact. The participants referred to it as the *crisis*: it was a mark in their history that coincided with general economic turmoil, as evidenced by the huge migration to the city and to foreign countries.

When the topic of potters owning land came up, Pepe Jornet told me why, at the end of the 1950s, his father bought land, a field of three and a half hectares in Benissanet, a neighbouring village. Previously, people had cultivated land primarily for their own consumption, and a potter household that had land would first dedicate itself to cultivating it, and then to producing pottery. But when Pepe's father bought this land, times had changed. Subsistence agriculture was disappearing, and he himself had one of the most economically successful pottery workshops. Pepe Jornet: 'Well, my father saw that when plastic was introduced there was going to be a recession, a decline in the demand for pottery. There were so many potters and there were piles of pottery. My father was the one who always had the largest production. My father feared a crisis and when he had the money he bought that land. He said that he bought it as an extra support, that instead of making eight or ten kiln contents annually, he wanted to see if he could make ends meet with four and with working the land. But on the contrary, everything went up. A lot of potters stopped because of the crisis. However, sales increased: instead of being seasonal we sold pottery year round. It was no longer necessary to store pottery.' Sales went up because a demand for new types of pottery developed: eventually Pepe's father sold the land and returned to working full time as a potter.

Flowerpots had always been produced. These pots usually had a simple conical shape, but their design could be varied. The shape could be a bit rounded. The rim could be convex, flat or nicked. And one could decorate the pot by, while still sitting on a slowly revolving wheel, allowing the rim to slip through two fingers thus creating a waving shape; engraving waving lines with a stick; or attaching horizontal *bordons*, 'ribs. These pots had always been produced, but only in small quantities or per order to be peddled in nearby regions: now the demand for this ware increased tremendously. To meet this growing demand, the workshops of Fabregat-Sales, Jornet, Ventura and Avante bought jiggers to produce flowerpots mechanically, and in large quantities. These pots, locally called *parra* or *torreta*,[9] had a conical shape, a thick flat rim at the top and a hole in the centre of the bottom: the generic flowerpot. The mechanically thrown pots were usually decorated in a quick and easy manner by painting, for instance with an air compressor that resulted in a decorative spotted texture. Initially the potters produced this ware in a few sizes, but in time they bought additional moulds so that eventually they produced it in six sizes, which, with the exception of the largest, nested into each other. Potters who threw these pots by hand followed in producing their flowerpots in similar sizes and also nesting.

The importance of the flowerpot in the production of a workshop varied. For potters owning a jigger and machinery for preparing clay, flowerpots became their most important product. During the 1960s the Fabregat-Sales workshop mainly produced flowerpots, and the Avante workshop exclusively produced flowerpots for six to eight years.[10] Potters not equipped with this machinery also produced a large volume of flowerpots, in addition to other pottery.[11]

Decorative types of pottery had always been made, but only in very small quantities, most likely as unique, singular pieces, never intended for the market. Around 1960 many new types – most of which were copied from elsewhere – were produced, and the renewal of decorative types has been ongoing ever since. These are pieces like money banks (spherical objects on a foot, often with a knob on top), vases, miniatures and pitchers for *sangría*, the Spanish punch. As this type of ware was so varied, and constantly varying, each potter developed his own repertoire of particular decorative pottery.

One type of decorative pottery stood out by virtue of being painted. This type of pottery was never labelled. But since – apart from the mechanically moulded flowerpots – this type was the only pottery which was painted, in interviews with potters I used as reference the ad hoc label *gènere pintat*, 'painted ware', or *terrisseria pintada*, 'painted pottery'. All participants clearly understood what this label referred to, though they did not seem to have used this label outside my interviews. The 'painted ware' was painted with a latex paint in a copper green or a yellowish metal colour, and patined with black bitumen. These were objects for decorative use, like vases, jugs, figurines, reliefs, planters, columns, amphorae and umbrella stands. Typical of this pottery was that it had fantastic, elaborated shapes and that it was decorated in relief with floral, animal and human motifs. The paint and patina brought out the decoration in relief. This type looked unlike any other pottery and made, as one participant put it, a 'plastic impression'.

Only four pottery workshops, those of Pedrola, Papaseit, Avante and Jornet, produced 'painted ware' as an addition to their regular production. As Pepe Papaseit told me, 'It was only produced in small quantities. It was not the basis of the production, no. But in those bad times when no vessels were sold, when no bowls were sold, of course you had to take an order for anything. It was just one more thing to make.'

In contrast to the previous open attitudes regarding techniques and designs, the potters who made painted pottery were not willing to share their methods and designs. Pepe Jornet learned the techniques of painting and applying patina from an outsider. He applied this technique in the loft of his house so that no other potter could discover his procedure. The designs were derived from pieces made outside of Miravet, but some potters like Pepe Jornet developed their own. Pepe Jornet also dried this pottery in the loft: he found out that another potter had secretly made drawings of his pottery when it had been standing outside to dry, as was common. Other participants did not confirm this degree of secrecy surrounding these particular techniques and designs, but they did acknowledge that one was not as open about them as usual.

By introducing new techniques into their production, potters also differentiated their production in interrelation with new products. The economically stronger potters employed machinery and wage labourers, and focused on the large-scale production of flowerpots. Other potters shifted to producing decorative pottery, each with a different repertoire, and were unwilling to share their designs and techniques.

New clients

During the 1950s the potters shifted to marketing their products to wholesalers and shop-owners, and although the pottery changed in certain respects due to this shift, it still consisted of the types already sold to muleteers. At the end of the 1950s and especially during the early 1960s, as new products were taken into production, these were sold to new clients.

When visiting wholesalers and shop-owners, Josep Fabregat came across the mechanically moulded flowerpots that his clients were selling, and recognized its commercial potential. The most important advantage of mechanically shaped flowerpots was that they were shaped much faster and therefore were much cheaper. Another advantage was that the sizes of the mechanically moulded flowerpots were more standardized and therefore nested better, something important when it came to transportation and storage. Josep Fabregat recognized the commercial potential of new products, but clients also ordered new types of pottery. For instance, a client of Pepe Papaseit had obtained a catalogue of 'painted ware' and he placed an order for copies of this pottery in addition to regularly ordered pottery. New commercial relationships could be formed on the basis of new products. A few wholesalers originating from Bailén (Andalusia) and Salvatierra de los Barros (Badajoz) and relocated in Barcelona as part of the enormous migration to that city, searched out Jornet, Ventura and Avante to order copies of pottery from their home towns, mostly small decorative pieces and many miniatures. As potters produced new types of pottery, they could contact new clients on the basis of these types. When Josep Fabregat focused on the production of flowerpots, he produced more than his regular clients could turn over, and so had to look for new clients. The most important client he acquired was one who bought a truckload of just flowerpots about every 20 days.

These shifts could be gradual. Potters had clients who bought the newly introduced pottery in addition to the regular types. When the demand for these regular types declined, a client might only order the newer types. These shifts could also be more abrupt, as when potters acquired new clients on the basis of new products. As these shifts occurred, types of pottery tended to become specific to a few clients, or even custom made for one particular client. Potters who produced the mechanically moulded flowerpots only sold to a few clients. Potters who produced the copies of the Bailén and Salvatierra de los Barros ware only supplied wholesalers originating from these towns. Jornet supplied his 'painted pottery' mainly to one client.

Potters and muleteers had been intimately familiar with the pottery. All potters had produced the same repertoire, but all potters and muleteers could readily recognize pieces by their individual maker. Serving the new clients, potters might still recognize pieces made by fellow villagers, but less so as potters began to be secretive about certain decorative pieces. Even if they were not secretive, potters produced at times highly customized pieces for individual clients. Their fellow villagers may not have been aware of these orders. The new clients did not come from the village, and usually only knew their supplier and his products, not those of others.

Potters and muleteers had also been familiar with the consumption of pottery and shared the geography of consumption. As potters no longer produced the same repertoire, this knowledge diminished. This surfaced clearly as Pepe Papaseit, his wife and I leafed through bill books of this period. In the accounts, he had specified several pieces of decorative pottery with labels of his own making (see also Appendix F). These were descriptions only used between him and particular clients that even his wife, who had at that time assisted him frequently in the workshop, did not know. Apart from very generic terms, during the 1960s potters began to refer to their products with labels that were no longer shared by the community, even labels whose use could be limited to dealings with one particular client.

Hence, as potters introduced new techniques and new products, they also changed the nature of their marketing relationships with existing clients, and more often contacted altogether new clients to supply with these new products. With these new clients, they no longer formed a community that recognized the pottery by its maker and knew of its consumption. As the sense of community with clients lessened, so the community among the potters lessened as well.

New consumers

Pepe Jornet recalled that Joan Avante Senr. wished to explore the production and marketing of 'painted pottery', but had no idea where to seek clients. Pepe's father had been producing and marketing this kind of pottery, and hence could advise Avante on this. The fact that Pepe Jornet actually recalled Avante asking for this advice is significant as it indicates a change. When marketing to muleteers, potters largely shared their knowledge about consumers and their consumption of pottery. This knowledge was collective. When they marketed to wholesalers and shop-owners, potters expanded their knowledge by directly observing the contexts formed by other merchandise in which their pottery was offered, but still to the same consumers. As Pepe Jornet's recollection indicates, potters no longer shared this knowledge.

As potters introduced new techniques and differentiated their forms of production, they produced different repertoires of new products. These new products were sold to new clients, and these products became particular to certain clients. Thus the potters' knowledge of consumers became fragmented: a potter knew of the consumers of his particular products, but far less, or even nothing of the consumers of products of other potters, hence Avante's request. When interviewed about the previous consumers and their forms of consumption, the participants differed in details of their knowledge of these consumers, but their different stories confirmed and complemented each other. In the following description of consumers, I have patched their stories together.

Potters began selling flowerpots to wholesalers and shop-owners who already purchased their other pottery, and to clients in towns and cities. For instance, the Fabregat-Sales workshop sold mechanically thrown flowerpots to one of their wholesalers in Saragossa, and to their clients in Barbastro, Huesca, Mollerusa, Lleida and Villafranca de Penedès. Hence the flowerpots were sold in contexts in which vessels and bowls were sold: basically hardware, kitchen utensils, flatware, cutlery, glass and fine pottery from Manises. When sold in shops, the merchandise usually also included food.[12] Similarly, potters who mainly sold handthrown flowerpots did so to their regular clients, like Papaseit to his clients in el Reino and the southern Catalan coastal region. Some clients who initially bought vessels and bowls, and later added flowerpots, shifted to flowerpots more and more at the expense of the first types. Vicen Sales recalled as an example a client in Huesca who ended up only buying flowerpots.

Those potters who mechanically produced flowerpots also acquired clients on the basis of these pots, clients who bought in large quantities. These clients were all located in Barcelona or its suburbs. For instance, the Fabregat-Sales workshop had a client in Barcelona, one who traded only in pottery and ceramics. This client also obtained merchandise from other places in Spain like Manises (Valencia) and even from Italy. The merchandise ranged from mechanically shaped flowerpots and mechanically shaped bowls to small vessels from Manises and hand-painted decorative earthenware plates from Italy.

Initially, potters sold decorative pottery to urban clients who also bought other pottery, but later they gradually targeted their marketing to clients located in the southern coastal region of Catalonia. After leafing through a bill book of Papaseit, it appeared that all decorative pottery was sold to their clients in that particular region.

The decorative type called 'painted pottery' was typically sold in the southern coastal region of Catalonia. The Jornet workshop already had clients in Saragossa, Lleida and Barcelona to whom they began to sell this pottery in addition to other ware, but when they acquired a client specifically for it, a florist who sold only this pottery next to his usual merchandise, this client was located in the southern Catalan coastal region, in Tortosa. When Avante desired to produce this type of pottery, Josep Jornet advised him to look for clients in the coastal area, especially in and around Salou. The Pedrola workshop acquired their most important client for the 'painted pottery', a shop specializing in pottery, glass and flatware located in the city of Tarragona. (Potters acquired clients for this pottery in the southern coastal region of Catalonia: the northern coastal region was dominated by the potters of Quart and La Bisbal d'Empordà, who had initially developed the 'painted pottery' that was subsequently copied by the potters of Miravet.)

The other decorative pottery, the small pieces and miniatures copied from southern pottery, was ordered by clients in Barcelona, who only traded in pottery. They sold the small decorative pieces in addition to small bowls, small vessels, tubs, flowerpots and decorative objects like fantasy water jugs. These clients had warehouses in Barcelona, Sabadell (a town close to Barcelona) and in San Feliú de Guixols on the Costa Brava. Joan Avante added that in winter he delivered mostly small bowls and vessels to the warehouse in Sabadell, in summer mostly decorative pottery to San Feliú de Guixols. These clients had employees peddling with vans and donkeys. These donkeys carried the pottery on their back, very clearly displayed, and were outfitted with colourful halters adorned with fringes and copper plates (much like the one I encountered in my youth in the Costa Brava). As the participants emphasized, the peddlers transported the pottery on donkeys with an eye towards the tourists. Colourful beasts of burden packed and adorned with pottery appealed to tourists and became a favourite theme for postcards.

Although the consumption and consumers of these flowerpots and decorative pottery differed from potter to potter, they coincided in three dimensions.

First, pottery was no longer offered in the context of household utensils and hardware, or in the more extended context that also included food and everyday products. At the time of muleteers, flowerpots were produced infrequently and in small quantities They were considered a small luxury, not a household necessity. Then (as might still happen today) a woman would take what she would consider an old worthless container and recycle it as a planter. To buy something for potting a plant would have been considered an indulgence. Muleteers peddled few flowerpots, only nearby, and only when ordered: Vicen Sales recounted that his grandfather would carry at the most half a dozen on a trip to Les Garigues. Now that flowerpots were produced in large quantities, these tended to be sold in contexts of other pottery and ceramics, no longer as part of household necessities. The other newly introduced pottery, the decorative pottery, was previously never made for the market. Initially this pottery was sold to shops also carrying household utensils and hardware, but soon it became available as a product for wholesalers and shops specializing in pottery, ceramics and glass. The participants said that the flowerpots and decorative pottery were bought as *capritxes*, 'caprices', or 'on a whim'.

This shift of context is correlated with another shift. As far as the participants could recall, bowls, vessels, tubs and cántirs had always been produced in Miravet. When they introduced jiggers to mould flowerpots, they perceived this as a technical development to supply more economically a demand that had always existed, but in the late 1950s was growing. However, when they introduced decorative pottery, it was related to a qualitatively new demand. In relation to this new demand, new types of decorative pottery were developed, and potters have since continued to produce decorative pottery, continuously replacing the specific types of decorative pottery as the demand for this pottery also changed. Pottery bought on a whim also changed with the whims of fashion. The decorative pottery that stood out by being painted, also stood out as being most typical for the fashion of the 1960s. As a participant noticed, with its copper green or yellowish metal colour and its plastic impression, it fit in well with that era's taste, characterized by formica and bright artificial colours.

The second dimension was that this pottery was no longer mainly marketed and consumed in rural areas, especially the western part of Catalonia and Aragon, but was now marketed to customers in cities, especially Barcelona, and on the coast. In Miravet, as everywhere in Spain, people have always loved to display flowers on their balconies, on terraces, on the pavements in front of their homes, wherever possible; yet the people of Miravet perceived a difference in this love of flowers between countrymen such as themselves and urbanites. Though they loved flowers, and they presumed that this was a general norm among villagers, they were not willing to spend as much time in taking care of flowers as city dwellers. As one woman explained, countrymen spent enough time working the land, and did not want to care for plants in their free time. As she had remarked retrospectively, the contrast between countrymen and urbanites had been stronger in the 1960s. Similar to the greater attention paid to gardens, urbanites paid more attention to interior decoration. They therefore might use pottery to liven up their homes. The flowerpots and the decorative pottery were urban consumption goods.

But the decorative pottery was also sold on the coast. These were the regions increasingly visited by tourists. With respect to the small and especially miniature decorative pottery, the participants did not doubt that these pieces were specifically sold to tourists. Tourists wanted to buy something that was cheap and could easily be carried home.

This spatial shift of the location of consumers from western regions to eastern coastal ones, and from rural areas to urban ones, was complemented by a temporal shift. The flowerpot was a product mainly for the spring and summer, the decorative pottery mainly for summer. These seasonal cycles affected the potters in relation to their involvement in the production of these types of pottery. For instance, the Papaseit workshop continued to sell vessels and bowls in addition to flowerpots. Since the vessels and bowls were products for the winter, the flowerpots were products for the summer. Although for several years, the Fabregat-Sales workshop mainly produced flowerpots year round.

The third dimension in which the consumption and consumers of the newly introduced pottery coincided was the economic realm. This occurred as the purchasing power of Spaniards increased. Due to the Spanish miracle, many Spaniards could now afford the luxury of new consumer goods. On the one hand, they consumed less of the previous types of pottery, but on the other hand they consumed new types of pottery. However this prosperity was unevenly distributed over the Spanish population: urbanites in general were more prosperous than the rural population, and apart from this, prosperity was also unequally distributed among the different Spanish regions. Catalonia was by and large a prosperous region, and Barcelona a very prosperous city, but the county of La Ribera d'Ebre was less fortunate; its per capita income fell below even the Spanish average. As tourism exploded, inhabitants of north Western Europe and of the United States also experienced an increase in their purchasing power, an increase that made it possible to vacation in a relatively cheap Spain. Hence, the consumption of the newly introduced pottery became one of luxury, as opposed to household necessity. It was both an urban and a tourist phenomenon, and the consumers of the newly introduced types of pottery were more affluent than the potters and the people in Miravet generally.

* * *

In the 1950s, potter workshops might differ with respect to the workforce formed by male, adult household members, and hence by volume of products and market regions targeted, but all potters employed basically the same techniques and equipment, produced the same types of pottery and had the same kind of client. In the late 1950s, but especially during the 1960s, potters began to shift their production, repertoires of products and their clients. Each of these shifts occurred gradually and was only slightly discontinuous with previous production, repertoires and clients. Examples are the hiring of an employee, or the ordering of 'painted ware'. Because of their individual insignificance, these shifts were hardly noticed.

However, these apparently insignificant shifts were conjoined in a particular structure. For instance, the Societat Fabregat-Sales invested in equipment and a labour force to produce mechanically moulded flowerpots to be sold to new clients. These new clients were wholesalers in ceramics, pottery and glass located in and around Barcelona and catering to new, urbanite consumers. Or, another instance, the Papaseit workshop lagged in the acquisition of new equipment – and never bought equipment like a jigger – and had a labour force consisting only of father and son. They produced the older types of pottery (though in diminishing quantities), but in addition new ones like handmade flowerpots and decorative pottery. They marketed their products to clients primarily located in the southern coastal region of Catalonia, and often these products were particular to particular clients. In that region a new demand for pottery had emerged: tourists.

These shifts in production, repertoires of products and clients considered separately were gradual, but by their structure of conjunction they were pivotal. Their conjoined results were that at the end of the 1960s potters had become individualized in their production, repertoires and clients.

---

1 In 1960 the Spanish government eased and adapted the tariff system; in 1962 it applied for membership in the EEC – an application that was rejected because of the dictatorial nature of its government; and in 1963 it joined the General Agreement on Tariffs and Trade (GATT) (Tamames 1986: 141–3).

2 In 1965 60% of the population was employed in agriculture and 22% in industry; the next time that numbers were offered, 1975, 41%, was employed in agriculture and 38% in industry (Anon., *La Ribera d'Ebre* 1989: 82).

3 Only two companies had between 51 and 500 workers – both textile factories – and the large chemical plant in Flix had more than 1,500 workers (Anon., *La Ribera d'Ebre* 1989: 135).

4 The number of landholdings decreased by a little more than a tenth, but this decrease was quite evenly distributed over holdings of all sizes. Land was not being concentrated (Anon., *La Ribera d'Ebre* 1989: 101, 104, 107)

5 In: Anon., *Censo y Classificación de vehiculos de motos y bicicletas*, for the fourth trimester 1963, Municipal archive of Miravet.

6 After the Civil War an Andalusian potter who had to leave his hometown for political reasons, came to Miravet and was hired by Fabregat: this was the first time that a potter hired a person who had mastered the potter's craft. In the early 1950s this Andalusian potter started his own workshop, as potters were still better off self-employed than working for a wage.

7 Although a few others villagers, working in other types of business, had.

8 Wedging is the mixing by kneading of clays into a consistent mass ready for throwing.

9 Another locally used term is *clavel.let*, a derivative of *clavell*, carnation, one of the most popular flowers. This term seems to be an old fashioned one. When Vicen Sales spoke about the trade of muleteers, he would use this term, but when talking about the ones he was presently producing, he applied the other labels. He, as well as other participants, pointed out that these labels were also used in Aragon. He conjectured that the people in Miravet had taken over these labels from the Aragonese. The use of these labels was in contrast with the use of *tiesto* and *maceta* in northern parts of Catalonia, especially Barcelona.

10 I have no data on production of flowerpots in the Ventura workshop.

11 The Jornet workshop is an anomaly. They did own a jigger, but for unclear reasons did not use it much. Still, they made kiln loads consisting only of hand-thrown flowerpots. According to Pepe Jornet, a hand-thrown flowerpot was stronger. Other participants did not mention this advantage, or regarded it as of no importance.

12 I have no data for the Ventura workshop with a large production of mechanically shaped flowerpots.

Chapter 6

Potters and their clients – wholesalers and shop-owners

The national and international context of the 1970s and 1980s

In 1973 the OPEC countries raised the price of oil significantly; this rise caused a global economic crisis. In contrast to the Great Depression, which had hardly any impact on the isolated Spanish economy of the 1930s, this global crisis hit Spain hard (Harrison 1985: 171, 175), not only directly, but, as the Spanish economy had become dependent on the general European prosperity, also indirectly: tourism, foreign investment and the demand for Spanish immigrant workers declined. The general result of the oil crisis was increasing inflation, unemployment, trade imbalance and capital flight (Tamames 1986: 228).

During the 1960s and the early 1970s the Spanish economy became modern, industrial and capitalist, but it still contained problems inherent to the Spanish political regime. Though the Francoist ideologists claimed the economic bloom was a result of the suppression of political debate, Spaniards increasingly felt that democracy was compatible with a flourishing economy. Because of its political structure, Spain was still rejected as a fully fledged member of the European Community. After the fall of the dictatorial regimes of both Portugal and Greece in 1974, the Spanish government was more than ever an anomaly within Western Europe. When the Spanish dictator died in 1975 during this economic crisis, Spain also had to undergo a political restructuring (Carr 1982: 749 and 760–1, Carr and Fusi 1979, Tamames 1986: 228).

In 1975 Juan Carlos, chosen and prepared for this role by Franco, became king of Spain. However, he turned out to be a closet democrat who initiated the transition towards democracy. Political parties and workers' unions became legalized, free speech became a right, democratic elections were held, and in 1978 a new constitution was approved. While democracy was still very fragile – as late as 1981 there was an attempt at a *coup d'etat* by Francoists – Spain had to cope with an economic crisis. One of the actions the new government took was the devaluation of the peseta, a boost for exports and tourism. 1978 was a year of record growth in earnings from tourism (Harrison 1985: 175). When the Socialists won in 1982, Spain's request to join the European Community gained momentum, and by the end of 1983 it was decided that Spain would join by 1986 (Harrison 1985: 182–3).

The socialist government succeeded, though slowly, in achieving economic recovery and in 1986–7, the period immediately following the joining of the European Community, Spain, together with the other new member, Portugal, experienced the highest economic growth in the whole European Community. Though this government had been concerned with the regional distribution of economic growth, it was still unevenly spread. Catalonia achieved the highest economic growth, its economy prospering through high investments and internationalization, i.e. trade, foreign investments and capital transfers (Gasòliba i Böhm 1991).

In the county of La Ribera d'Ebre employment in industry and the tertiary sector increased (though most businesses were still family run), especially in and around the largest towns. The types of industry remained the same, with the exception of a newly constructed nuclear power plant that attracted some metal industry.[1] The result was a split within the county between places where the main part of the population was employed in industry and services, and the villages, where the main portion of the population employed in agriculture (Anon., *La Ribera d'Ebre* 1987: 81).

The modernization of agriculture progressed even more rapidly than before. Land holdings became concentrated, the usage of machinery sharply increased, as did the area of irrigated land, the main cultivars of which became almonds, apples and peaches. With respect to peaches – an export product – this county became one of the three most important in Catalonia (Anon., *La Ribera d'Ebre* 1989: 104 and 114–15). Agriculture became fully market-oriented. A farmer, in his seventies, explained to me that he had once cultivated everything, but now only cherries and peaches. I commented that it would be nice to have a small kitchen garden for fresh vegetables. He responded with laughter to that. He made more money with fruit and was up-to-date on the latest techniques. The vegetables which could be bought at the store were just as fresh. They were all grown nearby, by other people specializing in their cultivation.

The main effects of these developments in the national and international context for the potters of Miravet were beneficial to those potters who remained. During the 1960s and 1970s many potters in Miravet, but especially in the rest of Spain, gave up their profession. During the 1980s, when the potters of Miravet became integrated into the international market, those remaining potters also turned out to be scarce in the international context.[2] At the same time, the economic changes of the 1970s and 1980s effected an enormous growth in demand for pottery. Thus the Spanish economic crisis of the latter 1970s and the early 1980s scarcely affected the demand for pottery. Many consumers of pottery were foreigners – tourists and retirees. Moreover, the demand also increased due to the closer international trade ties: around 1980 foreign demand came into being and proceeded to grow enormously. Therefore, during the 1970s and 1980s, potters gradually raised their prices and increased their production volume by investing in mechanization and the purchase of raw materials, and as a result they made more money.

Potters as members of their village community

During the 1970s and 1980s Miravet followed the general socio-economic trends of the county. The village population declined from 1,014 inhabitants in 1970 to 785 in 1991. (This last number becomes less accurate as people work and study outside the village and return only during weekends, or even less frequently, though they are still registered as inhabitants.) Although some small industries were established – a biscuit factory and one producing plastic articles for agriculture (like hoses) – and the two local construction companies expanded, Miravet remained basically an agricultural village. In agriculture there was the enormous increase, almost three times, in the area of irrigated land due to the cultivation of a river bank (see Appendix C). In addition, a large nursery garden for decorative plants and fruit trees was opened.

From 1970 to the early 1990s, Miravet was increasingly integrated into the region. Transportation was now widely available: every household owned at least one car, and households in which every adult had one car were not unusual. Primary education was provided in the village, though some children attended primary schools in Mora d'Ebre. But secondary education, which had become obligatory, was not provided and adolescents had to commute daily to Mora d'Ebre or even further away to Reus and Tarragona. Daily work outside the village had become usual: several people worked in Mora d'Ebre in places like the leather factory, shops, county administration and the Educational Centre, but many people also worked even further from home, like construction workers in Valls, while still commuting daily.

The village shops offered a limited stock and many people only frequented them to complement their weekly shopping in supermarkets in Mora d'Ebre. In 1992 an enormous supermarket and department store near Reus was opened, and villagers went there to shop on a regular basis. For clothing, a shopping trip to Reus, Tarragona or Barcelona was common. The village doctor, assisted by a nurse, provided only basic health care, for more complicated illnesses, villagers were treated by the hospital in Mora d'Ebre (opened in 1988) or hospitals and clinics in Reus and Barcelona. Cultural events in the region and in Barcelona were regularly attended. During summer, people invited me almost every weekend to attend a rock, folk or classical music concert, held in places varying from neighbouring villages to Barcelona. The parish of Miravet was no longer a separate one: the pastor of Benissanet was also in charge of services in Miravet. Family and friendship tied the villagers closely to Barcelona: many people who had migrated to Barcelona still had second homes in Miravet, where they spent many weekends and vacations, and villagers often visited family and friends in Barcelona. Vacation trips had become an annual or more frequent phenomenon, especially among the younger generation, and villagers now travelled far and wide, including the Americas. Thus villagers not only integrated themselves into the outside world, outsiders also came to Miravet. Catalans, Spanish and foreigners included Miravet, with its picturesque setting and its potter workshops, in their tours. During summer it was hard to find a vacant terrace seat, and some visitors even purchased second homes in Miravet.

Though the tendency of the last decades had been the increasing integration of Miravet into the outside world, the villagers also increased their community integration, most notably through municipal politics, especially since 1978 when democracy returned. In the 1980s two local magazines began publishing; and in the mid 1980s a communal dinner in the school yard became an annual event.

During the 1950s and 1960s subsistence agriculture began to disappear, and by the early 1990s it had done so completely. All agriculture, all economic activities, became market-oriented. At the same time, all economic transactions became articulated by money. The participants now judged economic activities and the resulting living standards basically by one value, and one value alone, money. The villagers would choose a livelihood, within or outside the village, that provided them with a good income. Like any other economic activity, the participants valued pottery production primarily with respect to the money one made with it: the more the better. Thus, in contrast to the previous prevalence of acquiring land and working it, potters became motivated to invest in their own production, and to increase the productivity of their workshops by modernization.[3]

Previously, during the period of mixed market oriented and subsistence agriculture, potters had been in one of the least desirable social and economic positions of the village – only wage workers were worse off. At the end of the 1950s and in the early 1960s a number of potters gave up working in their profession and migrated to the city. The ones who stayed modernized their production and increased their productivity. This increased productivity, combined with the demand that developed during the 1970s and 1980s, improved potters' economic positions. For instance, in the early 1990s, every potter was engaged in one of the most expensive activities: building houses and/or workshops. The only exception was Vicen Sales: he was about to retire and on his retirement would shut down the Societat Fabregat-Sales. Furthermore, his only offspring, a daughter, had left his home, so he did not feel the need to build. All potters owned their workshops and the houses in which they lived, and all sons of potters who had married or soon were about to be married were provided by their parents with a house or apartment. (There was only one recently married potter daughter who lived in an apartment provided by her parents-in-law.) Potters still differed in their incomes, but being a potter no longer implied poverty.

Ferran Segarra's life history is a success story. In 1980 Ferran Segarra and his wife Teresa Pedrola started their own workshop from scratch. They made enough to own a large workshop with a guest apartment and exhibition space for their private pottery and antique collection on the second floor. They lived in an apartment that they owned, and had built a house within the village as an investment (and hence not yet finished but used as storage space). They had also bought an apartment in Barcelona for their student daughter. In 1995 they built a new house to live in just outside the village, complete with swimming pool and antique doors. Of all the potters, Ferran

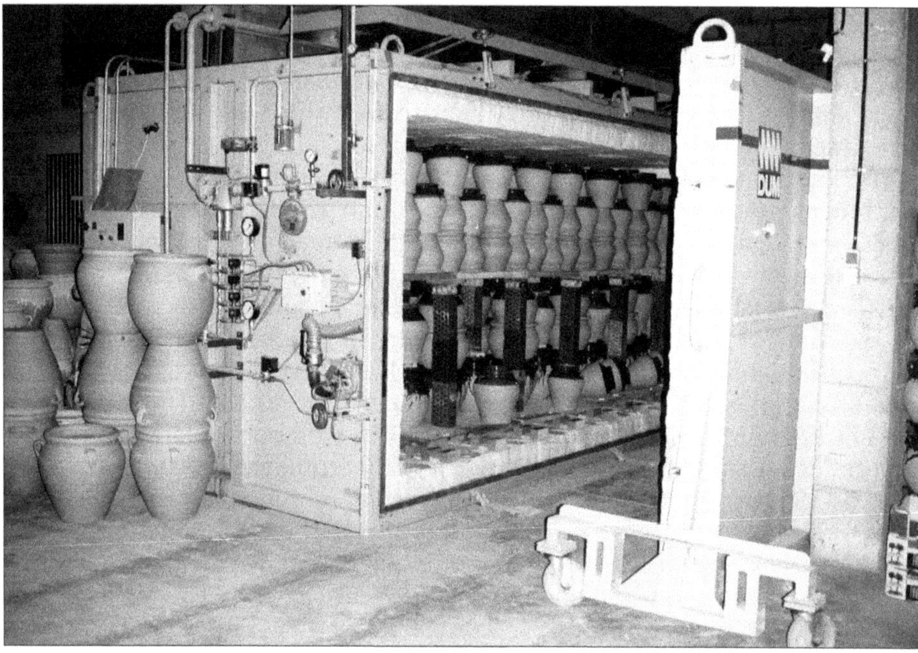

Plate 6.1 ***The gas kiln of the Pedrola workshop.*** Author photo, 1992.

Segarra became the most visibly successful, but in general all potters did well. This contrasted with their fellow farming villagers, who faced problems due to European Community competition.

Further differentiation of production

Since the early 1950s potters had been investing in labour-saving equipment, and they continued doing so during the 1970s and 1980s. Subsequently they made investments in improving the quality of the pottery.

During the 1960s potters had shifted to the use of ponds in which they mixed the two components of clay by employing human or animal power. In the early 1970s they bought electric mixers to do that job. The capacity of these ponds differed in relation to the size of the pottery produced: large pottery needed larger ponds, small pottery, smaller ones. Traditionally potters had used kick-wheels, rather strenuous tools. In 1975 three potters jointly ordered electric potters' wheels, and within a few years all potters used such wheels. In 1981 Ferran Segarra bought the first gas oil kiln in Miravet. The firing of a wood kiln required at least three people more or less continuously engaged, while the firing of a gas oil kiln required only one person who occasionally checked the process. Furthermore, one could better control the firing in a gas oil kiln – something especially important for the firing of glazes. However, a gas oil kiln was relatively small in comparison to a wood kiln. Potters producing small, delicately glazed pottery used gas oil kilns; potters producing large pottery, often unglazed or only roughly glazed, continued using wood kilns. In 1987 Ferran Segarra replaced his gas oil kiln with a fully automatic gas kiln: the firing of this kiln required only one person to switch it on, and the kiln automatically switched off. This kiln was relatively small but produced a perfect glazing. Four years later Segarra's brothers-in-law, the Pedrolas, followed him by buying a similar kiln. Although focused on producing larger pottery, they bought this kiln for their small pieces that demanded flawless glazing.

During the 1980s and early 1990s potters bought all kinds of supplementary equipment, some directly related to their craft, such as an electric grinding-mill, a spraygun for glazing, a spraybooth and a machine to apply bat wash to the bottoms of vessels, and some only indirectly, like a forklift truck, an emergency electricity generator and air heater. Of course, potters also bought this equipment in relation to their repertoire of products. For instance, the Pedrolas bought a forklift truck to transport their large pottery; Pepe Papaseit bought the spraygun and spraybooth for glazing fine pottery. Though potters specialized their productive capacities mainly for producing large pottery or for producing small pottery, their mutual differentiation did not stop with this distinction. They could 'personalize' their productive capacities with recently acquired equipment. The Pedrola's were the only potters who had a forklift truck and shrink-wrap equipment, and Pepe Papaseit was the only potter who had a spray gun and spray booth.

Potters also invested in their buildings. Since the weather is changeable, potters need indoor space for drying freshly thrown pots, for storing them until the firing and for storing finished pottery. They had been used to moving unfinished and finished pots around in their workshops and houses, often even using lofts and living areas, including the space underneath beds, in order to make the maximum use of available space. To save all the extra work of moving pieces around, and also because they needed space for the new equipment and the expanding production, they extended their workshops or built new, more spacious ones. They did this in relation to the size of the pottery they produced: potters producing large pottery extended their workshops more extensively than the others. Moreover, the large pottery could be piled up when drying and stored, in contrast to the small pieces, which needed to be dried and stored on shelves. Potters producing large pottery did not need to furnish their drying and storing areas with shelves as the others did.

Potters used a clay combined of two components, terra refractaria and *llac* or 'mud'. Initially they extracted their own materials, but by the early 1950s they had shifted to buying terra refractaria. The other component, llac, is a residue deposited by the river after its annual flooding. In the dry season of summer potters used to dig it by hand and transport it by cart and mule. In the 1970s they employed machines and tractors to do this. In the early 1980s in a village upstream, a peddle washery was established and potters could now buy llac from this washery. But around that same time, potters also started to buy packaged ready-to-use clay; they mixed this clay with their regular clay in order to improve the clay, but also to save themselves work. Again potters differed according to the kind of pottery they produced. Potters producing large pottery consumed a lot of clay, and in order to work cost-effectively, preferred to prepare their clay themselves; potters who produced smaller pottery consumed a smaller volume and could more easily buy the more expensive ready-to-use clay. Moreover, for this smaller pottery it was necessary to improve clay by blending their own with the ready-to-use variety.

During the late 1950s potters had begun to add skilled and unskilled workers to the permanent labour force of their workshop, but this labour force was still basically formed by the male adult household members. Women had the primary task of housekeeping. In general, work in the pottery workshop was considered too heavy for them. Women were only called upon when many hands were needed for light tasks, for instance when packing and unpacking a kiln, or in periods of severe hardship, for instance to assist a thrower.[4] Continuing the trend of the 1970s and 1980s, some potters increasingly shifted to producing small types of pottery. By the early 1990s they had more light tasks – like applying spouts, forming handles and finishing the pottery – for which they employed their wives and daughters (who, by the way, had more spare time from housekeeping due to modern labour-saving household appliances and running water).[5]

* * *

Already in the 1960s, potters had differed with respect to introduction of new techniques and equipment in relation to their repertoire of products. This differentiation became even more pronounced in the subsequent decades. To exemplify how potters specialized their production, I describe the Papaseit and Pedrola workshops, the first producing small pottery, the latter planters.

The Papaseit workshop

The Papaseits lived and worked in their house according to the local custom. The ground floor was used as a working and/or storing area. For example, carpenters had their workshops on this floor, farmers used it for things such as peeling almonds and packing fruit, and people stored tools, vehicles, products and merchandise there. Thus, the Papaseits had their workshop on the ground floor. The second and/or third storeys were customarily used as living areas. The Papaseit house only had a second floor, where they had the kitchen, bathroom, bedrooms and living room. The top level, with at least one side open (a kind of loft), was used for domestic purposes like storing food and drying laundry.

Pepe Papaseit described the history of the productive capacity of his workshop. In 1944 his father built the house and equipped the workshop. He and his father had been adapting it since that time. Around 1958 they filled up the pit used for the barefoot mixing of clay and constructed a pond with a mixer moved by a donkey. In 1968–9 they demolished the pond and constructed a new one with an electric mixer. Since their lot was small, they had to squeeze this pond in between the house and the road, and therefore limited its size to approximately 2 × 2 m. The family had been using a communal wood kiln, but in 1985 they bought their own gas oil kiln, which required an extension to the workshop. Behind the actual workshop there was a back room. When they built the extension for the kiln, they

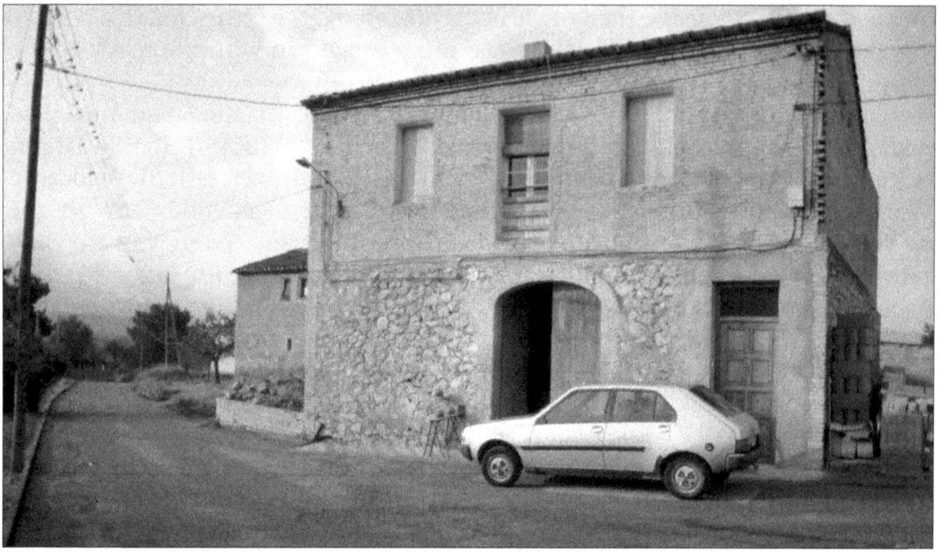

Plate 6.2 ***Papaseit workshop, exterior.*** Author photo, 1992.

also fitted this back room with metal warehouse shelves. Since the son had plans to get married, they built him a house next door with a workshop on the ground floor. Till his wedding, they used this new work space for storing new equipment. In 1990–1 they bought a small and very accurate gas kiln, a compressor to spray glaze, together with a booth in which to do so, and a machine used for removing the excess glaze from the feet of pottery before firing. None of the other potter's workshops had this kind of equipment: with it the Papaseits 'personalized' their workshop.

In the early 1990s the Papaseits had a workshop consisting of one big room of approximately 4 × 7 m, and a slightly smaller back room. They placed their two electric wheels against the back of that first room and in its left-hand corner. In the right-hand corner they placed their electric clay mill. The rest of the room was filled with shelves, trestles and stands for glazing, forming handles, finishing pottery and so on. They arranged these shelves and trestles according to their changing needs, but always kept a passage of about a metre free to get from the entrance to an open space in the back, in front of the wheels. They centred their work around that open space, having the tools they needed within reach of it, and used the rest of the room for drying, storing and small tasks. As they had gradually specialized in making small pottery, the tasks traditionally performed by women also increased. Pepe's wife had become more involved in the production. Usually she was working in this open space while her husband and son were throwing. She performed tasks like applying spouts, and also assisted her husband and son by handing them clay and taking thrown pieces away. The open space had a heater and a stool: visitors, like me, could sit there and talk as they continued with their work. They used the back room for storing finished products and the leftover space in the room with the kiln was used for drying and storing unfinished pottery.

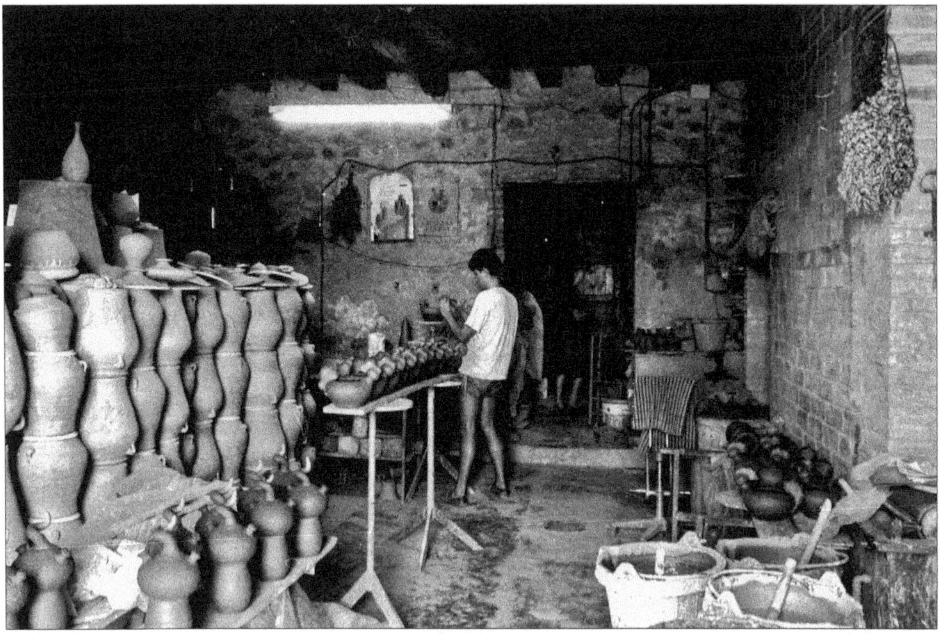

Plate 6.3 ***Papaseit workshop – interior 1.*** View from the entrance toward the back. Author photo, 1992.

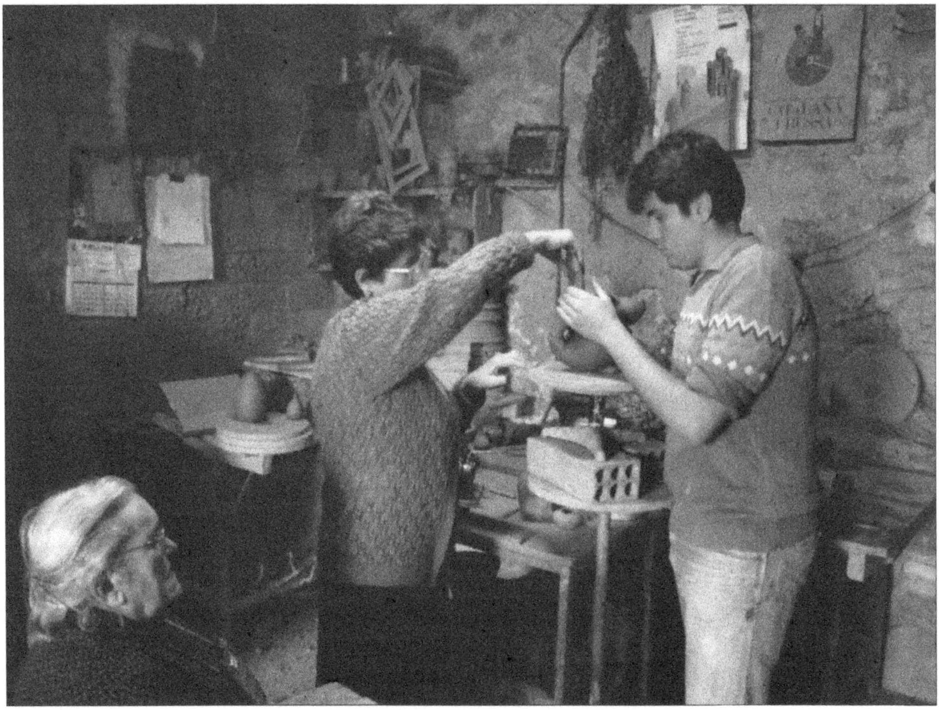

Plate 6.4 ***Papaseit workshop – interior 2.*** Josep Papaseit and his mother Dominica Fabregat working on a pitxell, while Josep's maternal grandmother watches. Author photo, 1992.

The Pedrola workshop

The brothers Pedrola also described the history of the productive capacity of their workshop. The Pedrolas used to live and work like the Papaseits but as they changed their workshop, they did things very differently. At the end of the 1950s they too wanted to construct a pond for mixing clay, yet they did not have space to do it near their house and workshop, so they did it slightly outside the Raval in their vegetable garden. They too had been firing in a communal wood kiln, and when, in the early 1970s, they no longer wanted to do that, they built a new workshop and their own wood kiln at the site where they had the garden. The workshop they built was about the same size as Papaseit's (then the average size workshop) with the kiln placed at a distance of about ten metres behind it. They soon decided to extend the workshop into the space between the workshop and the kiln. At that time Antonio Pedrola Senr. was working there with his two sons and making the usual types of pottery, including a variety of small pieces. As he grew older, his sons gradually took over and realized that the production of small pieces was too labour intensive, and gradually shifted their production to large types of pottery. In 1980 they bought a gas oil kiln similar to Papaseit's, but larger, and extended their workshop again with a room for this kiln. From then on they used the old wood kiln for the firing of unglazed pottery, mostly large types, and the gas oil kiln for glazed pottery. Sometime in the 1980s they extended their workshop again with a 10 × 15 m two-storey shed, and again in 1990 with a space of the same size. At the time of the fieldwork, they had far and away the largest workshop in Miravet. In the same period, they had demolished ponds for mixing clay and constructed new ones three times, each time constructing larger ponds. The last one made in 1990 measured approximately 5 × 10 m. Mixing clay in ponds usually took a couple of weeks, depending on weather conditions. Josep Pedrola told me that he was looking into further expanding their capacity for preparing clay by buying a machine which would mix clay instantly.

Josep Pedrola had come across an old machine used for moulding sewer pipes and had adjusted it for making cylinders. He and his brother had developed a throwing technique using these cylinders as the first step in the process. With this machine they saved themselves time and the strenuous work of the customary first phase of throwing large pots. None of the other pottery workshops employed such a machine. The last new piece of equipment they acquired was a large gas kiln, one that was fully automated.

One entered the Pedrola workshop through its oldest part. In that room, which was heated, sat three electric wheels; the centre of their work. Through two wide doors, this room lead to the first extension with the wood kiln. The rather narrow passageway along the wood kiln led to the room with the gas oil kiln, and from that room into the two last extensions. In the first one, they placed the newest gas kiln. These extensions were filled with dried but still unfired pots and also with finished pots, all neatly piled up. Since they produced large planters, they had the problem of moving these heavy objects around. Josep Pedrola devised two kinds of trolleys for moving freshly thrown pots. One was a composed of shelves on wheels, shelves

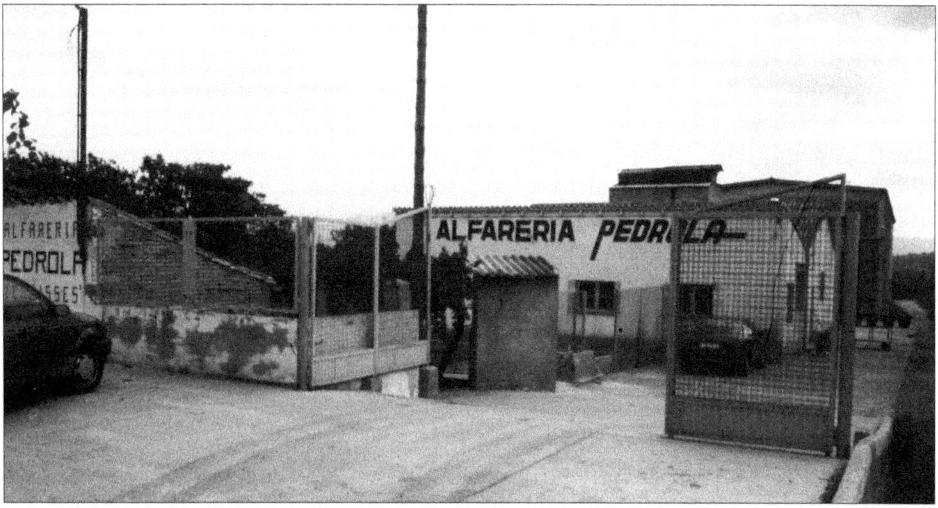

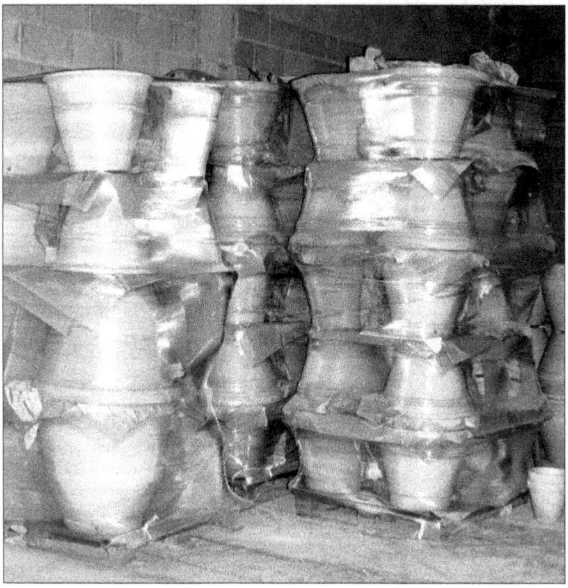

Plate 6.5 (above) ***Pedrola workshop, exterior.*** Author photo, 1991.

Plate 6.6 (left) ***Pedrola workshop, interior.*** These are pallets with pottery packed in shrink-wrap. At that time, only the Pedrolas made use of pallets and shrink-wrap, especially for exporting large quantities of large pieces. Author photo, 1991.

designed for larger pieces than those the Papaseits use. The other was a very low trolley on which very large pieces, up to a metre or so, could be moved. Other workshops might employ trolleys similar to the first type, but none employed anything similar to the second type. They arranged their workshop in such a spacious way that these trolleys could be moved around. Also, they had a small cart to push buckets carrying engobe and glaze to the pots which needed to be treated with these liquids. In contrast, the Papaseits produced small pieces that they carried to a particular place to be engobed and glazed. Depending on the work, the Pedrolas employed one to three unskilled strong male workers. Their tasks were to give the brothers, who were the throwers, the clay, take thrown pieces away, and move these pieces to places to dry.

Casa autonomy

In the early 1990s, in Miravet, the casa, 'household', was still ideally an autonomous unit consisting of a complex of persons and things, but it was no longer necessarily a unit of both production and consumption (see also Harding 1984). Households were always having to balance their actual labour force with their actual access to means of production. For instance, a household with too much labour force to work its own land would send the excess workers out to work for a wage. On the one hand, with the general development towards less labour-intensive forms of production and, on the other hand, the creation of new employment inside or outside the village, more and more people began to work outside the household.

Though a household was no longer necessarily a unit of production, it still formed a unit of consumption. For instance, a person might drive a particular car, but other household members would certainly feel free to make use of that car. When talking about owners of cars, people referred to persons as well as to households. A person who worked outside his or her household did not pay board – i.e. this person was not perceived as enjoying goods and services provided by the household for which he or she in return needed to pay. Such a person would either turn in all of his or her income – with the exception of some pocket money – or would contribute to occasional expenses. For instance, a young professional told me that he did not need to contribute to his parental household income on a regular basis, but when they installed central heating, he paid for it. If someone married and left the household, the parental household still helped financially with setting up a new household. For the majority of the villagers this previously meant buying furniture, but this has come to include an apartment or house. Hence the villagers still held on to the ideal of autonomy of the household, but this autonomy had become increasingly focused on consumption.[6]

However, potters continued to form households which were also units of production. During the fieldwork, only the daughters of Pepe Jornet and Joan Avante did not work in their households, all other sons and daughters did. By having their sons as potential successors working in their workshops, potters were especially motivated to invest in modernization. The Societat Fabregat-Sales was a case in point of a workshop as it had no successor and hence did not invest in modernization. In the late 1950s and 1960s, this was the most advanced workshop in terms of equipment and workspace. However, the son of Fabregat did not want to continue, and the son of Sales, who was a bit older, did not himself have a son. Therefore, after the 1970s, the Societat Fabregat-Sales barely invested in their workshop. In the early 1990s all potters' sons worked as potters in their parental workshops and presumably would take over these workshops in due course. Though people in Miravet did not necessarily perceive the household as a unit of production, potters modernized with the idea that at least one of their sons would join the household's labour force and continue the workshop.

Potters adhered to the cultural value of autonomy of the household by financing the modernization of their productions with their own financial resources and/or those of the parental household. At least this was the image they projected to their fellow villagers. Household autonomy was an important source of pride. Even if they took out a loan, they were not likely to admit it publicly, especially to a stranger like me.[7] Previously, potters were forced to maintain this autonomy, as they had few financial resources outside their own household. By the nature of their profession, potters lacked such assets as land for collateral, and only when they were doing economically well did bank credit become available to them. This was in stark contrast with farmers. The Franco government created easy loan conditions for farmers and significantly reduced risks by fixing agricultural prices. Potters were also acutely aware of agricultural and industrial subsidies given by the European Community to fellow villagers. They told me that they were not eligible for any such support. They did become eligible for governmental support in 1986 when the *Subdirecció General d'Artesania* (a subdivision of the Department of Industry and Energy of the Generalitat) was founded with the goal of supporting artisans.[8] Until the early 1990s, only Ferran Segarra received such a subsidy, a minor one, he said, as it only covered the costs for designing his new workshop. Note that Ferran Segarra, with his urban mentality, became the first potter to acquire such a subsidy. He knew how to play the bureaucracy to acquire it.[9] But also note that this was a subsidy and not a loan, hence was risk free and could not compromise the household autonomy.

By purchasing labour-saving equipment and by enlarging workspace, potters no longer needed to exchange and co-ordinate labour with each other, and could better realize the ideal of the household as an autonomous unit of production. Previously, not every potter's workshop owned its own kiln but rented in turn a communal one by paying a fee to the owner. This entailed co-ordinating with other potters using the same kiln. These potters constructed their own kilns during the 1970s or 1980s. Previously potters had been forced by their lack of space to glaze pottery outdoors, always on the same day that they packed the kiln. With this density of tasks, they required more labour than was available in their households, hence they were forced to exchange labour with other households. With enough workspace, they no longer needed to exchange labour.

Thus, as potters modernized and ceased to exchange labour, they lacked the direct, intimate knowledge of each other's production. In the early 1990s they hardly visited each other's workshops, but on occasion they could still observe each other's production. Workshops were not off-limits but semi-public. (For instance, one needed to be invited to enter someone's home, but not to enter his workshop. Still, a person who was not on good terms with a potter would not enter that potter's workshop.) Potters could not be secretive about certain aspects of their production. For instance, when I talked with Josep Pedrola about his adapted employment of the machine moulding sewer tubes, I asked him how other potters had responded to his use of this machine. He replied that he had not hidden this machine. It was sitting in his workshop easily observable. But then again, he had not advertized it either.

This reticence to share technical knowledge was most obvious when it came to knowledge of glazing. The colour of glazed pottery is the combined result of the colour of the engobe (or slip, a thin coating over the body of a pot) and the colour of the glaze (the top, a glass and transparent coating). Initially potters used a kaolin for their engobe, which they called *lapiz*, extracted from the nearby mountains. It was either quite pure and white, or mixed with iron oxide and red. They used this kaolin as they found it. They covered this with an almost colourless glaze, hence this glazed pottery appeared as the colour of the engobe: either yellowish white, or red.[10]

During the 1970s and 1980s, when potters developed their new repertoires of products, they added colours to the original range by mixing oxides into the engobe. As Pepe Papaseit pointed out: 'One could say that the period in which oxides began to be used pretty well coincided with the one when every person worked in his own workshop. The moment had arrived when everyone faced the problem of acquiring new skills as one no longer made just the usual pots. The secrets of the craft came from the moment when one worked with oxides and produced new types of pottery.'

The final colour of glazed pottery is the result of many factors including the composition of the engobe, glaze and clay, the kind of fuel used in the firing, and the duration of the firing. It is only predictable after a series of proofs in which these factors are observed and varied. Since the potters differed in the meticulousness with which they controlled these factors and in the extent that they examined proofs (and the directions these proofs took), the colours became very personal. For this reason the potters were reticent to discuss this aspect of their work with other potters.

Potters mainly used copper oxide to achieve green, manganese oxide for brown and cobalt oxide for blue. Potters learned the uses of oxides through different channels. Some potters had known of the use of oxides years before they actually applied them in their production. Out of curiosity, some had even made proofs with them. Other potters saw the products of their fellow villagers and also started to use oxides. I queried each potter about how they had learned the use of oxides. With the exception of two cases, each one told me that he had learned it from an outside source like a book or a company specializing in products for the ceramic industry. For instance, Pepe Papaseit had seen other potters in Miravet producing new colours and he knew that they did this by using oxides, but did not know which oxides and exactly how it was done. When he had to make a trip, he passed near a place where he knew there was a company specializing in products for the ceramic industry. There he obtained information about the uses of the different oxides. I probed why he had not asked the other potters. 'I do not know,' he replied. 'I have not asked anyone. Since these things are so private, if they do not tell you spontaneously, how can I ask?' He went on to say that if he had seen, for instance, a green achieved by another potter, he would not ask that potter the composition of the engobe, but rather would start investigating companies in order to obtain the desired result. The precise use of oxides, the composition of engobes, was not secret. Every potter could read about oxides and enquire about them, but a potter would not easily tell a fellow potter about his particular recipes. 'I hide my recipes a bit and others do that too. Because, it's not like it used to be. Nowadays you can show me something and there is the possibility that I will use it to compete with you.' He defended his reticence by saying that he had invested time in the development of his recipes.

There were exceptions to this reticence. In a demonstration of uncharacteristic forwardness, Vicen Sales explained that he had asked Milio Treig Senr. about the use of certain oxides, and had gained the desired information from him. Vicen immediately elaborated on this by listing others whom he might have asked and those whom he would not. The difference was that he had *més amistat*, 'more friendship', with some than with others. Other potters who lacked the characteristic secretiveness regarding colour were Ferran Segarra and Josep Pedrola: both are brothers-in-law, very close friends, and have developed their recipes together. Hence, unless friendship existed, potters felt uncomfortable asking for such information.

To ask a potter for certain technical knowledge was to ask him a favour; to give a potter certain technical knowledge was to grant him a favour. For potters who were good friends, this exchange of favours was an aspect of their friendship. For potters who were not such good friends, the exchange of technical knowledge functioned differently. Several potters explained that they had given particular colleagues information about the use of oxides. When I asked these colleagues about this, their answers did not confirm the stories of those who claimed to have shared this knowledge. Since the use of oxides is a complex process, the recollection may have been altered in the telling. One potter might very well have told another potter something about oxide use, but the second one might have learned other aspects of oxide use from an outside source. Hence, both sides of the story could be right, but the most interesting thing is that both told different stories and not complementary ones. By telling me that he had told another potter about the use of oxides, a potter showed his friendliness towards that other potter. This was an aspect of the general attitude of people of Miravet: everyone was everyone else's friend, or at least that was how one wanted to appear. Yet by denying that another potter of Miravet had given him information on the use of oxides, a potter refused to acknowledge that he was indebted to another. Such an indebtedness could only be reciprocated by the continued mutual exchange of favours as occurring between friends, and not all potters were such good friends. Therefore, potters' reticence to credit another potter of Miravet with providing information on the use of oxides characterized the general mutual relationships between potters as developed in relation with the differentiation of production. In addition to this technique of applying oxides, I have come across many other instances of newly introduced techniques, like the use of quartz, of a certain glue to fix flawed pottery, of chemicals to prevent pottery sticking on shelves in the kiln, or of chemicals to prevent the glaze from dripping. Potters were equally reticent about these techniques, or had been when such a technique was newly introduced.

A concomitant effect of this reticence to share techniques, but also one of the general changes in the production and repertoire of products, was that potters could not, even if they wanted to, exchange labour with each other. Previously, with their old equipment and workspaces, potters had been forced to exchange labour with each other, and since they all produced the same products using the same equipment and techniques, they could do so. At that time, potters did not know each others' techniques and were not familiar with each other's equipment; they no longer had the option to exchange labour. The last regular case of labour exchange I witnessed was two potters exchanging labour for the firing of a wood kiln, the oldest form of firing. But in 1991 one of these potters bought a gas oil kiln: though he continued to exchange labour with the other in the firing of the other's kiln, the symmetry in the exchange was lost and the continuation of their exchange relationship potentially endangered. Labour exchange, which had characterized the general mutual relationships among potters, disappeared.

Between the early and late twentieth century the potters' status within their village community no longer depended on their profession, but on their individual monetary income, which in some cases was relatively high. From the 1950s on, potters no longer marketed their products to muleteers – fellow villagers – but to wholesalers and shop-owners from elsewhere. Finally, the mutual relationships potters had formed by reciprocally exchanging labour between workshops had almost disappeared.

New repertoires of new products

During the 1960s some potters specialized in the production of flowerpots by investing in jiggers, machinery to prepare clay in large volumes and in the employment of workers. At the end of the 1960s a pottery factory was founded in Naverette (La Rioja), Fansa. This factory produced flowerpots on a much larger, industrial scale and eliminated the flowerpots of Miravet from the market. The potters ceased producing flowerpots and therefore employing jiggers, and had to redirect their production to a new type of pottery that then came into demand: the *jardineria*, a derivative from *jardí*, 'garden', which constituted a wide array of models.

This category of planters had some forerunners. The *clavellet* marketed by muleteers could be considered by its function jardineria, although interviewees never used the label jardineria in descriptions of pottery produced before 1960. Other pottery produced before 1970 might also occasionally be used as a planter. I had, for example, seen several old solseres with decorative wavy lines similar to those on flowerpots, but not glazed on the inside. When asked about these pots, Vicen Sales remembered that the Fabregat-Sales workshop had been producing these vessels in limited quantity for one of their clients in Saragossa during the 1950s. Since they were unglazed he surmized that these might be used for storing water, but since they were smaller than the vessels usually used for storing water, and had those decorative lines, they might also have been used as planters.

Vicen Sales surmized this use because at the end of the 1960s when potters began to produce the larger types of vessels as planters, they left the inside unglazed. The shoulders and rim of these were, however, still glazed.[11] The colour of the glaze had been reddish, but when used for potting plants, the colour was either reddish or green. Some potters explained that the green went better with plants. In the former use, the vessels had to be closed at the bottom or have an opening reinforced with a ring in order to put a bung in it. When used for potting plants, the vessels needed to drain and therefore have a simple opening at the bottom. When used as previously, the mouth of vessels had to be large enough to make it possible for oil or food to pass through, but not too large, so as to prevent debris from falling in. When used as a planter, the mouth was wider in order to give plants more space. When used for the storage of oil, vessels ideally had a smooth surface which made it easier to keep clean, although some potters striped vessels in order to camouflage flaws in the throwing. When used as planters, the vessels were always striped to camouflage flaws, but more importantly for decoration. Previously, all the vessels had handles, mainly to differentiate the several sizes. However, used as planters, these handles were occasionally left off. Furthermore, some potters made strawberry pots by adding three 'pockets', quarter spheres stuck just underneath the shoulder of the vessel which covered a hole that was made in the wall. Plants could be planted in these pockets and the hole in the wall allowed the plant to root into the vessel itself.

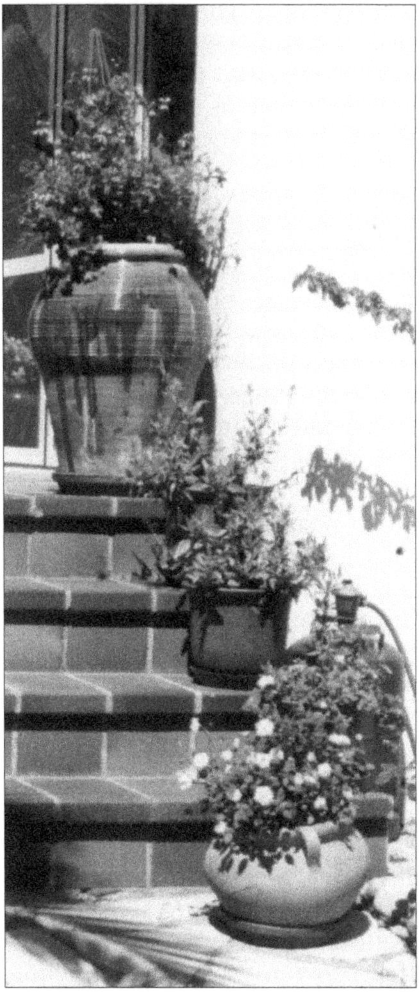

Plate 6.7 ***Planter in the shape of vessel.*** The top planter has a shape derived from the vessel of the first half of the twentieth century, and is produced by the Pedrola workshop. Author photo, 1997.

In the late 1960s and early 1970s many other types of pottery underwent similar changes as they became used as planters. The calandrin was a cossi used both as a washtub and for doing dishes. The wholesalers and shop-owners who bought flowerpots became interested in this type of pottery for planters as it is larger than the flowerpots. The calandrin originally had a horizontally flat rim that made it easier to place a board on it while doing the laundry. But this rim also chipped more easily. When the calandrin was produced as a planter, the rim was rounded, and therefore stronger and more visually appealing. Sometimes the rim was even vertically nicked with a stick or a small toothed roulette just like bordons, the horizontally attached ribs, were nicked. The calandrin could also be decorated with wavy lines and with bordons. Since the calandrin could be decorated in different ways, clients could order varieties. For instance, in one order they could order two or more variations, or from one order to the next they could chose another variety. Potters have also added more sizes than the initial two.

The last type of pottery that transformed into planters was the cossi de paret. Most participants had never made this type for its initially intended use as a washtub. They knew of its existence through their fathers and by seeing old pieces. In the 1970s they returned to producing it as a planter, again sometimes with some changes in its physical form. Like calandrins, cossis de paret originally had a horizontally flat rim that, as this pottery was produced as a planter, became rounded. This pottery was also decorated with such things as a lion's head.

In the early 1960s, when delivering the old types of pottery, Josep Fabregat had shown clients the mechanically moulded flowerpots in order to solicit orders for them. During the 1970s and 1980s potters did this not only with pottery that was actually in production, but also with pottery for which they had only produced a prototype. They derived ideas for new models of planters from different sources. Some potters bought pottery in order to copy it. For instance, Pepe Jornet did this with an old tub made in another potter village, as did Josep Pedrola with a planter decorated with garlands. Some potters copied more indirectly, like Pepe Jornet did a planter he had seen in, of all places, a soap opera on television. A few potters designed planters. Josep Pedrola had designed a series of formally similar planters.

Clients could also suggest new models. For instance, in the early 1970s, the Jornet workshop received an order from a client of theirs to copy what this client called a *bolo*, an onion-shaped planter that this customer, an amateur potter, had designed. Jornet added three more sizes, and incorporated this planter into his regular repertoire of products. Such an introduction of new types became commonplace, as clients made, drew, or described new models, or brought them from elsewhere. For instance, one client of Jornet ordered copies of the cossi as it was typically made in Quart (Girona). He considered the potters in Quart too expensive and therefore ordered these from Jornet.[12]

As orders for planters continuously shifted over time, the repertoire of planters produced at any given time also shifted. I kept coming across more varieties during fieldwork. Once, while cleaning the Pedrola workshop, I came upon a planter resting on three feet that I had not seen before. While

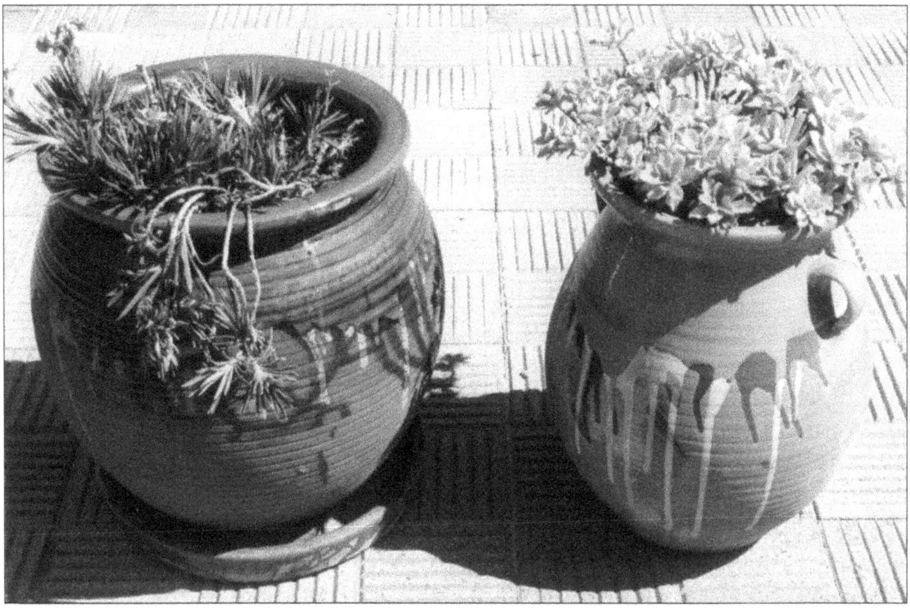

Plate 6.8 ***Planters produced by the Pedrola workshop.*** The left one is a bola. Author photo, 1997.

leafing through carbon copies of bills, I came across a planter described as '*cuenco palmera*', a variety I had never heard of. The potters had a seemingly inexhaustible repertoire: at the time of the fieldwork the Fabregat-Sales workshop, ironically, received an order for hand-thrown flowerpots, a planter they had not made since they had bought the jigger; so this time they threw them by hand.

These repertoires could overlap between potter workshops when planters were derived from old types of pottery. Previously potters and muleteers, i.e. the people who were intimately familiar with Miravet work, could recognize pottery as being produced by an individual potter. Formerly these differences were minor, such as the finish of a rim or the surface texture. With these planters, however, differences were much more apparent: I could recognize them easily even upon my first arrival in Miravet. The planters differed in final shape and details from potter to potter. For instance, only Pedrola and Avante decorated the cossi with lion's heads, though their heads had very different expressions. In other cases the repertoires of planters did not overlap. When a potter found an idea for a planter from a sample not made in Miravet, or when he developed a new design for a planter, no other potter copied that planter. Usually, when a potter received a new type of planter from a client, only he would produce it. For instance, Jornet was the only potter who produced the bolo and the cossi originally from Quart. The sole case in which potters took over planters produced in other workshops was the *bola*. This was an ovoid planter brought in and ordered by a client of the Fabregat-Sales workshp Later, the Pedrola and Papaseit workshops added this planter to their repertoires, although in clearly distinguishable versions.

* * *

In the early 1970s the production of what had been the main types of small pottery came to an end: the demand for the 'painted pottery' ceased, and the Barcelona traders who marketed their merchandise by employing colourfully decorated donkeys stopped ordering. As the demand for these types disappeared, potters responded by developing new repertoires of small pottery.[13]

These potters increased the production of pottery that until the early 1970s had been produced in small quantities, or even as single pieces. This was pottery that had been peddled in villages nearby. By the 1990s this small pottery consisted of pieces like the *baci* (the chamber pot), the measurements for wine, complete with funnels and drain bowl, the bedwarmer, the colander, the bottle for olive oil and the simple pitxell. This was also pottery that was not peddled at all but made for use in one's own household or to be sold at the workshop to fellow villagers or to people from nearby villages. The latter consisted of pieces like the mortar, vases, the *caduf* (a ladle for a Moorish irrigation machinery) and three types of pitxell. These pitxells were the *pitxell de tronc* – a pitxell resembling a tree-trunk; the *pitxell de barral* – the 'pitxell of the barrel', a pitxell specially designed for use by muleteers; and the *gargola de moxó* – a piece always classified as a pitxell but called gargola (similar but smaller than the canteret) which has a small bird, *moxó*, on top. This last piece is an *engany*, a 'tricky piece': it looked like a pitxell but the opening for water was hidden in the bottom and it had six spouts, only one of which water could be drunk from.

To the kitchen utensils already produced, new ones were added. As with the planters, this happened because clients ordered copies of kitchen utensils made elsewhere, or potters, on their own initiative, copied such pottery from elsewhere, or potters designed new kitchen utensils. Included were the following types: several types of cruet developed by reducing the size of the old bottle for olive oil, and by copying and designing other models of cruets; the *tomba-truita* – literally 'turn omelet' – which was a large plate with a knob at the bottom used to turn an omelette in order to fry its other side; the *fregall* – the 'pan-scrub' – a small ball with a side opening to keep pan sponges and so on out of sight in the kitchen sink; small vessels with a sieve as lid for storing oil that had already been used for frying (the lid sifted out crumbs of food and thus cleaned the oil; these small vessels had 'meat' or 'fish' written on them in Catalan in order to distinguish the oil used); a vessel with holes in it to be used for storing garlic; several models of canister – some with shapes clearly derived from traditional vessels – which had written on them, in Catalan, 'pasta', 'sugar', 'honey', 'sweets' and the like; several types of salt-cellar, including one to be hung on the wall; vessels to be hung on the wall with holes in the bottom for holding cutlery; and finally flatware like dinner plates, bowls and also coffee and tea sets. All this pottery was glazed; the only new unglazed ware developed was the *pitxell de nevera* – the 'pitxell of the refrigerator'. This was a pitxell low enough to fit on the shelves in a refrigerator and was used to keep water cool.

The repertoires of these potters also expanded through the introduction of many new objects with a wide variety of functions. These items consisted

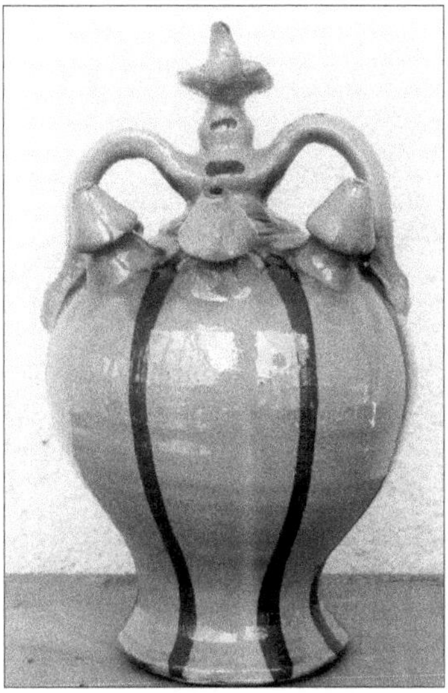

Plate 6.9 ***Gargola de moxo.*** This piece is produced by the Segarra workshop. Author photo, 1992.

of ashtrays, pencil holders, small planters in the shape of roosters or cadufs (the cadufs now glazed), candlesticks, lanterns, a tray for the 12 grapes to be eaten, as is the Spanish custom, at every stroke of the clock on New Year's Eve and, above all, pitxells. Decorative pitxells had already been made, like the pitxell de tronc and the gargola de moxó, but in the 1980s the production of pitxells increased and many new decorative pitxells were developed (about 80 in all according to Pepe Papaseit and Milio Treig). In the early 1990s one could find pitxells resembling vegetable forms like the aubergine, tomato, melon, acorn, pumpkin and pineapple. There were pitxells resembling animals like the rooster, elephant, turtle, owl, snail, squirrel and peacock. Others resembled objects like a certain pastry or a bell. Some pitxells had geometrical forms like the rectangle and the circle. A variety of pitxells were decorated with patterns of spouts and handles, some very intricately. One pitxell had a spout in the shape of an erect phallus. Pitxells were also copied from those of other places, like the *doll* from the Empordá and a Greek pitxell. The pitxell de barral was previously unglazed but, like all the other decorative pitxells, it became glazed, often with a spiral design. Though not actually a pitxell, the gargola was then glazed. In addition to the usual size, the pitxells were produced in miniature and medium sizes. And, finally, small containers in the shape of baskets, eggs, cups and the like began to be produced: these are meant to be filled with sweets and to be given as souvenirs at weddings and baptisms, and hence were inscribed with the dates of these events and the names of the bride and groom or newborn. The one type of unglazed pottery was the *cargolera,* a vessel with round holes in it designed to be used to fasten and thus clean snails (*cargoles* in Catalan), but which was actually used as a lamp shade. (The Papaseit workshop produced small pottery and fortunately carbon copies of their bill books for a part of 1971 and 1972 still exist. These gave the most accurate possible cross-section of the repertoire of this workshop. See Appendix F.)

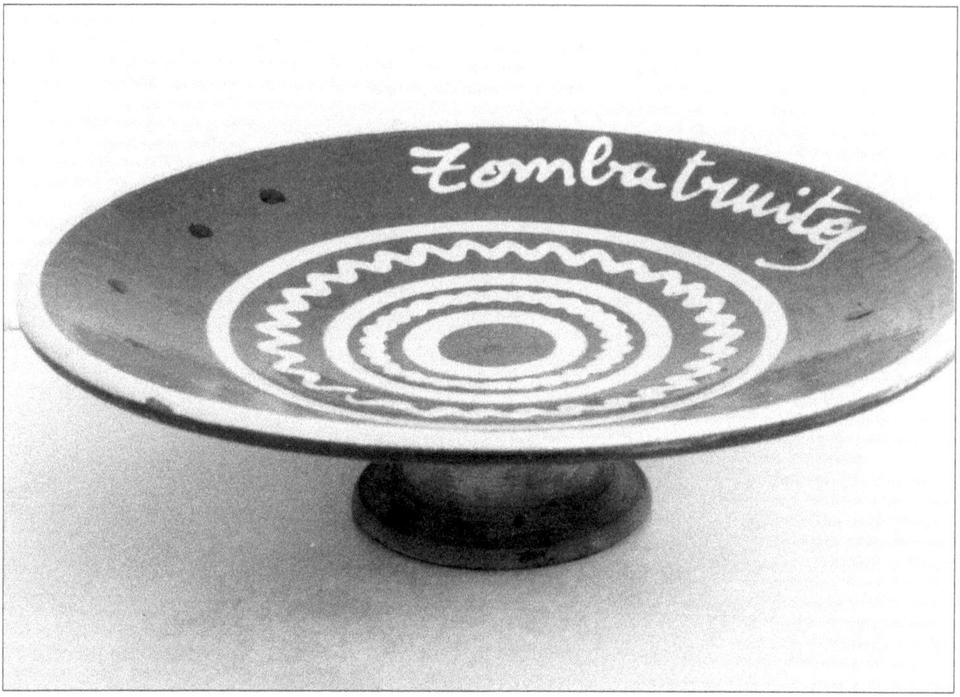

Plate 6.10 ***Tomba truites.*** This piece is produced by the Papaseit workshop. Author photo, 1997.

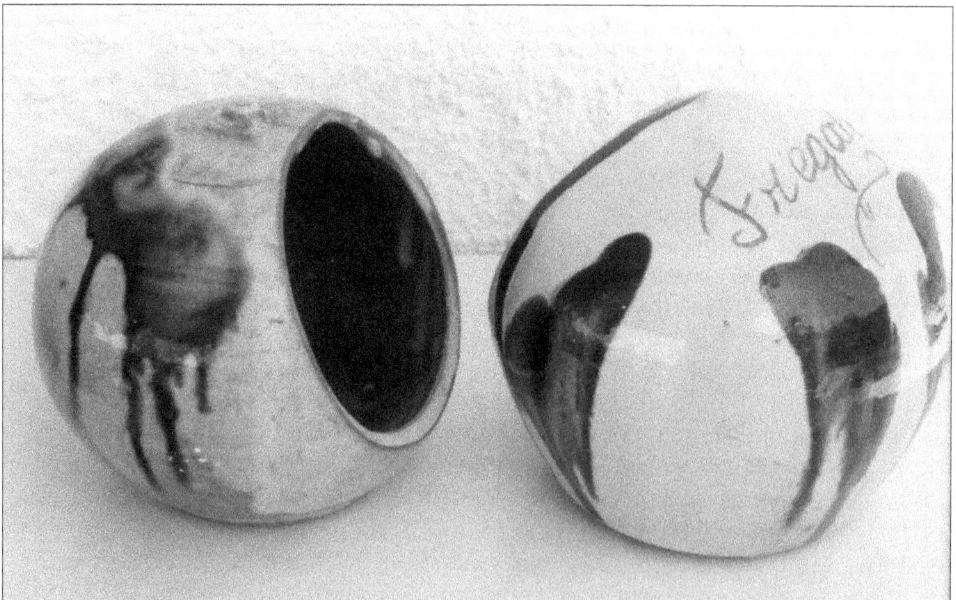

Plate 6.11 ***Fregalls.*** These pieces are used to keep sponges on kitchen counters and are approximately 10 cm high. Left: Segarra workshop; right: Papaseit workshop. Author photo, 1997.

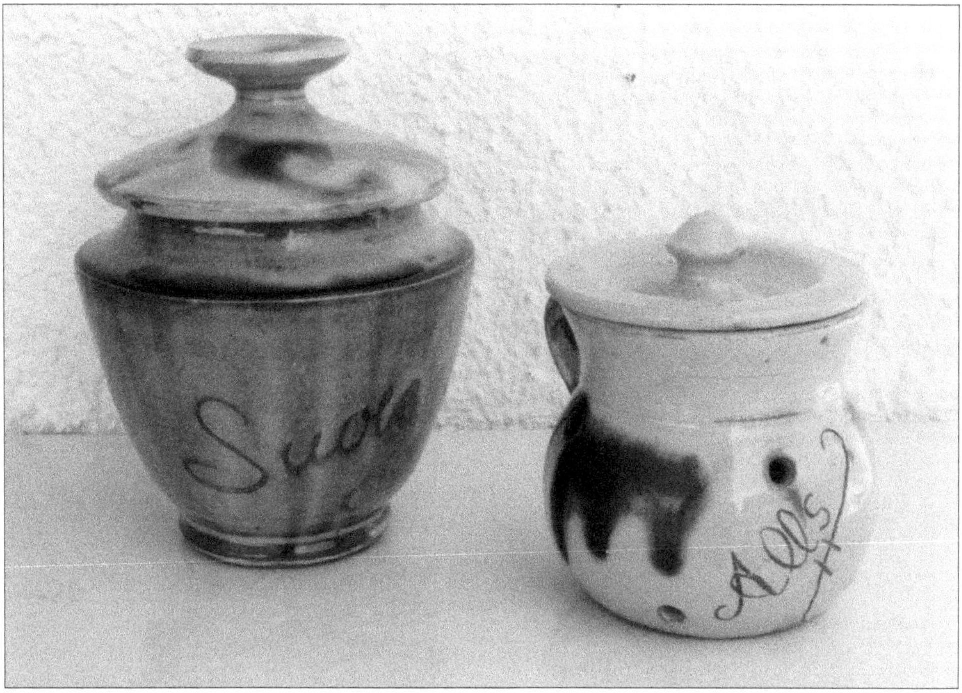

Plate 6.12 ***Containers.*** The one on the left is used for sugar, the one on the right for garlic. Papaseit workshop. Author photo, 1997.

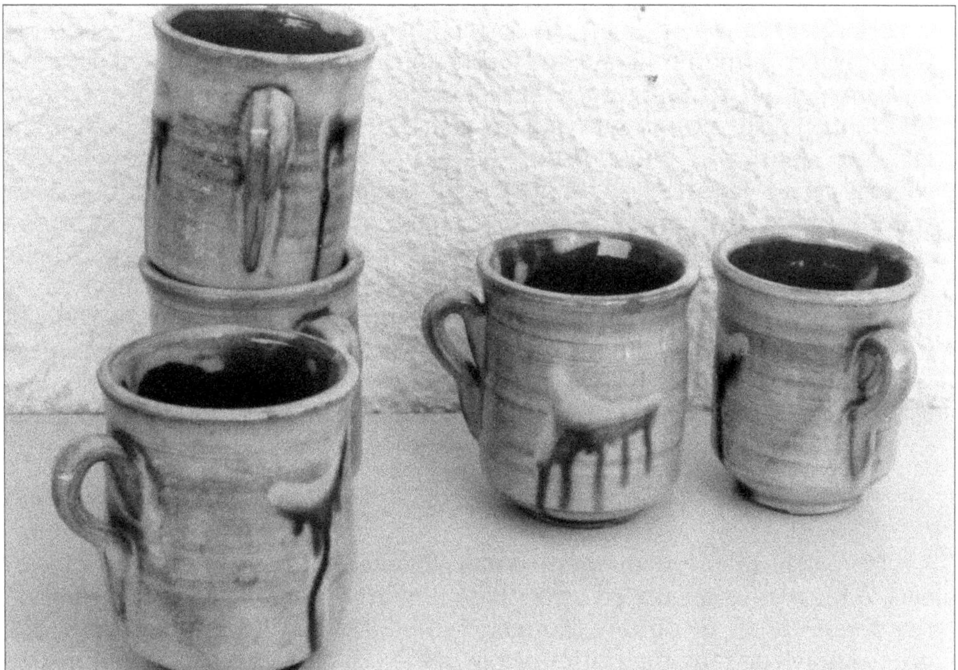

Plate 6.13 ***Coffee mugs.*** Segarra workshop. Author photo, 1997..

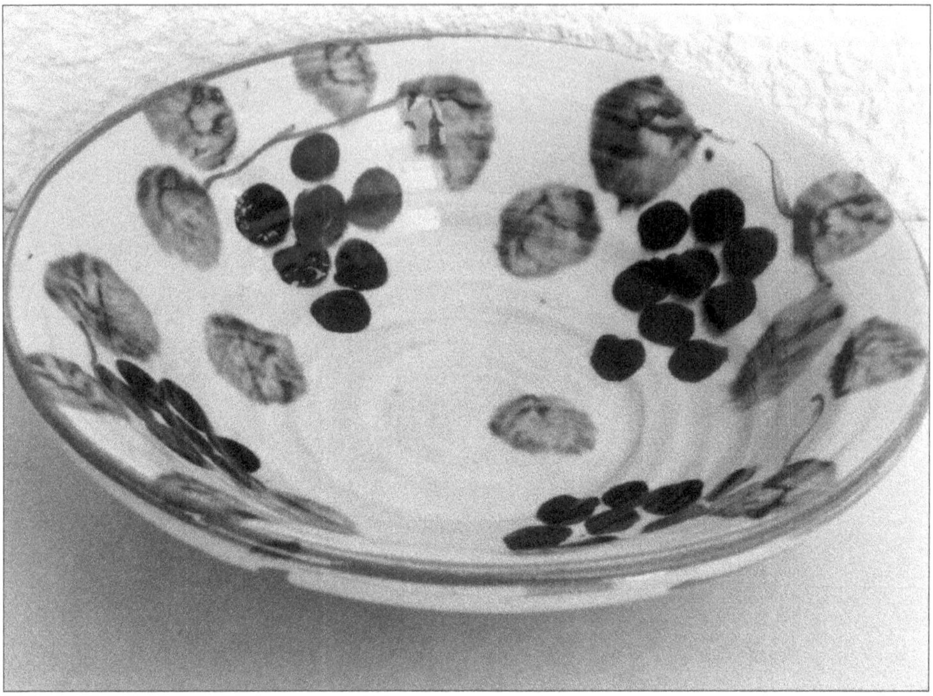

Plate 6.14 ***Bowl with grape design.*** Segarra workshop. Author photo, 1997.

As with the planters, potters stopped producing several old models of small pottery and developed new ones. Potters overlapped their repertoires with respect to pottery that had been previously produced, like the wine measurements that were quite standardized. Potters overlapped also with other previously produced types in that they produced pieces that they labelled identically, though they varied shape and glaze between the workshops. For instance, one potter made the gargola de moxó with a more spherical body than another might, or one potter made the bird on top of the gargola de moxó with its wings spread out while another made it with the wings folded. Potters considered other types, especially the decorative pitxells, as highly individual. In some cases when two potters made similar items one accused the other of plagiarism. By contrast, no such accusation would ever be made against a potter copying from an outside source. Especially with respect to these decorative pitxells, some potters attempted to design five to ten new models annually.

* * *

As previously mentioned, when potters marketed their products to muleteers, both parties calculated prices based on an arithmetical organization of the several types of pottery. Potters contested this calculation since it did not acknowledge the quantities of raw materials and labour invested in the different types of pottery. When potters shifted to marketing to wholesalers and shop-owners, they changed the price calculation to acknowledge these different quantities. When asked how they calculated their prices, potters invariably answered that it was based on input of raw materials and labour, and perhaps such additional factors as space occupied in the kiln and inherent risk of coming out flawed, but they could not further specify their calculations. To illustrate, Josep Pedrola told me that one day Pepe Papaseit had come to ask him for the price of a certain type of pottery. Pepe had not produced this type for a while, had received an order for it, and was not sure what he could charge. Josep Pedrola had told him the price he charged, and Pepe then asked if it was worthwhile to produce that pottery for that price. How, Josep Pedrola asked me, could he measure the quantity of clay needed, and the fuel, and the labour, that is the preparation of the clay, the throwing, the drying, the glazing and finally the firing? Josep used this anecdote as an example that even among potters the calculation of prices was hard to explain. One really just asked a price that seemed to take all these factors into account without actually calculating them. The price should only give the impression that it was calculated according to these factors. The importance of this impression was made clear in an anecdote told by Ferran Segarra. As a translator, he had helped another potter in negotiations with a French wholesaler. This wholesaler had ordered the same planter with and without decoration, and the other potter had charged different prices. Yet, Ferran emphasized, the difference between those prices had not been enough to convince the wholesaler that it was based on different labour and material input and he had been reluctant to accept these prices.

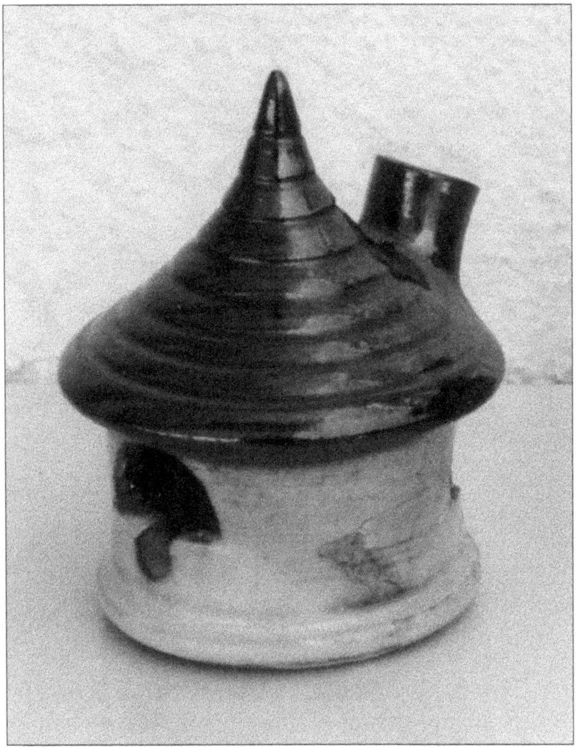

Plate 6.15 *Fantasy ash tray*. Author photo, 1997..

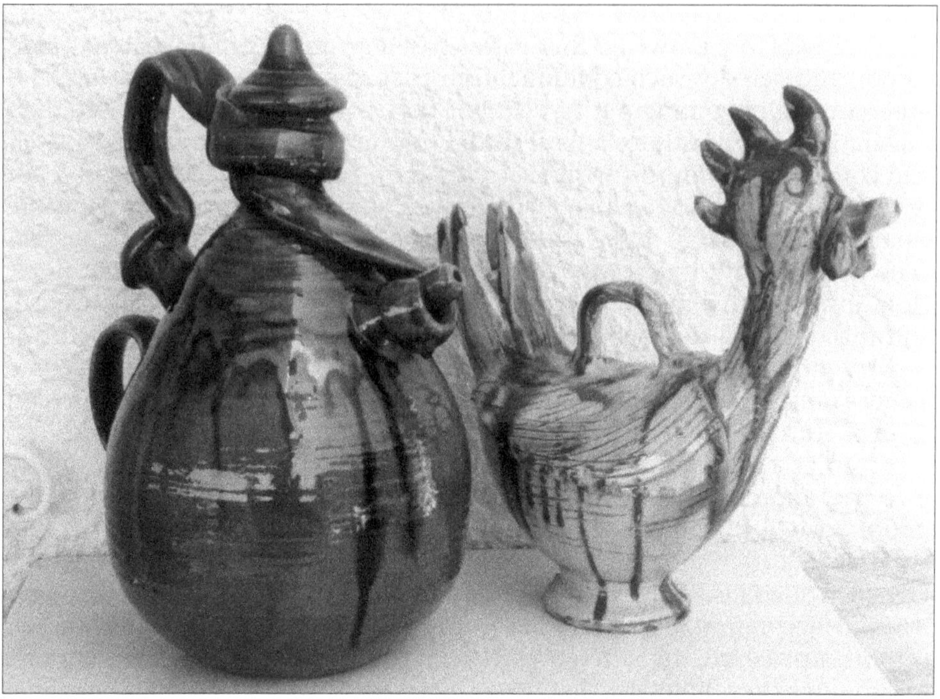

Plate 6.16 *Pitxells 1.* Left: Segarra workshop. Right: rooster, Jornet workshop. Author photo, 1997.

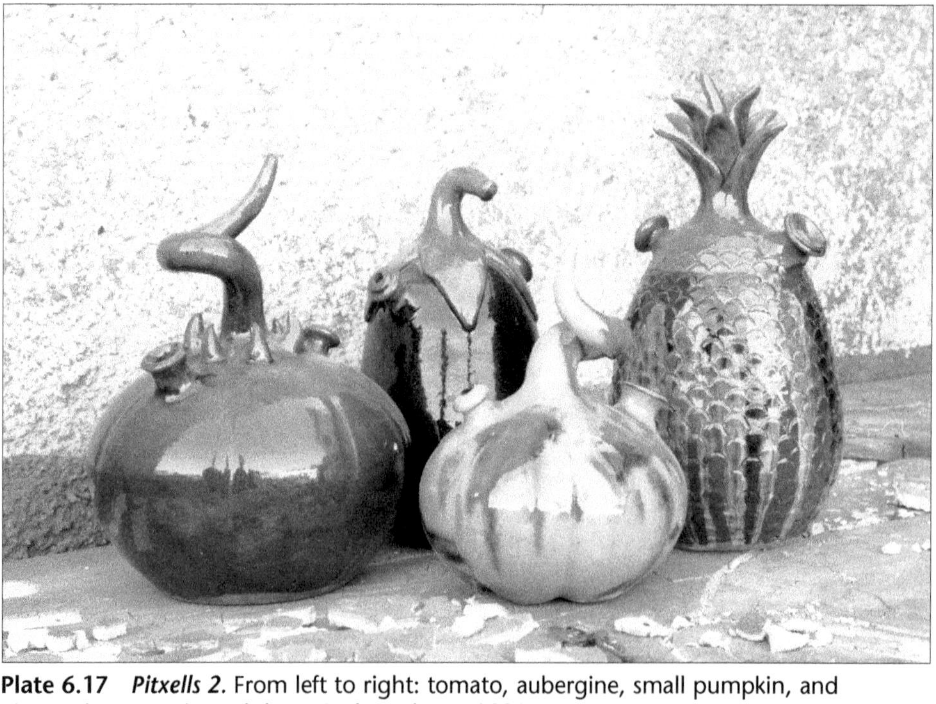

Plate 6.17 *Pitxells 2.* From left to right: tomato, aubergine, small pumpkin, and pineapple. Papaseit workshop. Author photo, 1992.

The calculation of price based on labour input influenced the differences between potters. Previously, when marketing to muleteers, potters had also differed with respect to the quality of their products, but these differences were only a factor in the personal relationships between potters and muleteers, not in the price calculation. During the 1950s, price had become a private matter as opposed to the previous period of collective prices. As potters took stronger market positions, they no longer suspected colleagues of undercutting prices. By the 1990s the personal quality differences had become a factor in the calculations. Potters declared, at times they even bragged, that they charged higher prices than others. Potters who charged higher prices did so because they claimed that their products were of better quality. As potters modernized their production they perfected their products in accordance with the old quality criteria, but also developed new ones.

During the age of the muleteers, potters had valued pottery that was well fired and did not have cracks and holes. In 1949 Josep Fabregat used new construction techniques to build a kiln. This new type of kiln provided greater control over the firing and eventually all potters built such kilns. In addition, they devised a new technique for packing kilns. These changes resulted in pottery that came out evenly and well fired, and cracks and holes in pottery became rare.[14] Previously potters prepared clay with a technique that left grains of chalk in it. After the firing, these grains might cause pottery to chip. Upon receiving the pottery and bringing it home, muleteers had to *remullar*, 'soak', it: they immersed it in water until the chalk diluted. This extra work was inevitable and therefore accepted. When potters switched over to preparing their clay in ponds, they filtered out these grains of chalk, and hence improved the quality of their clay.[15]

Gradually potters developed new criteria for evaluating pottery. They did so in relation to the modernization and subsequent individualization of their production and their repertoires, and, of course, in relation to their new marketing practices and, indirectly to new consumers and consumption of their products. Potters in general worked towards a perfection of their products according to new criteria, but at the same time potters also differed in their criteria and the attention they paid to them.

The potters and muleteers had judged pots by their *bona línia*, 'good line'. At that time the potters threw pottery in conventional shapes: the 'good line' referred to the clarity of the contour of these shapes. Shape was then of no concern. As the potters started to produce new types of pottery (i.e. non-conventional ones), not only was the clarity of the contour a concern, but also the shape. A client brought the first bola to the Fabregat-Sales workshop to be copied. Eventually, the Papaseit and Pedrola workshops also copied this type of planter. But due to the concern with shape, these three workshops had, at the time of the fieldwork, distinctly different versions of this planter. From that first model, each had developed their own variations on the ovoid shape. (Yet at the Pedrola workshop, both brothers were throwing these boles exactly the same. Hence the differences were not between individual potters but between workshops.) Josep Pedrola, Ferran Segarra and Josep Papaseit were especially concerned with the perfection of shapes. They often explained how they had been working very consciously on these and they could relate histories of shapes of specific types of pottery. Not all potters had this same concern. For instance, Pepe Jornet made the bolo and the cossi from Quart the same as the planters that were brought to him. He explained that if he designed a new type he did not develop it further. Of course, he was concerned with the 'good line' of his products, with the clarity of the contour, but he did not explore the possibilities of improvement.

With respect to details of the shapes, potters became more strict. During one of our discussions on the shape of pottery, Josep Pedrola pointed out how he took care that the rim of his vessels and planters were well *pasat*, 'passed'. He explained that since the rim was the last part thrown, a potter could do it hastily and therefore not distribute the clay equally. The rim might not form a pleasing ring, and that was something that he wanted to prevent. (I have in fact observed old pottery with both nicely shaped and badly shaped rims. During the fieldwork, I observed all the potters taking care to 'pass' the rims well.) Josep Pedrola pointed out that he cut the foot of vessels and planters with a rib at a diagonal angle. The pot then had an angular transition from the flat bottom to the rising wall that prevented chipping the edge. (Having observed old pottery which had such an edge and old pottery that had not, it was obvious that all the potters took special care to cut the foot similarly.)

The general increased concern with details of shapes applied to aspects of pots which were developed out of the old repertoire (like the planters developed out of vessels and tubs), but also applied to those planters which shared physical characteristics with vessels and tubs, planters with similar rims and feet. When it came to newly introduced types, potters differed in the attention they paid to details. For instance, Pepe Papaseit pointed out how he and his son had improved some of their pitxells. They had made the pitxell in the shape of a tomato with a straight stalk, but since real tomato stalks often curl, they began to make the stalk curled. They had made the pitxell in the shape of a pineapple with five leaves forming the crown, but eventually made it with nine.

The potters and muleteers had also judged pottery by the even distribution and gloss of the glaze. Later, the criteria for glaze became more exacting. Where glazed objects touch each other in the kiln, their glaze melts together and on these spots scars occur.[16] Previously, potters and muleteers accepted these scars as inevitable, and only bowls were packed in the kiln in a way that the scars occurred on an inconspicuous place underneath the rim. In due time these scars became unacceptable and potters prevented these by firing pottery on shelves and taking care that pieces did not touch one another. This technique of firing pottery on shelves requires more space in the kiln than when the pieces are stacked on top of each other. For that reason, only potters producing small pieces fired them on shelves. The planters were still stacked but the scars were camouflaged. Potters removed these edges with a whetstone and subsequently painted them. Again, potters differed in the degree of attention paid to this process. Some potters painted the scars with a colour more or less the same as the surrounding glaze and used the same paint for all pottery of the same colour. But someone as meticulous as Josep Pedrola mixed the paint for each individual scar in order to more closely approach the colour of the surrounding glaze. Still, the Pedrolas, who specialized in the production of large pieces, began to fire their smallest planters on shelves in 1992. Perhaps in due time all glazed pottery, regardless of size, will be fired on shelves.

The small pottery had another problem. This pottery is almost always wholly glazed and during the firing the glaze drips from the foot of these objects on to the shelves. Upon breaking the pieces free from the shelves, razor sharp edges occur on the feet. For safety reasons, potters had always scraped these sharp edges off with a coarse metal blade. Yet they began to consider this insufficient and therefore to grind them off, or prevented the glaze from dripping onto the shelves. This grinding off of the edges required much labour. Once, Pepe Papaseit was doing it and commented to me that previously it would have taken perhaps an hour to scrape the edges off, while at that time he easily could lose a day on it.

Potters and muleteers had also judged the glaze by its gloss. Though the type of wood kiln introduced in 1948 provided more control over firing, the subsequently introduced gas oil, and especially gas kilns, provided even more control, and hence produced a glossier glaze. Moreover, wood kilns caused impurities due to whirling ashes: the latter kilns did not cause such impurities. As a result, potters producing small pottery invested in gas oil and gas kilns. When the Pedrolas bought a gas kiln in 1992, they too used it for firing their smallest products, planters with a glazed rim. As grains of clay can ruin the gloss, some, but not all, potters wiped off the smaller objects before applying the engobe. One such careful potter identified these grains as typical of badly finished work. A further step in perfecting glaze was taken by the Papaseits. They bought a compressor together with a chamber in order to blow glaze on pottery, and thereby distribute it evenly. Obviously, Pepe Papaseit was a potter very concerned with the gloss, as well as with perfection in general. I was once present when Pepe Papaseit unpacked a kiln. Pepe could not stop admiring his achievement: the glaze had come out very glossy.

As the quality of the pottery gradually improved through the interaction between potters and clients, participants could only recall a few specific instances in which clients explicitly required improvements. One such instance involved a case in which a potter was asked to eliminate tracks left by hands on pottery. (When a thrower is finished with a pot, he takes it from the wheel with his wet hands. After the firing the tracks left by these wet hands can still be seen.) One client of Joan Avante required that he wiped off a pot with a piece of cloth in order to prevent these tracks. Joan Avante did this only for this one client. Another such instance was a case in which a client insisted on a certain perfection, but refused to pay extra. An American placed an order with the Pedrolas for planters with glazed rims and insisted that the glaze should have no scars. This meant that these planters had to be fired on shelves, hence less of them could be packed in a kiln. For these extras, the Pedrolas needed to charge a higher price, but since this client would not pay the higher price, they declined his order.

Already during the 1950s, potters shifted to calculating their prices based on their investment of labour and materials. During the 1960s, but mostly during the 1970s and 1980s, they gradually perfected their products according to their old criteria as well as to newly developed ones. Previously the quality differences between products of different potters were not included in price calculations, but as marketing relationships changed, potters began to include them because these differences related to different labour input. Eventually potters differed in the degree to which they perfected their products, hence their prices differed.

New trade relationships

From the 1950s potters had been contacting new clients with whom they established new forms of trade relationships. As the numbers of potters in Miravet, but also elsewhere in Spain, declined, new demand, including international demand, developed. These trade relationships developed altogether differently.

When potters started to market to wholesalers and shop-owners, they had to travel and visit potential clients. Even when potters had established regular relationships with clients, they still had to visit them from time to time to request orders. Pepe Papaseit and Josep Jornet said they were ill at ease with this situation. They used excuses to explain to their clients why they had come, rather than admit they were there to solicit new orders. The wholesalers from the south who had relocated to Barcelona were among the first clients who came to Miravet to look for potters to take orders from them. By the early 1990s this had become the usual modus operandi. Wholesalers and shop-owners needed to come to Miravet to find suppliers. Sometimes they even had difficulty finding a potter willing to accept their orders: potters had so many orders that they could not accept more.[17] Moreover, if clients had placed an order, they often had to accept long waiting periods for delivery, a year being quite a usual interval. Often they had to accept many months of additional delay.

Potters accepted orders only for the kind of ware in which they specialized. They also refused orders for pottery for which they felt they could not charge the full price. During fieldwork, only twice did I observe potters producing gerrots, the largest vessels, and then only in small numbers. Potters argued that as a gerrot took up a lot of space in a kiln, was strenuous to throw, and carried a relatively high risk of coming out flawed, they needed to charge a client a commensurable price, which they might not be willing to pay. For these reasons, they did not seem to like to produce gerrots.

When potters began to market their products to wholesalers and shop-owners, they delivered the merchandise to their clients' warehouses or shops and charged a price that included transportation. The interviewees used the expression that the pottery 'was sold there', i.e. at their clients' warehouse or shop. As Pepe Jornet had said when talking about the competition in the Saragossa market shortly after the demise of the Co-operative, his father was at that time exceptional in that he charged for transportation. By the early 1990s potters no longer dealt with transportation. The few potters who still occasionally delivered, and charged for transportation, did so only for long-term clients and would not accept new clients on such conditions. To refer to the new practice and to contrast it with the previous one, the interviewees used the expression that the pottery 'is sold in Miravet' or 'at the workshop'.

Potters still loaded the trucks for clients, a job that required skill, but they charged for packing material. The potters explained this reversal simply: they would rather use time making pottery than delivering.

Though the potters were in a stronger position, they knew their limits. Once I was at the Pedrola workshop and the two brothers were fantasizing about how much money they could make if only they were smarter. They were playing with the idea of printing a catalogue and sending it to prospective clients, such as those they had contacted at a fair in Barcelona. They would double the prices in their catalogue: if these prospective clients accepted these prices that would be great, if not, nothing would be lost since they would keep on working for their old clients. But after they had had all kinds of wild ideas of the wealth they could amass, Josep came down to earth and said that they would never do such a thing. They had their clients, some for a considerable period of time, and it would be safer to continue supplying them and raising prices in an acceptable way. I had heard this kind of argument more often: potters were not inclined to give up clients with whom they had been doing satisfactory business and on whom they could rely, not even when lured by big money. Similarly, they were not too stubborn when clients wanted a certain type of pottery that the potters did not like to produce. The two cases observed when gerrots were actually produced, were for good clients. The potters considered they were doing them a favour. In another case, the Pedrolas produced a strawberry planter, a larger vessel with attached pockets in the shape of quarter spheres. I was able to help Josep Pedrola with attaching these pockets and I learned how much work was involved. Such a strawberry planter took twice as long to make as a regular vessel. Josep Pedrola said that he had to charge a client a higher price for such a planter but not the commensurate double price, which would be far too much for what a client would expect to pay. Yet Pedrola only made these pots for clients who ordered a large number of other ones, never for clients who would only order these pots. Also, he usually made a smaller number of these than a client had ordered. He explained that in this way he kept his client satisfied, made money on the order, and did not lose too much time and money on the making of strawberry planters. Hence, though the potters were aware that they might fetch higher prices, and that the production of certain types of pottery was even unprofitable, they were not too calculating and respected other aspects of their relationships with clients as well.

However, there were exceptions. For example, one potter may have gone too far by accepting orders which were advantageous only for himself. He required that clients place orders of whole kiln contents that allowed him to work in neat cycles of producing one order at a time, and freed him from the problem of storing finished pottery. Since his kiln was extremely large, these orders also had to be large. During the fieldwork he was the only potter of Miravet who occasionally had to worry about orders, probably, as other potters pointed out to me, because of the conditions he set.[18]

In earlier times, when in the weaker position against muleteers, potters had attempted to confront the muleteers as a group. Potters were aware of opposing interests and felt that their interests were best protected by sticking together. This was epitomized by 'the price of Miravet'. They established their co-operative to enforce, among other things, this collective price. After the demise of the co-operative, the potters assessed collective prices for the last time. This was during the period that Josep Jornet insisted that Miquel Borell be present when he settled accounts with a wholesaler so that Borell could testify to the other potters that he, Jornet, had adhered to these prices. By the early 1990s potters had totally given up on collective prices. They did not even assess a price on the most basic communal level, between two potters. Quim Borell and Toní Vives[19] had an order from a certain wholesaler for the same type of pottery. Toní Vives had approached Quim Borell to ask what he charged for this particular type. Later on Quim Borell found out that Vives had sold it cheaper. He was angry: 'Why did he come to ask the price if he was going to sell it cheaper?', he asked me. He was not annoyed that Vives had sold it cheaper as he could ask whatever he wanted, but that he had pretended a kind of solidarity. *Cada caragol la seva casa,* 'every snail [in] his own house', ended the recounting of this story.

When selling to muleteers, potters deplored others selling cheaper than the collective prices. By the early 1990s some potters rated their products higher, as more perfect and of better quality, than those of their fellow potters, and therefore charged a higher price. They stated explicitly, not only to me but also to clients, that they sold at higher prices because their products were of higher quality.[20] At one time Ferran Segarra even bragged that he was the most expensive potter of the village and related that one other particular potter was selling petxins at 450 pesetes while he was already asking 500 pesetes for the much smaller pot for storing garlic. For some potters at least, selling cheaper was no longer an act of non-collegiality, but a sign of poor craftsmanship.

As the potters gradually differentiated their production, especially in planters and small objects, they also began to differentiate their clients accordingly. For instance, once when Antoni Pedrola and a client of his were loading this client's van, I gave a helping hand during which we discussed the history of the marketing relationship. I learned that they had been doing business with each other for almost 20 years. During that period, this client had bought what the Pedrolas were producing, planters and small pottery, but as the Pedrolas increasingly specialized in the production of planters, they became less willing to accept orders for small pottery and eventually this client had to order this from a potter who specialized in such production.

It was then typical that wholesalers and shop-owners who carried large and small pottery obtained their merchandise from two different potters. Muleteers had often obtained their merchandise from more than one potter, but that was because of the volume of the production of individual potters. That could still happen in cases when clients needed very large orders, but it also happened because potters specialized.

During the 1950s and 1960s potters had sold to wholesalers who distributed the pottery to shops, but during the 1970s and 1980s many shop-owners purchased vans and station wagons, especially those interested in small pottery who had begun to obtain their merchandise directly from the potters. For instance, during the 1960s Pepe Papaseit had a client wholesaler who bought the major part of Pepe's production. Gradually this client bought less, so shop-owners who had been served by the wholesaler then came directly to the workshop. Hence, during these decades, the proportion of shop-owner clients increased. In the 1990s, these clients came at least once a year, but often more frequently, to place and to pick up orders. During these occasions potters and these clients did not limit their interactions strictly to business, but also engaged in amiable conversation about business, soccer matches, etc. Over these years, often 10 or 15, potters and clients came to know each other well. On one occasion, as I helped Antoni Pedrola and his client, we talked about their 20-year marketing relationship, and they included in their account the history of the Pedrola workshop, how it grew, the acquisition of machinery, the history of the repertoire of produced pottery and who had been working there. The client knew this history very well since he had witnessed it on his visits and he and Antoni Pedrola obviously enjoyed recalling these memories. These visits might also include family members: some clients regarded their trips to Miravet as an outing, or combined these trips with a vacation, and were accompanied by their wives and children. For instance, the Papaseit workshop had a client from Madrid who annually went to the Catalan coast during Holy Week. He then came with his family to place and pick up an order. Potters had developed closer relationships with these shop-owners than with wholesalers, with whom they usually had contact through employees.

Foreign clients were a different matter. There were two accounts of the first foreign client who came to Miravet. According to Vicen Sales and Santiago Benaiges, in the mid 1970s, a Dutch wholesaler contacted the Fabregat-Sales workshop through a representative in Barcelona and was consequently supplied several times. According to Ferran Segarra and the Pedrolas, around 1980 a Frenchman came to Miravet to look for potters who would supply him with planters. All the potters refused to accept an order from him: they found a stranger too risky. (Note that Fabregat and Sales did not react with distrust of their first foreign client; but then, they were contacted through a representative in Barcelona.) Ferran Segarra was at that time an apprentice in the workshop of his in-laws, the Pedrolas, but shortly afterwards started his own workshop. He did not see any problem with supplying foreigners, and when he started to work on his own he was eager to acquire such orders. He called this Frenchman by phone and together with his in-laws he accepted this order, the first foreign order.

Foreign clients had become very common, especially after Spain's joining the European Community. Potters lost their distrust, and even become very appreciative, since these clients, French, Belgian and Dutch, and occasionally American, Canadian, British and German, boosted production. By the early 1990s, all potters producing planters did so primarily for foreign clients.

In the marketing of planters to foreigners, the identification of specific types of planters with specific clients became an explicit factor, as potters understood that wholesalers within one national market could become each other's competitors. This competition clearly surfaced in the following event. Pepe Jornet was visited by a Dutch wholesaler who presented himself as a friend of a good Dutch client of Jornet. Since this wholesaler presented himself as a friend, Jornet saw no problem in accepting an order of the types of planters he was also supplying to his first Dutch client. But unknown to Jornet, these wholesalers were fierce competitors in the Dutch market. When the other wholesaler found out about this order, he immediately proposed to Jornet that he would buy it outright, and urged Jornet never to sell any of the types of planters he was buying to other Dutch wholesalers. Pepe Jornet was very apologetic to his regular client and explained to me that he would never normally have accepted that order but that he was tricked. He would not only have refused that order to keep his good client, but he understood that competition in one national market eventually might hurt him as a producer. Therefore, of their own accord, potters began to limit the types of planters they sold to a client in one particular national market, and if the potters did not do this, their foreign client, as happened with this Dutch one, might require it.

Normally, foreign clients stipulated that certain types of planters were sold only to them and to no one else in their national market. Such types were models that these clients had designed or models that they brought to the potters, or models they were ordering. For instance, a French wholesaler who had stayed in Miravet and had developed a series of planters in co-operation with Joan Avante, owned the monopoly on these planters. Only when he could no longer buy a significant quantity of these, did he give permission to Avante to sell them to other clients. This particular wholesaler went even further than requiring exclusivity of models. He required that the pottery produced for him be stamped with his initials, as he had also been doing with potters at other sites.[21] He wanted this in order to have his products marked primarily in the French market, but also in the Belgian, Dutch, German and American markets as well, as he had been selling to wholesalers from these countries.[22]

In general, planters were sold in large quantities, but to these foreign wholesalers the quantities were enormous. With very few exceptions, foreign clients were all wholesalers who covered their national markets or substantial parts of them. To make cost efficient use of transportation, they used huge trucks with trailers that had to be completely filled. Often one potter could not produce enough to do so, and another potter or potters might receive an order at the same time as well.

Related to the difference in volume of orders, was a difference in the method of payment. Previously, when potters had delivered their products to the client's warehouse or shop, they had been paid immediately. When the clients had to come or sent someone to Miravet to pick up their orders, they paid potters after having loaded the products. However, these foreign clients never transported their merchandise themselves. Also, since they placed such large orders, potters invested a lot of work and material in producing those orders. Potters were not very open to me on how they solved these problems – money matters were very private – but at times they complained about long waiting periods before receiving their payments. Josep Pedrola explained that he considered these waiting periods the result of a discrepancy in scale. These wholesalers were accustomed to work with large sums of money and with large companies who accepted long waiting periods. Potters worked with smaller sums of money on which they were directly dependent, therefore they could not wait long for payments. The Pedrola's had an important French client who ordered a trailer load of pottery about twice a year, but because they had problems with the payment, they stopped supplying him. When he agreed to pay upon loading the truck, they started to supply him again. Similarly, some potters required a down payment upon accepting a large order, half or at least a substantial part of the value of the order, and required the remainder upon loading. In such a case, the truck driver usually had a cheque with him.

The last aspect in which marketing relationships with foreigners differed from those with Catalans and Spaniards was in their degree of integration with other forms of contact. Potters' contact with their foreign clients was usually restricted to business transactions. Almost all these clients had their merchandise shipped by transport companies. They came to Miravet to place orders only. However, they often did this by mail, phone or fax, or they accompanied the transporter they hired. Hence potters had infrequent and only occasionally direct contact with them. If potters interacted with these clients directly, language was often a severe limitation, as very few foreign clients spoke Castillian, and then often very badly, and none spoke Catalan. For instance, Pepe Jornet had several times met and received orders from a Belgian couple. Only after I filled him in on some detail, that was obscured by the language barrier, did he start to understand the dynamic between this couple. It had actually been the wife who selected pottery and placed orders. She was the entrepreneur and her husband was just accompanying her. Potters also formed an impression of their clients in other ways: in their description of these clients, potters often mentioned the cars their clients drove, invariably expensive ones, more expensive than the cars that potters or their fellow villagers would dream of driving.[23] At times, some potters showed, by off-hand remarks, a poor understanding of where their foreign clients came from. For instance, upon hearing radio news on the Bosnian war, a potter asked me if Bosnia was near the Netherlands, my home country! Furthermore, these restricted contacts were of a relatively short duration. It was only in the last decade that potters had been obtaining foreign clients. These clients generally only placed orders for a few years. Hence, potters and foreign clients had hardly any contact beside the most necessary exchange of information, often in a hodge-podge of languages, numbers and drawings.[24]

Some foreign clients seemed to feel awkward with the language barrier and made up for this by bringing small gifts for the potters: tulip bulbs, cheese, chocolates or bottles of alcohol, often gifts which were associated with their countries. As mentioned, some foreign clients also combined their visits to Miravet with a holiday in which they were accompanied by their wives and children: though potters could hardly talk with their clients and their families, they still met them.

* * *

During the 1970s and 1980s Spain, and especially Catalonia, continued to develop economically, most importantly by strengthening economic ties with the European Community. As an aspect of this economic development, the demand for pottery grew. The potters of Miravet made good money and invested in the modernization of their production. Thus they improved their economic positions within the village community and in relation to their clients. In this period they continued to differentiate production and repertoires from each other, and these eventually became individualized. They perfected their products, and invested labour in the perfection, but differed to the extent that they did so. Pottery became differentiated in yet another dimension, that is, with respect to its quality. During the 1960s potters began to differentiate their clients in conjunction with the differentiation of production and repertoires. In the early 1990s, they marketed all types of pottery to Catalan and Spanish wholesalers and shop-owners in long-term close relationships. They only marketed planters to foreign wholesalers in very restricted, purely business relationships.

---

1 In 1983, the main industries (in descending importance of number of people employed) were chemical products, wood (furniture and coffins), textile, food products, metal and construction ceramics (Anon., *La Ribera d'Ebre* 1989: 138).

2 As seen below, the potters of Miravet, like most Spanish potters, exported mostly planters. These were scarce in the international context. From the offering of pottery on the Dutch market it appeared that of the EC countries, only Greece has an extensive manual pottery production, mostly larger planters. There were some French and Italian handmade planters on this market, but this pottery was far more expensive than the Spanish. Most mechanically produced planters (except for the generic flowerpots that do not have a clear origin) came from Italy. There variety was very limited: it was always red and came in only three or four different shapes. (A lot of glazed pottery like bowls and plates were imported from Portugal, but hardly any planters.) Apart from the Spanish pottery, mostly Thai pottery was sold in the Netherlands. This pottery was always glazed and looked very different from the terracotta from Spain. I made these observations by visiting shops in the Netherlands and trade fairs, and by my contacts with Dutch pottery importers. From what I saw on occasional visits to shops in Germany, Belgium and France, and at international trade fairs in Barcelona and Frankfurt, the offering in other EC markets appeared to be similar. (This was also corroborated by pottery exporters whom I met in Miravet.) In the US far fewer pottery planters, apart from the generic flowerpots, were offered. Here too the mechanically produced Italian ones were ubiquitous, and there were some Thai pots. Spanish handmade planters were only offered in a catalogue, and for a few years in a shop in Manhattan owned by a Dutch importer.

3 Land ceased to be the dominant means of production that it once had been. People were still identified by their casa and casa still included land, but in contrast to the previous situation in which land was always worked to provide the necessary produce, not all plots were worked. Farmers had given up working plots that could not be farmed with modern means of production. People who were not farmers but still owned some land had also given up working the land as they could buy produce in a shop more cheaply. Some worked it for sentimental reasons. For instance, a woman told me that her household still worked the little land they had, that they were losing time and money on it, but that they did it because the land had always been part of the family and it would be a pity to let it lay fallow. Yet she added, if costs should go up and if her elderly father would be too old and her son unwilling to work it, they still might have no choice but to stop.

4 Although Josep Papaseit agreed that heavy tasks were not done by women, he pointed at one exception. Previously women sometimes put handles on vessels that required them to bend

over, a very strenuous job. Furthermore, all of the participants mentioned the case of one woman who had been working as a potter's assistant, doing exactly the same tasks a male one was doing. Because of the developments in marketing, bookkeeping had become more laborious. Josep Pedrola's wife took care of the bookkeeping for the Pedrola workshop. Though not a full-time job, it still required many hours a week.

5 But this relationship between the labour force of a workshop and the type of production is not very determined. There are some exceptions, i.e. potters producing small pottery and not employing their wives and daughters. Clearly, other factors are also in play, like the interest of wives and daughters in this work.

6 A recent development was that members of one casa were living in two apartments, on different floors, but in the same house: the family with children in one apartment, the parental family in the other. People felt free to walk in and out of each others apartment and usually all casa members had their meals together, this arrangement though created some distinction between the two families.

7 Also, they financed much with 'black' money, something they also did not publicize to outsiders. Both pride in household autonomy and financing by black money were made very clear to me on one occasion. A potter with whom I had good rapport invited me for a short car trip. He had just ordered a very expensive piece of equipment and was excited about it. In the intimacy of his car (and fueled by his excitement), he told me that probably everyone in the village would be guessing about how he financed his latest purchase, but he was proud to tell me that he did it all with his own money. Of course, he said with a wink, no one knew that he had that money, nor was that something he was going to tell.

8 This is confirmed in a conversation and letter with Mr Ferrer i Llobet, head of this subdivision. The next chapter elaborates on this organization.

9 People commented to me that Ferran Segarra received the subsidy because he had *tios*, 'connections', with the implications of patronage. Indeed, especially during the Franco era, such subsidies were favours granted through patrons, and still were, at least at the village level. During the elections for mayor in 1991 I heard many examples of patrons providing favours in exchange for votes. For many villagers it was inconceivable that one could obtain, as Ferran Segarra in fact had done, such subsidies purely through bureaucratic means.

10 The only exception being the Catalan bowl on which a chemical used in the viticulture by fellow villagers was splashed. This chemical contained copper sulphate and therefore resulted in a green colour.

11 By 1972 Papaseit was already producing vessels as planters. When I was leafing through carbon copies of bills with Dominica, Pepe Papaseit's wife, and her son Josep, we came across an entry in which vessels were specified with 'moro'. *Moro* is the term used by the potters to refer to the ring of glaze on the upper part of the outside of a vessel, the glazed rim and shoulder. She explained that this specification referred to vessels that were unglazed on the inside and only had this moro: these were planters. Josep added that in March 1972 when this bill was written, making vessels unglazed on the inside was something new and thus this term was marked on these. He implied that since this planter/vessel had become 'normal', that is, had ceased to be a marked category, he no longer differentiated these vessels from the ones with the glazed insides in this way.

12 An exceptional case is one of a French wholesaler who stayed several months in Miravet and in co-operation with Joan Avante developed a series of planters.

13 In contrast to the planters, the potters did not have one label for this small pottery. In my interviews we referred to it with such descriptions as 'the small pottery' or the 'small ware'. These descriptions were used very loosely and were very dependent on their contexts. One of these contexts, and a very important one, was that these two types of pottery, planters and small pottery, correlated with a differentiation of the production of the workshops: talking about these repertoires of products implied talking about the different forms of production.

14 In addition, a new material for repairing cracks and holes invisibly came on the market.

15 The Petit household was the first one to employ ponds for preparing clay for a few years shortly after the Civil War. They stamped their pottery with their name and the name of Miravet, and with *barro colado*, 'strained clay'. The latter referred to the fact that their clay was prepared in ponds, thus had the chalk grains filtered out, and hence was of better quality, which was emphasized with this stamp. Due to the history of this particular workshop, these ponds were only used briefly, and it was not until a few decades later that other potters shifted to this preparation technique.

16 In Miravet such a scar is called *tocadura*, a derivative from the verb *tocar*, to touch.

17 Once, in 1982, Ferran Segarra and the brothers Pedrola had an offer of a free stand at a commercial fair in Barcelona, where they were contacted by many interested wholesalers and shop-owners. However, since they already had enough orders, they did not repond to any of them.

18 This potter was very commonly considered the best thrower of Miravet: his problems were definitely not caused by poor quality.

19 Pseudonyms.

20 One former client told me that this was already the case in the 1970s.

21 Furthermore, in addition to the fact that the French wholesaler had developed his own series of planters and had these stamped, he was also concerned with the perfection of his merchandise. He was also the one who required that the freshly thrown planters were wiped off.

22 When he stopped ordering from Joan Avante, Avante started to stamp his products with his own initials. He was the only potter at that time who stamped his products with only initials and not, as some had previously done, with their full name and address. He could not explain to me why he did this.

23 Only one villager, an important contractor, drove an expensive new Peugeot. In general the people of Miravet seemed to have a very practical attitude towards cars and were reluctant to express status by cars, yet they were sensitive to status expressed by cars.

24 Those potters who did speak French – that is, Ferran Segarra and the sons of Joan Avante – had naturally more immediate contact with their clients. These potters made vacation trips to the Netherlands and Belgium, where they visited their clients. Ferran Segarra had especially close relationships with his foreign clients. When they visited Miravet, he often took them out for lunches and dinners, and sometimes they stayed overnight in his guest apartment.

Chapter 7

The exchanges at the workshops

The potters first marketed the bulk of their production to muleteers, and then later on to wholesalers and shop-owners, distributors who subsequently sold the pottery to consumers. Yet, at the same time, potters sold their products directly to consumers, people who came directly to the potters' workshops. Potters sold only relatively small quantities to these people (with the exception of Ferran Segarra, who had been selling his whole production in this manner) and their relationships with people coming to their workshops were therefore not economically significant. Still, these exchanges had great social importance for the potters.

These exchanges differed in nature. Though the potters themselves did not make an explicit distinction – as for instance expressed with different terms – between these exchanges, they expressed a distinction by conducting and perceiving them differently; they therefore articulated and constituted different kinds of relationships through these exchanges. The basic distinction was between what I call locals – that is, members of the same community as the potters; and visitors – people coming from afar to visit the potter workshops.

The exchanges with locals

When pottery was marketed to distributors it was transported from Miravet to other places; people from Miravet and neighbouring villages – a radius which has been greatly enlarged by the increase of cars – went directly to a potter's workshop to acquire pottery. At this local level, exchanges of pottery, as well as of other objects and services, always took place between people who related to each other in additional ways. A local always went to a potter to whom he/she was related (and/or befriended) and these exchanges would become occasions for conversation. These exchanges were always aspects of more complex relationships constituting the community in which the potters were integrated. How were these exchanges with locals articulated?

The participants invariably mentioned that if a local entered a potter's workshop then this person expected a bargain, either a free piece of pottery or one for a reduced price. As Pepe Papaseit explained: 'I still have farmers who come here for a gargola and ask me "how much is it?"' As he obtained fruit and vegetables from these farmers (not in direct and immediate exchange, but on other occasions), he refused any payment for a gargola. 'Perhaps what they give you costs more, or even if it does not cost that much, it's not important, it's friendship.' A gargola is one of the cheapest pieces of pottery, and Pepe continued elaborating on more expensive pieces. 'Well now, if someone comes for a vessel that costs 5,000 pesetes, I would say "give me 4,000" ... because I can't give it to him. Well now, if a relative or a friend comes, I might give it.'

Like the exchanges between potters and muleteers, these exchanges can also be analysed applying Sahlins' continuum in which social distance is related to a specific form of reciprocity (see Chapter 2). For Pepe Papaseit, the difference between giving away a vessel costing 5,000 pesetes or charging for it, related to the social distance between the person and himself. The first exchange of giving away the vessel for nothing is located at the extreme end of 'generalized reciprocity', and the second exchange of selling it for a reduced price is located closer to the mid-point of 'balanced reciprocity'.

As a critique of an approach to exchanges in which actors are perceived as 'economizing' (Sahlins 1972: 230) to maximize their economic positions, Sahlins develops this continuum to analyse exchanges as expressing, creating, and/or strengthening social relationships. And indeed in these two exchanges Pepe Papaseit assessed two forms of reciprocity as expressing – and probably strengthening – two different, but rather close, social relationships. Yet Sahlins develops this continuum with respect to 'primitive societies', while the villagers of Miravet were embedded in the Spanish political economy, a political economy based on 'economizing' individuals and on commodities with exchange-values created in the market. Pepe Papaseit gave a gargola for free to the same person he would charge a reduced price for a large piece. Although in these exchanges the social relationship was constant – it was between the same people – the form of reciprocity varied. The difference between the forms of reciprocity was caused by the prices of these pieces, by their exchange or market value. The market value of 5,000 pesetes was the price Pepe Papaseit charged a wholesaler or shop-owner, that is, the price based on economizing as he makes his living by selling the bulk of his production for these prices. Yet Pepe Papaseit did not charge that price to a friend but a reduced one. Hence, he was acknowledging the relationship he had with this friend, as well as the market value, i.e. an economizing attitude. Sahlins' continuum of interplay between social distance and reciprocity needs to be adapted for the analysis of these exchanges embedded in a market economy by adding the factor of exchange-value, the value created in the market.

To analyse the form of reciprocity assessed in exchanges between locals, it is necessary to look at the relationships in which these exchanges take place (that is, at the social distance between the potter and the local), as well as at other relationships – the relationships in which objects acquire an exchange-value, a value which 'appears to them [i.e. potter and local] as an objective character stamped upon the product of [men's] labour' (Marx 1967: 77). Therefore, these exchanges are ambivalent: the reciprocity is a compromise between the assessment of the social distance or relationship between the people involved, and the objective value of the exchanged object. This ambivalence is not only apparent in the assessed reciprocity, but also in the interaction between the exchangers. For instance, Pepe Papaseit told me about a person who, when he came to his workshop for a piece of pottery, always asked Papaseit the price, but in the end, Pepe Papaseit always gave it for free. Pepe Papaseit: 'And he always said "I do not come any more because you do not want to charge me anything."' Pepe Papaseit explained that because this man was a farmer and at times gave him fruit, and because they were such good friends, he never charged him a peseta. This man knew this, even probably expected it, still he always asked the price and threatened not to return when not charged. Hence, in this exchange (and many others which were recounted and observed) the farmer pretended to hold on to the exchange-value, to a value perceived to be objective and independent from his relationship with Pepe Papaseit, while simultaneously both knew that the reciprocity assessed was a function of their relationship.

Due to the ambivalence in these exchanges between locals, the reciprocity assessed can be the exchange or market value, while the other aspects of these exchanges – i.e. the exchange as expressing and strengthening a relationship as opposed to maximizing one's economic position – were also maintained. This was the case if a household economically depended on its production and subsequent marketing to local consumers – something that the potters no longer did since they sold the bulk of their products to distributors. For instance, Pepe Jornet explained that his household alternatively bought bread at one of the two bakeries in Miravet: one baker was his brother-in-law, the other a close friend. His choice of baker was not motivated by the quality of the bread, the price or the location of the bakery, but by the relationship with the baker: he felt that he had to patronize both his brother-in-law and his close friend because of the closeness of his respective relationships. Pepe Jornet would pay for his bread, while if one of the bakers needed a piece of pottery, he would give it to him, or charge a reduced price. The difference is that a baker was economically dependent on the patronage of villagers, while Pepe Jornet was not. Potters and muleteers had adhered to market values for the same reason. Josep Benaiges and Llop were close friends and traded their whole lives with each other, but Llop paid Josep Benaiges the full market value for his pottery, and Josep Benaiges the full price for food Llop would bring from a trip (see Chapter 2).

The market value was not necessarily only quantitative, but could also be 'negative'. Potters always had pieces which came out so badly that they could not fix them up and sell them for a regular price. Because of their poor quality these pieces were excluded from the exchanges for which they were intended and in which they would have a market value. These pieces had a negative market value, as opposed to objects that were never sold or intended to be sold, which had neither a positive nor a negative market value, but no market value at all. Locals could also explicitly ask a potter for such a piece. Knowing that such a piece had a negative market value, potters were more inclined, actually implicitly forced, to give such a piece for free. They anticipated this: once when unpacking the kiln, Vicen Sales was putting some flawed pottery aside and explained that these were the pieces he was going to give away were a local to come by asking for one.

Similarly, the vegetables and fruit mentioned in the situations above might have been vegetables and fruit not suitable for sale. Vegetable and fruit wholesalers required a certain quality standard. For instance, when a peach had ripened with a leaf sitting on it, it had a mark on that spot. A wholesaler rejected that peach, though such a peach was equally good for consumption. Such vegetables and fruit also had a negative market value and were like the flawed pottery given away to friends and/or kin. The vegetables and fruits mentioned in those instances above might also have been surplus produce. Villagers still produced some vegetables and fruit for their own consumption, but certainly not much compared to what was previously necessary. They did it more, as one put it, as *passatemps*, a 'hobby'. For instance, one had orange or fig trees – cultivations that did not require much attention. Such produce was available on the market, but since the particular cultivator did not produce for the market and did not have the contacts to sell it, the surplus he or she gave away also had a negative market value.

These exchanges of pottery and of produce differed due to their respective forms of production, marketing and consumption. Pottery was more regularly produced than vegetables and fruit: it was easier to predict that a potter had a certain piece of pottery than one could predict a farmer having rejected/surplus produce. (If a farmer did not have rejected produce, then one had to pay the regular price; if he did not have surplus produce, then he could not give anything away.) In Miravet, pottery was only marketed to its consumers at potter workshops, while produce was also marketed at shops and at the market place. One had to go to a pottery workshop in order to acquire a piece of pottery, while one had several alternatives for acquiring produce. One needed a piece of pottery at irregular, infrequent intervals, while one needed produce on a daily basis. A potter could never predict if a give-away piece of pottery might be needed, while a farmer knew that vegetables and fruit were always welcome. Hence, pottery was requested, while vegetables and fruit were offered, and therefore a potter had to assess the reciprocity in a specific exchange, while a farmer – since he or she volunteered the exchange – never needed to.

In all these exchanges, the exchange-value (i.e. the one created in the market) entered these exchanges in another significant way: all these exchanges of pottery and produce were of objects produced by the person who exchanged them. However, objects bought outside of Miravet to be sold on the village were always paid for. When at the end of the 1960s the flowerpot produced in Miravet competed with pottery produced in a factory in Navarette, some potters began to sell flowerpots produced in Navarette in their workshops. These potters delivered their products to the warehouses of wholesalers who traded in flowerpots, and upon delivery bought flowerpots to be sold in Miravet. Later on, when they no longer delivered, wholesalers sent them flowerpots. Similarly, when Pepe Papaseit acquired a wholesale client in Breda, a centre of casserole production, he bought casseroles to sell in his workshop. These casseroles had always been sold in shops or at the market place in Miravet, but never produced there since the local clay is not of the right quality for the production of these casseroles. These objects acquired in market relationships did not lose their exchange-values and were hardly ever sold for a reduced price. (Note that this is in accordance with the payment of the full market value for objects or services upon which a household economically depends.[1])

The exchanges with locals were of importance to the potters as the potters expressed their integration in the local community through them. In these exchanges, the pottery was ambivalent. The pottery had an exchange or market value, i.e. one created in the wider market. At the same time, the pottery also expressed the particular relationship between potter and local. This ambivalence underlied the articulation of these exchanges between potters and locals, that is, how these people decided to exchange which goods in which manner. Manner also included customary practices around these exchanges. Thus, in these exchanges, the pottery was differentiated and contrasted with those other goods, like fruit and vegetables, produced by other locals. As the pottery was differentiated and contrasted with these other goods, so were the potters integrated within the local community by the specifics of their production and marketing, as well as by the specifics of the production and marketing of those other goods by the locals.

The exchanges with visitors

On one of the first days of the fieldwork I was in the Pedrola workshop, chatting with and watching the brothers Pedrola and their workers, when unannounced visitors arrived. Their cars and clothing indicated to me that they were not from Miravet, nor quite probably from a village. They drove a fancy new car, certainly not the type of car people of Miravet would drive to visit a workshop. These visitors wore fashionable leisure clothing, the type of clothing the people of Miravet would not wear during a regular workday. They were warmly welcomed by the brothers Pedrola, who asked where they were from. The Pedrolas clearly did not know them. As the visitors approached the wheel where Antoni was throwing, Antoni interrupted his work and as a demonstration threw a small candlestick. While watching Antoni attentively, these visitors asked questions about the throwing. Antoni seemed to be flattered by this interest in his craft and responded enthusiastically. When the candlestick was thrown, Antoni left the wheel and took these people around the workshop, mainly showing them the kilns and the storerooms, and explaining the use of several pieces. He also pointed out a room upstairs where there was more ware for sale. The visitors were then left on their own to peruse in order to choose pottery pieces. All items sold to these visitors was at the listed prices.

The first time this happened, it seemed to me as an unusual work situation. Here people could drop in unannounced. The potter would interrupt his work to welcome these visitors. When travelling through Spain with my Dutch friend, who exported pottery, we had made similar visits and were similarly received, but we had a business interest. It seemed that the Pedrolas were not making much profit on the pottery bought by these people. Why then were they so well received?

I observed all kinds of visitors: people fascinated by throwing, some even videotaping it, and people clearly not interested in the process. I noted visitors asking all types of involved questions, even making notes on the answers, and visitors only engaging in short polite conversations. Some people purchased a lot, others nothing. Some visitors came by car and others came by bus. I noted and was told that these visitors came especially during the summer months, August was busiest, on weekends, and on holidays. What were these relationships potters constituted with these visitors? How did these differ from the ones constituted with locals?

In contrast to the exchanges with locals, potters invariably charged these visitors a fixed price, usually indicated on the products themselves or on a list. This price was slightly higher than the exchange-value as charged to wholesalers and shop-owners. Generally potters justified the slightly higher price by pointing at the proportionally longer time they needed to attend to these visitors. They charged a higher price, because the price was based on the labour invested in a particular piece of pottery, in this case in the selling of it. The price charged appeared to them to reflect the 'objective character' of their products.

Very clearly – and in sharp contrast with their exchanges with locals – potters expected to receive these prices. Indeed, visitors seemed to expect to have to pay the full price. I twice witnessed an attempt of a visitor to bargain for a lower price, and twice the potter in question displayed unmistakable indignation. Of course, locals were required by the etiquette of their exchanges to offer to pay the full market value, and it was up to the potter to charge a reduced price or not to charge at all. But once when I asked Antoni Pedrola why he exhibited such indignation during one such encounter, he answered that the bargainer had insinuated that he was unfair. In Antoni's opinion, to charge one person a lower price for the same thing as another person was unfair.[2] This answer surprised me: I had seen him selling pottery to locals, including myself, for reduced prices. But Antoni replied that this was something totally different. What surfaced in Antoni Pedrola's explanation of his behaviour was that potters perceived relationships with visitors as being all equal and hence they charged an 'objective' price, as Marx put it (Marx 1984: 77).

Complementary to this and also in sharp contrast to their exchanges with locals, the particular pieces sold to visitors were always the works of perfect quality. To visitors, potters sold for a fixed price – the market value – and therefore only pieces that fetched that value – perfect pieces.

Another difference was the content of conversations. With locals, usually, personal topics were discussed, topics also discussed independently from the context of the potter's workshop; with visitors, topics pertained to the potter's craft. In conversations it appeared that for locals pottery was something familiar, that did not require any attention apart from its utility. Pottery, though, needed to be explained to visitors in terms of its manufacture and use, and they demonstrated stages of the production and showed them their workshops.

In contrast to the exchanges with locals, potters sold their pottery to visitors unambiguously, charging the market value. Thus these exchanges express the potters' integration in the capitalist political economy of the wider society. The exchanges with locals were articulated in a manner that differentiated and contrasted pottery with the products of the locals: the practices and customs surrounding the exchanges with visitors, i.e. the explanations and demonstrations, placed pottery in very different categories. To explore these practices and customs and their accompanying categorizations further, the exchanges of pottery with locals and with visitors are placed in the wider context of different forms of consumption and encompassed production.

The differences in the consumption

The differences between exchanges with locals and with visitors relate to differences in pottery consumption – roughly a rural form of consumption versus an urban one. These forms evolved as aspects of national and international developments of the 1960s, 70s and 80s. During the 1960s and the early 1970s, the Spanish economy transformed into a modern, industrial and capitalist one. By the time of the fieldwork in the early 1990s, Spain had been integrated in the European Community and Catalonia benefited from this development with the highest economic growth in Spain. From the 1950s on, Spain transformed into a predominantly urban country and lost its rural character. For instance, in the first decades of the twentieth century, roughly a quarter of the Spanish population lived in villages of 2,000 or fewer inhabitants, and less than a tenth in cities of 100,000 or more. In 1980 less than a tenth lived in villages of 2,000 inhabitants or less, and more than four tenths lived in cities of 100,000 or more (Nadal 1984: 230). During the same period, Miravet's population declined from hovering around 1,500 to about 800 inhabitants. From one of many rural villages, Miravet had become one of the few remaining. Parallel to this statistical shift was a shift in the cultural meanings of urban versus rural life affecting villagers and urbanites differently, which in turn encompassed the different forms of pottery consumption. What were these different forms of pottery consumption, and how were these forms embedded in the cultural meanings of urban and rural life?

Consumption by locals

Miravet remained an agricultural village, but it fully profited from the economic development by modernizing its agriculture and enlarging the area of irrigated land. In the 1960s, villagers had migrated to cities not only for employment, but also for the more expansive consumption opportunities. In the next decades, these consumption opportunities came within reach of villagers as shops opened in nearby towns and villagers made shopping trips to cities. The village became in general more integrated with the outside world. During the fieldwork, not one person living in Miravet made a negative comment on village life. On the contrary, people told me how they enjoyed living in a place free from stress, air pollution, violence, drugs, crime, high rents and a lack of community spirit. Especially during relaxing moments such as dinner parties and terrace visits, they asked me if I agreed that in Miravet one lived well. They urged me to stay.

During the 1960s, as everywhere else, the local demand for pottery declined sharply as modern conveniences, like running water, butchers and food storage in bottles and cans came within reach of the people of Miravet. One exception was the large vessels. Villagers widely used Miravet products, mainly the solseres, for storing olive oil as they regarded this the best storage method for locally produced olive oil. Occasionally people still used gargoles for drinking water when working the land.

Even though the people of Miravet loved potted plants, they hardly used planters produced in their village. They would recycle any old container as a planter and indeed, considered the buying of a planter an indulgence. Walking through the streets of Miravet, I spotted plants on balconies, terraces and pavements. The plants were potted in cans, barrels and re-used pottery, like old vessels and cossis. Plants were mostly potted in the mechanically shaped generic flowerpots and, in a few cases, pottery planters that were not considered fit for sale. I hardly noticed a difference between the potters' houses and those of their fellow villagers. Even some of the potters' houses were adorned with the mechanically shaped flowerpots even though, as I knew, those potters produced a wide array of planters. Though the people of Miravet had more disposable income, they spent little on outdoor planters.

However, the people of Miravet did often spend money on decorative ceramics and pottery indoors. In their living rooms they usually had sideboards and open cupboards on which ceramic figurines and animals, like horses, cats and swans were displayed. It was the kind of pottery that was referred to by interviewees as *fina,* or 'fine', pottery, in contrast to *basta*, 'coarse', as made in Miravet. It was industrially produced, neatly moulded and glazed. These trinkets were often in the Lladró style – very saccharine elongated shapes and soft colours. In addition, they displayed small keepsakes, which were given on occasions like First Communions and weddings (a few made by the potters, but more often keepsakes were of 'finer' materials than the Miravet pottery). The only pottery produced in Miravet I saw thus displayed was the miniature pitchers made in the Avante workshops, which were finished with a fine texture. Painted plates, of the kind made in Talavera de la Reina (Toledo), were displayed on the walls of living rooms and hallways, and on tables, fruit bowls of the same kind. The few indoor plants were almost invariably potted in industrially produced, smooth white glazed planters, often with a slight relief decoration. The only other indoor pots I saw in use were the ones previously cited as 'painted pottery'.[3]

In the few kitchens I visited, I noted villagers using the fregal, the 'pan-scrubs', sugar bowls and other canisters then made in Miravet, but also elsewhere. People sometimes placed pitxells, usually the newly designed types in regular or miniature sizes on top of kitchen cabinets. Although I had not been in many kitchens, I did get the impression that these pieces were not widely used. Though villagers definitely appreciated the aesthetics of kitchens, like modern kitchen cabinets, kitchens did not get the aesthetic attention that the outsides of their homes, and the entrances and living rooms received.

Hence, most of the people of Miravet indulged in their newly expanded consumption possibilities, but seldom used the pottery produced in their own village. Their consumption of locally produced pottery declined, and although the participants could not recall changes in the practices around the exchanges of pottery in the workshops, due to the decline of local pottery consumption, one can infer that these exchanges must have decreased since the 1960s. Previously, potters produced certain pottery specifically for local consumption, pieces made to order received from nearby regions and never marketed by muleteers but always directly to consumers in the potter workshops. This production also dwindled.

When the exchanges of pottery with the locals were placed in the context of the local consumption, the practices and customs surrounding these exchanges can be more fully understood: the locals either consumed pottery in a very familiar way or as simple, insignificant decoration. Because of these forms of consumption, locals asked a potter for the particular piece that was needed. In order not to appear presumptuous, they then engaged with the potter in courtesies surrounding the assessment of the value of that piece as expressing their particular relationship, and then moved on to discuss their personal lives and the latest village news. In the contexts of these forms of consumption by locals and its subsequent exchange, production simply declined. However, seen in the following, it increased and transformed in relation to outside visitors.

Consumption by visitors

Visitors came from the cities and their consumption of pottery was an aspect of the cultural meanings rural life had for them. These visitors were both tourists and urbanites owning second homes in the countryside.

During the 1960s international tourism exploded and tourist resorts mushroomed all along the Mediterranean coast. Not all tourists stayed at the beach, and participants remembered how sometime in the middle of that decade foreign tourists began to visit Miravet on day trips. As the Spanish economy itself boomed, Spaniards also acquired the means to vacation and to make such trips, although slightly later than their foreign counterparts, Spanish tourists also began to visit Miravet. Initially these tourists came by car, but soon travel agents organized trips by bus. During the 1970s and 1980s, day-trippers from nearby cities – families travelling by car, or school children, social clubs and senior citizens travelling by bus – visited Miravet.

As in many other places in Spain, when villagers left for the city they did not sell their old homes: indeed, no one was interested in them. These people frequently returned to their village of origin and turned their unsold homes into vacation homes.[4] This trend accelerated, especially during the 1960s when they acquired cars. From the 1970s to the early 1990s, foreign and Spanish urbanites – mostly Catalan – bought houses in the countryside as second homes. The result was that in Miravet's county La Ribera d'Ebre in 1981 there was a ratio of about 1,400 second homes to 7,500 primary ones – a proportion of one to five (Anon., *La Ribera d'Ebre* 1989: 165). This trend continues today.

These tourists and second-home owners not only spent their weekends and vacations in the countryside, but also showed interest in things rural, like locally produced food such as olive oil, cheese and sausages, the latter two called *artisanal*, 'handmade'. The aesthetic side of their appreciation showed in their taste for decorating their interiors, especially the second homes, with such things as antique peasant furniture, and their exteriors with pottery planters and old pots (see also Prat 1983: 11–13).

The majority of the inhabitants of Miravet indulged in the new forms of consumption without much ado, but a minority shared the appreciation of rural life with the urbanite visitors. For instance, in 1990 a new pavement was constructed linking the village with the Raval. No one in Miravet disputed the necessity of this pavement, but what they disputed was its aesthetic aspects. Many found it beautiful because it functioned well. They did not give its style a second thought. A few others – among them Ferran Segarra – argued that it looked like the typical pavement one would find in coastal tourist resorts and in cities, and that it should have been constructed in a more humble manner using materials and techniques which had always been used in the village. People like Ferran wanted it to fit into a rural environment, or at least what they understood a rural environment to be. Those who liked the pavement were people who enjoyed rural life in a casual way. Those who did not like the pavement were people appreciating the countryside in a self-aware manner, an appreciation they shared with the urban visitors.[5]

This minority also showed another aesthetic appreciation of pottery and stood out among their neighbours by decorating their interiors with pottery of Miravet. They had pitxells and other small pottery on display in their living rooms and hallways. Several of them regularly used flatware, coffee sets and small bowls made by the potters. Often, they displayed antique pottery, made in Miravet or elsewhere.[6] In addition to the pottery, they often had antique and 'traditional rural' furniture. People who decorated the inside of their houses this way clearly bought outdoor pottery planters and/or old pots for aesthetic appeal.

As the visitors purchased pottery as an aspect of their appreciation of rural life, which was shared with a minority of the locals, one cannot assume that the potters themselves appreciated rural life either as the urban visitors did or as the local majority did. Therefore I will present three individual potters and their different understanding of the consumption of pottery by their visitors. But before moving on to discuss the appreciation by these individual potters, the urbanite appreciation of pottery from the perspective of these visitors is exemplified. To do this, I will analyse a number of texts, including television programmes and exhibitions, with potters and pottery as topics.[7]

During the 1960s, Spanish general production became industrial and mechanized and although during that period potters of Miravet also mechanized their productions,[8] they still relied on manual skills, most significantly throwing, a skill which takes much dedication and training to acquire. Hence, by the 1970s, due to the technical changes in Spanish production, the artisanal pottery production in Miravet transformed from the prevalent form of production to a highly exceptional one. With Spain's integration into Western Europe, this artisanal production also became highly exceptional within an international context.

In Spain, from roughly 1970 onwards, a number of books and guides on pottery were published emphasizing this exceptional form of production.[9] The three most influential ones are Llorens Artigas and Corredor-Matheos's *Cerámica popular española*, 'Spanish Popular Ceramics', published in 1970, Vossen, Seseña and Köpke's *Guía de los alfares de España*, 'Guide to the Potteries of Spain', published in 1975, and Sempere's *Rutas a los alfares*, 'Routes to the Potteries', published in 1982. This last, most frequently mentioned by the participants as being seen in the hands of their visitors, consists of two parts.[10] The first part describes the materials, the techniques of modelling and the kilns used by the Spanish and Portuguese potters, all geographically differentiated. The second part describes sites on the Iberian Peninsula in which pottery production takes place. This part is organized by regions; for each region a route leading to villages and towns with pottery production is given, hence the title of the book. The villages and towns are entries in this book: for every place the names and addresses of potters are given, the techniques and repertoire of products (and often the use of the several products) characteristic of that place are described, and sometimes an anecdote illustrates the text. In this book, Sempere focuses on what he called *tradicional*, 'traditional' techniques, 'traditional' types, etc. The meaning of this term is never made explicit, yet it is linked to *sobrevivir*, 'survival' in the competition with industrial production because '... the new products [i.e. industrial ones] have not been able to replace the traditional products derived from earth [*tierra* which also has connotations of land and country] and the agile hands of craftsmen who fulfil a social function' (Sempere 1982: 11, my translation).[11] Traditional pottery is furthermore appreciated because it is *productos naturales*, 'natural products',[12] developed in harmony with the natural environment;[13] it is produced applying techniques developed from *tiempos inmemoriales*, 'immemorial times';[14] it is maintained by people 'living practically isolated from the world';[15] it provides us with a bond with past civilizations, which have left their marks on ours;[16] and it provides an *auténtica relación uso-acción-forma*, 'authentic relationship between use-action-form'.[17/18]

Potters and pottery were not only topics of books and articles, but also of television programmes. For instance, in the summer of 1993, Catalan television broadcast a series following a hot air balloon travelling through Catalonia, stopping at places where the guide showed typical things. One of the subjects shown in Miravet was Vicen Sales working at his wheel.

Potters and pottery were also topics of exhibitions. For instance, in 1978, the *Obra Cultural* of the *Caixa de Pensions*, a Catalan bank,[19] organized the exhibition *Art Popular Català, Terrisses de Miravet*, 'Popular Catalan Art, the Potteries of Miravet'. This exhibition consisted of pieces of contemporary-produced pottery arranged according to workshop in which they were produced, accompanied by very large photographs of the heads of that workshop at work, usually throwing, and a handout that concluded with: 'The potter tradition continues. The times modernize but the craftsmen of Miravet, as artists of clay, resolutely continue to use their potter's wheels. Praise the neighbourhood called "the Potters Workshops" [i.e. the official name of the Raval].'[20]

Within the Catalan context, the highly exceptional artisanal production took on political meanings. After the end of the Franco government, when all Spaniards called for democracy, Catalans also called for a return of the autonomy they had been granted in 1932 but lost under the Franco dictatorship. In 1977 in Barcelona, after Franco's death and the first democratic elections in about 40 years, 1.5 million Catalans – a quarter of the total Catalan population – celebrated La Diada, the Catalan national holiday. Teresa Pedrola, who was then living in Barcelona, told me what a great emotional impact that day had made on her: it had been one of the most moving days of her life. The whole city was red and yellow – the colours of the Catalan flag – and everywhere one heard *sardanes*, the Catalan national dance. In the Spanish constitution of 1978, the possibility of autonomy was given and in a referendum in 1979 the Catalans voted for the Statute of Autonomy. The political scientist Daniele Conversi writes about this political restructuring: 'Thanks to nationalist pressures, Spain has undergone administrative transformations unparalleled in the rest of Western Europe and which are tantamount to a revolution. Its map has been thoroughly redrawn according to regionalist and nationalist principles, as the country was transformed from one of the most centralized regimes in the west to one of the most decentralized.' (Conversi 1993: 264).

This political restructuring of the relationship between Spain and Catalonia has also effected a re-evaluation of cultural symbols. The Franco government had attempted to rebuild Spain as a culturally homogenous country. Initially it had prohibited, and later severely restricted, the use of Catalan. With the Statute of Autonomy of 1979 Catalan was declared to be Catalonia's 'own' language and an official language on an equal footing with Castillian. In 1983 the Generalitat passed the law of linguistic normalization. Since then, the government has actively promoted the use of Catalan (Laitin 1987). As part of this normalization, the Generalitat published a popular poster with a map of Catalonia and all the places where pottery was produced with a few of their 'typical' products and their Catalan names (publication n.d., probably mid 1980s). The Franco government had projected the image of Spain as the country of flamenco and bullfights. It had oppressed performances of the *sardana*; now the Generalitat subsidizes its teaching. The Franco government had prohibited the celebration of Carnestoltes – the Catalan Carnival; after the return of democracy Catalans more than ever celebrate this festival. Other festivals became freed from Franco restrictions on imagery and they too gained in popularity. All these festivals have a strong local flavour. Historians study local archives and anthropologists study folklore and oral sources in order to 'recover' even more festivals. New festivals are created, festivals celebrating local activities from days past like reaping and threshing, and sheep dog training (Noyes 1992: 44–8, 55, Prat 1983: 13–14). These studies and revivals were promoted by an enormous congress in Barcelona on the *Cultura Tradicional i Popular* with the theme '*A la recerca de les nostres arrels*' – 'in search for our roots' – organized by the Generalitat in 1980 (Prat 1983: 14), the *Institut Catalá d'Antropologia,* founded in 1978, and what is presently called the *Centre de Promoció de la Cultura Popular i Tradicional Catalana* – 'Centre for the Promotion of Catalan popular and traditional culture', a subdivision of the Department of Culture of the Generalitat (Prat 1983: 14, 15, Calvo 1995: 24).[21] And certain schools with the basic objective of educating 'young Catalans', included in their curriculum traditional stories, traditional celebrations and the learning of traditional and typical artisanal techniques (Prat 1983: 15). With the return of autonomy, Catalans delight in everything Catalan.[22]

One of the clearest examples of how pottery acquired Catalanist connotations can be found in the catalogue of a Barcelona mail order company called *Xarxa Cultural*, 'the cultural net.' In a foreword, the goals of this organization were explained: 'Xarxa is an organization which disseminates the Catalan culture which all Catalans should know' (n.d.: 3).[23] In addition to this goal, it also had a political one: it contains information on the Statute of the Catalan Autonomy as a 'participat[ion] ... in the wide popular campaign to achieve an authentic Statute' (n.d.: 3).[24] the implication being that the Statute of Catalan Autonomy grants Catalonia autonomy within the Spanish state, but that the 'authentic Statute' would grant Catalonia independence from that state. The section *Artesania* opens with a quote from Ramon Violant very similar to Sempere's ideas but now placed within the Catalanist agenda: 'To know the folk art of our people gives us aesthetic pleasure, but in addition helps us to understand history, to deepen our roots, and to shape our personality in the most genuine sense of this word' (n.d.: 4).[25] Within this section, pottery is presented by location of production. Other merchandise offered ranges from suspenders with the Catalan flag, to posters with photographs of important Catalan historical and political figures, to books on Catalan culture and of Catalan literature.

Within Catalan politics, this re-evaluation of rural life had a class dimension. At the end of the 1950s and during the 1960s many Spaniards from poor regions of Spain had emigrated to Barcelona. In 1970, of the people living in Catalonia, 40 per cent were born elsewhere (Saez, quoted in Woolard 1989: 33). As they belonged in their regions of origin to the poorest part of the population, they formed the working class, while the Catalans tended to form the middle and upper classes. Being based on an electorate of non-Catalan origin, the leftist parties now had to moderate their Catalanism (see DiGiacomo 1985). On the other hand, the rightist party with its electorate of Catalan origin is strongly Catalanist. This party exploits the Catalan countryside as '... the source of authenticity upon which they must draw in order to safeguard an endangered national identity' (DiGiacomo 1987: 164, see also Prat 1983: 13–14).[26] Of course, these middle and upper classes are the ones with the second homes in the countryside, and the means to collect pottery.

From the perspective of the visitors, potters had become linked to their products by their production techniques – especially throwing. Pottery had become linked to the localities where potters lived and worked. Potters and pottery had also become linked with the past, with 'tradition', though this is not necessarily an exclusive link as there is also an appreciation for 'non-tradition'. Potters and pottery had become linked with Catalan identity.[27] Pottery workshops had become stops or destinations for trips, and pottery had become a collectible.[28] When urbanites came to a pottery workshop, they came to purchase a piece of pottery that had these connotations: the practices and customs surrounding these purchases expressed these connotations.

Consumption mediating production

All potters knew that if locals came to get a piece of pottery, it was needed for utilitarian usage, like a vessel for storing olive oil. They even interpreted purchases of decorative pottery by locals as motivated by this rationale: locals need to give someone (usually an outsider) a present. And they were very clear why these locals went to their workshop: it was the cheapest way to acquire pottery. But potters exchanged their products in their workshops very differently with visitors, just as the visitors consumed this pottery very differently. To place these exchanges in the wider contexts of production and consumption as mutually encompassing each other, the focus is now on how, in Marx's words, production and consumption 'mediate' each other (Marx 1972: esp. 229–30), how the potters, by interacting with visitors, especially by their understanding of the urbanite consumption of pottery, specified their production, as well as their marketing to these visitors. The inhabitants of Miravet, including the potters, were not uniform in their appreciation of rural life and pottery. Therefore three individual potters are presented as different in their exchanges, especially the practices and customs surrounding the sale to visitors, and the resulting specification of production.[29]

Vicen Sales on visitors

Vicen Sales explained to me the reasons visitors came to his workshop: 'Tourists cannot be on the beach everyday. They would get bored or they have days with bad weather. One day they might say "let's take a trip", and they come here.' He elaborated that these trips also included visits to the castle of Miravet and lunch in a nearby restaurant where one could sit outside and enjoy the landscape. I probed him on why the potters' workshops were included in these trips. Vicen Sales conjectured that the foreign tourists come to Miravet '... to take a small souvenir back home to France, Belgium, Holland, or Germany'. Though the pottery of Miravet is sold in the tourists resorts, these people might come to Miravet to buy souvenirs which are genuine, souvenirs made in the region where they have vacationed, and not, for example, in Japan. He gave similar reasons as to why Spanish and Catalan day-trippers came. I was left somewhat unconvinced – perhaps since I had been probing him and he had not given the issue much thought. However, he mentioned that visitors were interested in fet a ma, *'handmade'. I queried why he thought that was. Vicen Sales: 'Because it's* artesania, *handmade. When it's machine made it isn't any longer artesania ... it's not because it should be made out of clay. It can be made out of clay but if you make it with, for example, a machine, it isn't artesania ... When it's made on the wheel, it's artesania.' Again upon probing he could not elaborate on the visitors' interests in handmade pottery, though he was keenly aware that it attracted them: they could buy similar machine-made pottery much cheaper.*[30]

From all the visitors, Vicen Sales could only single out one group which he called 'people who dedicated themselves to ceramics'. These people showed their dedication by being informed on pottery, by carrying the guides on pottery on their visits, and by asking pertinent questions (as he said, he considered me among this group). These were questions about the techniques of production and the use of pottery. I asked Vicen if these people enquired about the use of a certain piece of pottery, did he think then that they were going to use it as such. Vicen was clear: these people might want to know how a sabonet was used – that is, a small vessel used for carrying soap when doing the laundry – but they themselves, he assumed, would use it, for example, as a vase.[31] *These visitors came especially for the pottery and sought pitxells and pieces that were 'typical' of Miravet. Initially he specified this category of pottery as pottery he had seen produced in Miravet all his life, but later narrowed it down to the small decorative pottery, especially de gargola de moxó. It seemed that for him 'typical' primarily referred to a demand. For instance, the gargola de moxó was most in demand as a 'typical piece', while other pieces, like the big bowls used in pig slaughter, were never bought by these visitors, hence were not 'typical', though they have been produced as long as he can recall. These dedicated-to-pottery people collected different types of pottery from different places. Again, when I asked him why these people dedicated themselves to pottery, he could only reply that it was something these people liked to do. Generally he understood visitors as people who can afford the luxury of taking a trip and buying a present for themselves and though he could single out this group of visitors, all visitors were eventually unpredictable in what they bought. Everyone just bought something he or she liked.*

Sales had specialized in the production of planters, largely for export, but as visitors mainly bought small pieces, he had begun to make these in response to demand. He did not produce much small pottery since he had not developed the required techniques nor acquired modern equipment to produce these pieces. He would just do it when he had a few hours to spare. For instance, one afternoon I dropped in at his workshop and he was throwing 20 gargoles de moxó. He had just finished packing a kiln that they were going to fire the next day. He could not work on orders and to spend that afternoon productively, he threw the gargoles.

Though he had hardly adjusted his production, he had adjusted his marketing to the visitors by also selling pottery made by others elsewhere. It was handmade and mechanically produced pottery – any pottery he could possibly sell. He had begun to offer pottery made elsewhere by offering mechanically produced flowerpots and casseroles to locals, and had expanded this offering with small decorative pottery – ashtrays, bowls, small planters, coffee sets, decorated plates etc. – in the same price range as he himself was producing with an eye to the visitors. If, for instance, people came looking for a coffee set, he would indifferently sell them one made somewhere else as he would one made by himself. When asked he would never lie about the place of production of a piece, but he had no problems selling others' pottery. 'Why not,' he asked me, 'if I can make a few nickels? The customers who know that a particular piece is not made here, that is, the ones looking for the typical pottery, will not buy it and the others will be happy with it.' And indeed, he explained, a few visitors recognized pottery made in Miravet, but most visitors could not differentiate between that pottery and the pottery made in other places. And even though they seemed to be attracted to handmade pottery, on the whole they could not distinguish that from industrially made pottery.

With regards to these visitors, Vicen had dedicated a small room in the back of his workshop to the storage and display of the pottery that they might buy. On these shelves – simple metal ones also used in local grocery and hardware shops – he piled up the pottery, some pieces half wrapped in order not to chip, and usually made the most of the space underneath and in front of the shelves by also piling pottery there. Often everything was covered with dust: it gave me the impression of being for storage rather than for display. He presented the pottery without much ado, just as other merchandise that was locally offered.

Most other potters shared Vicen Sales's understanding of the urbanite consumption of pottery without sharing with these visitors their appreciation of pottery. These potters had made similar adjustments in their production and marketing like producing pottery that visitors might buy, and offering in their workshops pottery made elsewhere in an intended room or corner of their workshop. There were two potters, however, who diverged due to their sharing with their urbanite visitors the urbanite appreciation for rural life and pottery.

The Segarras on visitors

On several occasions, I interviewed Ferran Segarra and his wife Teresa Pedrola about the visitors. They also told me that visitors came to buy pottery on 'a whim'. But in contrast to Vicen Sales, they did not need to conjecture about the visitors' motivations. They knew why people came, often already knew them by the manner in which they entered their workshop and greeted them. The Segarras categorized these visitors and their respective purchases consistently into three kinds of visitors.

The first kind of visitor was taking an outing which, in the course of a trip to the castle of Miravet and a local outdoor restaurant, happened to include a visit to a potter's workshop. The most annoying of these visitors were the ones who came by bus: they were on a trip organized by some travel agent and were brought to his workshop as part of a group package. Hence this kind of visitor was more often than not uninterested in pottery per se. Sometimes the travel agents tried to arrange in advance the arrival of their buses, and Ferran would then try to prevent it. But when Ferran could not, he suddenly found himself with a busload of people in his workshop, a place that was certainly not prepared for such a mass of people. Fortunately these tours did not happen very often. The visitors coming by bus were Spanish senior citizens and schoolchildren. Arriving both by bus or by car were tourists whom Ferran described as interested in 'beach, barbecue and a bit of folklore'. All these different visitors watched the throwing a bit, and they might buy a piece of pottery – any piece of pottery – but usually they bought nothing and always had to use his bathroom. Segarra was very sceptical about tourism and did not support the municipality of Miravet in its promotions of it. Since these people were neither commercially interesting nor very interested in his craft and his products, Segarra did not like to spend much time with them.

The second kind of visitor came to buy a piece of pottery, in Ferran's words, 'for the pleasure of buying it'. They were by far the largest group and they had money to spend. They came to buy – bowls, kitchen utensils, ashtrays, candlesticks, etc. – to decorate their vacation homes in the fashionable country style, or to buy jugs, the type of pottery that had become a collectible. They preferred these pieces to be handmade since that carried connotations appropriate to the country style or appropriate as a collectible. Ferran held these people in low esteem, especially the collectors of pitxells, whom he described as 'ladies who have little work and don't know what to do. They keep themselves amused with nonsense like that.' Some potters designed new pitxells for these collectors. As Ferran said, it could have been any such absurdity with a hole through which water entered and one through which it left. Ferran was reluctant to make these types. He had only on rare occasions designed new pitxell which these collectors would buy: 'That with the pitxells, it's all invented, and to make bullshit, I don't like that. To make the Olympic pitxell, no!'[32] *By this phrase he referred to the craze then building up for the Olympic Games in Barcelona. He exaggerated: none of the potters made an Olympic pitxell, but his point was that he preferred to make pottery developed from traditional types.*[33] *He was, however, commercially interested in them.*

The third kind of visitors were, as Ferran called them, people with an interès cultural, *a 'cultural interest'. They came to buy pottery, as well as to talk with the potter and to see his workshop and watch him work. Ferran talked with them '... about the world of pottery ... about everything a bit, but above all about the use, the shapes, the technique, how one prepares clay, how one fires. People were really interested in this.' (From observing Ferran, I know that he engaged in dialogues with these people, not just passively answering questions as other potters did.) They were Spaniards, especially Catalans, and came from the cities but were '... looking for a re-encounter with their origins, with their grandparents, something that they had lost ...' These people were interested in their past, in* 'tradicion' *as Ferran called it, and especially in the localization of that past, that is, in the locally specific types, production techniques and uses of pottery. Since they were well informed on this by books and guides and by their conversations with potters, Ferran called these people 'intellectuals'. These people typified a period from the end of the 1960s through the 1970s. At that time these intellectuals were in their 20s. Because they were already coming before Franco died, Ferran did not perceive a link between these visitors and the end of dictatorship, but he did acknowledge a link with the granting of autonomy to Catalonia because the interest of the Catalan visitors in Catalan pottery corresponded with their revived interest in and appreciation of their culture and language. These visitors were interested in purchasing* ceràmica popular, *'popular or folk ceramics' – a term he thought was introduced by the above mentioned books on pottery – that was 'typical' for Miravet, but, of course, they were actually looking for the smaller pieces (such as the simple pitxell, de gargola de moxó, the 'trunk pitxell') since they lived in relatively small urban apartments. As examples of pieces they were explicitly not interested in, Ferran mentioned flatware and the pitxell resembling an aubergine. Although these visitors in general had far less money to spend on pottery than the previously mentioned kind, Ferran liked them: 'we are not salespeople, we are artisans and therefore I like that kind of person. They are interested and you can talk with them... It's a relationship I have with these people, very different from the commercial one, that is, with people not culturally interested in pottery.'*[34]

As the Segarras, like all other potters shifted to a capital-intensive form of production and specialized in the production of small pottery, they specified their production in response to these urban visitors. As all other potters at that time, they initially mainly marketed their products to wholesalers and shop-owners and only a bit in addition to visitors. But by investing in the most modern equipment, they got to produce the most perfect – in terms of shape, firing, glaze and finish – pottery, the ware in which the visitors were mostly interested and willing to pay the highest prices. Wholesalers and shop-owners were far less interested in these more expensive pieces and the Segarras thus shifted to producing for visitors. They preferred to do so because they also made more money on their products when they sold these in their own workshop. In addition, they could produce the types of pottery they liked to produce.

Ferran had also directed his repertoire of products in response to these urban visitors. Initially he had produced both planters and small pieces, but he quickly specialized in the production of the latter. He returned to produce pottery that potters had ceased to produce during the 1960s. He referred to this return as 'reviving traditional' pottery. For instance, he again produced middle and small sized bowls with the same shape and glaze as had been done before and he called them 'typical bowls of Miravet'. In addition he developed a wide range of new types of small pottery. He emphasized the 'handmade' character by making every piece clearly distinguishable from other pieces. For instance, if he threw a series of one particular type, he glazed them in different colours and patterns.[35] The handmade character was made explicit by stickers attached to these products saying 'Ferran Segarra, handmade. Miravet'. The relatively large pieces, he usually signed with his initial F and 'Miravet'.

Initially Ferran Segarra started by renting an old workshop, but in the mid 1980s he built his own. Potters usually worked on the ground floor of their house or, since the early 1960s, in sheds outwardly similar to the ones in use by the local factories (factories that produced plastic irrigation material and cookies). When Segarra had to build a workshop in the mid 1980s, he did this with an eye toward the visitors. He had situated it at the entrance to Miravet, had built it in a retro-tradition style and had constructed a parking lot in front. The result was very inviting to the passer-by: some visitors told me that their first impression of the building had been that of a restaurant. In gothic letters it said on the walls Ferran Segarra, Canterer – canterer being the local expression for potter – and Terrisseria Popular, 'popular or folk pottery'. On the first floor there was one big room. On the left stood potter wheels and shelves for drying freshly thrown and recently glazed pottery; on the right merchandise was displayed. The merchandise was displayed on antique tables and wooden shelves. All was neatly lit with spotlights and regularly dusted off. Moreover, the pieces were displayed in small quantities of each type, and care was taken that the shelves and tables were always well filled but never overcrowded. On the second floor, free for every visitor to enjoy, he exhibited his own collection of antique pottery. Ferran Segarra also presented a kind of visiting card. The text on it linked the present pottery production in Miravet to that of the Iberians and the Moors. The pottery was described as a combination of 'the most traditional pieces' with 'new ornamental forms required by a changing society'.[36]

In addition to producing and marketing pottery, the Segarras gradually started to trade in old or antique pottery. They set up a network of gypsies who travelled in the countryside and bought old pottery, and more recently old furniture, which they then sold to him. This network covered a large part of Spain: sometimes gypsies came from as far as Madrid. The Segarras cleaned and restored this pottery and sold it in their workshop. They started to sell this pottery retail and subsequently also wholesale. At the time of the fieldwork, they exported truckloads of it to Belgium, the Netherlands and France.

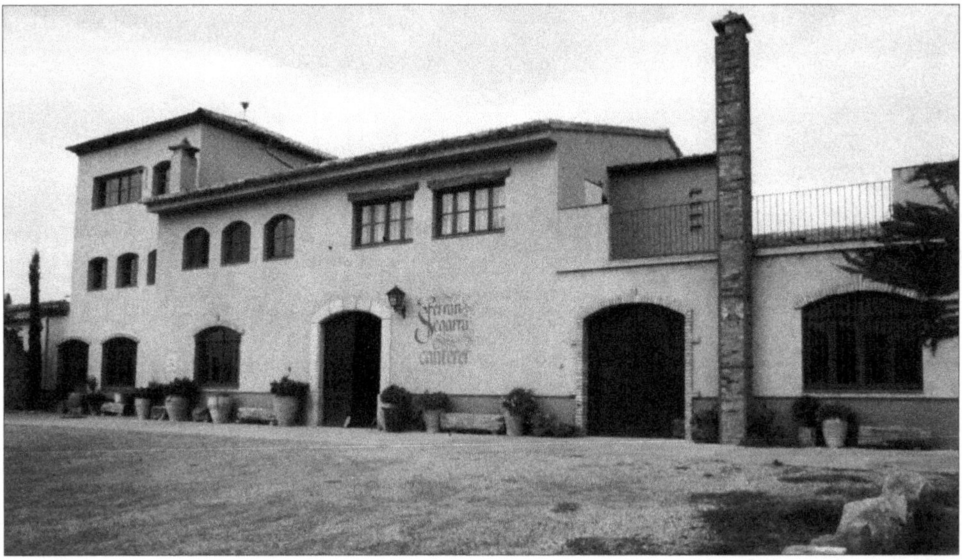

Plate 7.1 ***Segarra workshop, exterior.*** Author photo, 1992.

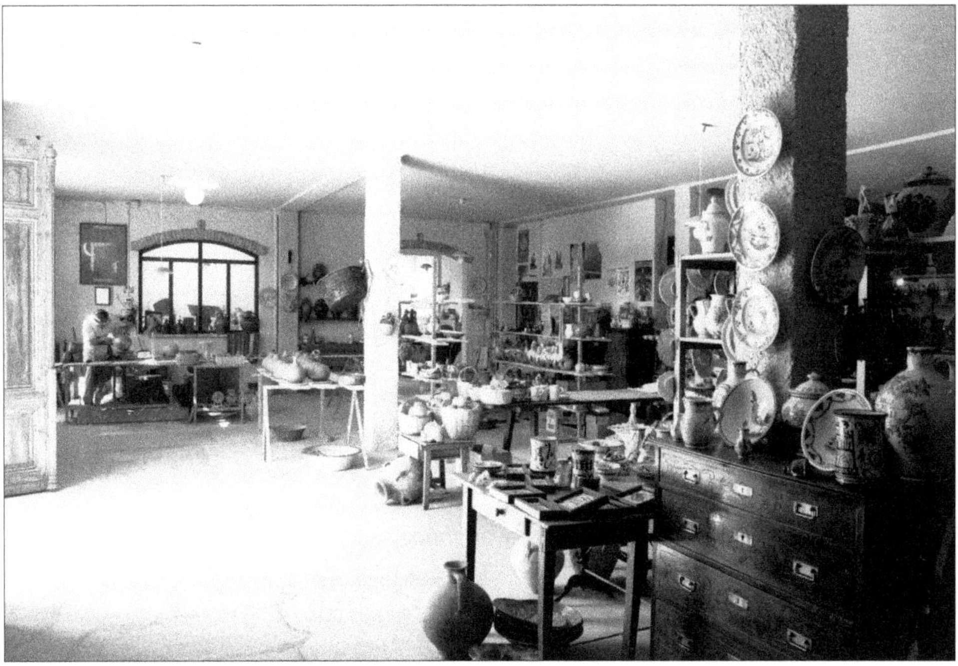

Plate 7.2 ***Segarra workshop, interior 1.*** On the left, view of the potter wheels, on the right of merchandise display. Author photo, 1992.

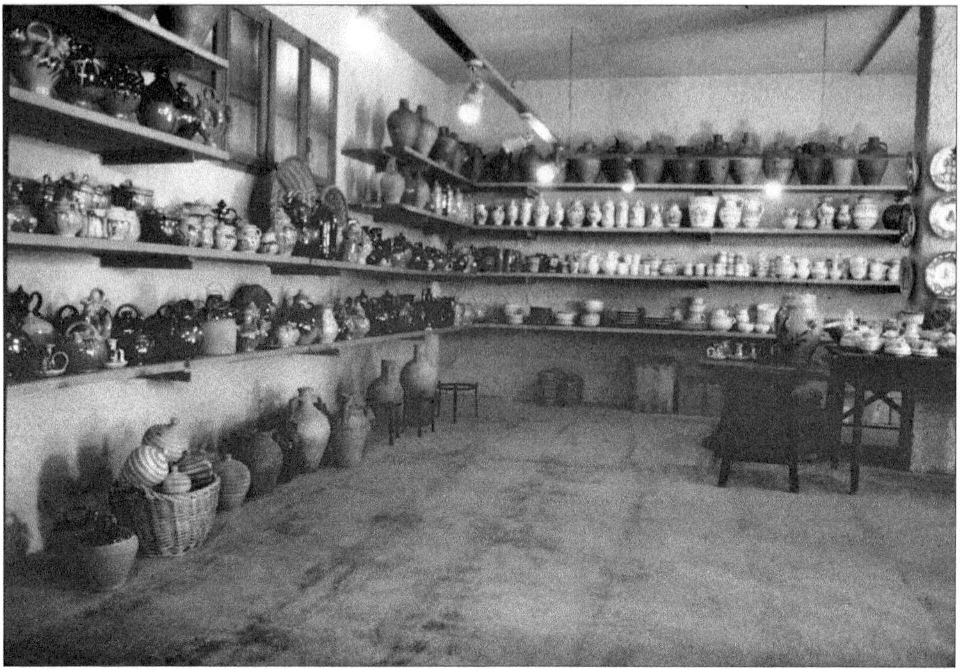

Plate 7.3 ***Segarra workshop, interior 2.*** Display of merchandise. Author photo, 1992.

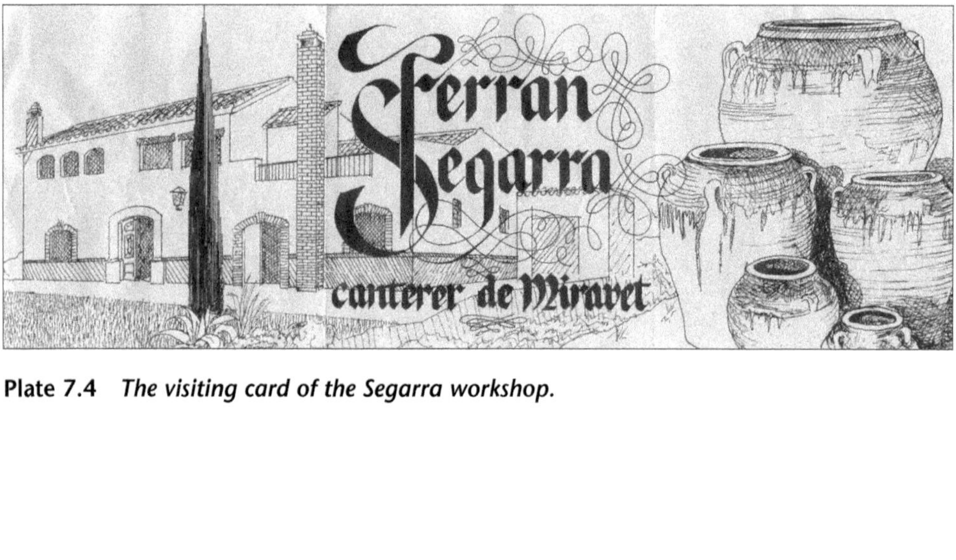

Plate 7.4 ***The visiting card of the Segarra workshop.***

Like other potters, they also sold products from others – pottery, glassware and baskets – but they selected their merchandise according to other explicit criteria. Since they claimed to be craftsmen they only wanted to sell handicrafts: the glassware was profitable but was also handmade; the baskets were not profitable but were regional handicrafts. The only pottery they sold that was not handmade were casseroles; indeed, nowhere in Spain were they handmade. In addition to these handicrafts they sold a few books on 'traditional', 'typical' Catalan pottery, the type of books mentioned earlier.

Segarra offered his merchandise to the customer as in an elegant urban shop, yet simultaneously emphasized its traditional handicraft character. To this new workshop, Ferran Segarra attracted the majority of the visitors of Miravet. In summer and on holidays it got very crowded in their workshop. Ferran Segarra and his wife both worked very hard selling their merchandise, and were occasionally assisted by their daughter. Though they employed people for other tasks, they alone attended to customers. Teresa: 'The people who visit our workshop want to talk with the potter or his family, not with an employee!' Besides, such a person could never adequately answer visitors' questions. Moreover, as she said, and as I often observed, they tried to establish personal contact with their visitors and Teresa prided herself in her recollection of people from previous visits and her good memory of personal details like who accompanied a visitor on a previous occasion.

The Segarras' sharing of the appreciation of rural life and pottery with their urban visitors related to their life histories. Ferran Segarra was born in Miravet, a son of a muleteer. Unlike Vicen's and all the other potters, his family emigrated to Barcelona in the 1950s when he was four. He only returned to Miravet during vacations, and grew up in Barcelona. In the 1970s he married a woman from Miravet, Teresa, the daughter of Antoni Pedrola Senr., and for several years they owned a bar in Barcelona. However, they preferred to live in Miravet and therefore returned. But even though, as he himself said, they were then living in a village, they continued to live an urban lifestyle. Upon their return, Ferran began, at the age of 30, to learn the craft of potter from his father-in-law. He choose to become a potter because he perceived that the new market for pottery was increasing, and he was attracted to this craft because of its associations with village life and making things with one's hands. When he decided to become a potter, he had been interested in pottery, an interest that had deepened over time. At the time of the fieldwork he owned an impressive collection of new and antique Spanish pottery, and a reasonably good collection of books on pottery,[37] and a nationwide network of colleague potters which he had visited at their workshops, or had met at the national annual pottery fair. In 1993 he even attended an international conference of potters in Istanbul. Moreover, he occasionally wrote short articles on pottery in the local magazine (Segarra 1990a, 1990b, 1990c and 1993) and has assisted Josep-Maria Saez, the former parish priest of Miravet who wrote articles on Miravet and its 'traditions', with the writing of a book on Miravet pottery (see Saez 1992). As my acknowledgements indicate, he was also very supportive and helpful to me with my study of potters and pottery.

The Papaseits on visitors.

Like Ferran Segarra and Teresa Pedrola, the Papaseits understood the visitors as more than just a commercial opportunity, and like the Segarras, they too identified themselves with their products, as for instance, they too had returned to producing pieces which had no longer been produced. Yet the content of this identification was different in a few interesting aspects.

Since their workshop was small and off the road, they hardly received visitors who would come to a potters' workshop like a tourist attraction. They distinguished clearly between visitors with an interest in traditional local potters and their pottery and those, in Ferran Segarra's words, 'buying for the pleasure of buying it'. But their attitude towards these latter differed importantly from Ferran Segarra's. Segarra considered the pitxells designed especially for these collectors nonsense and looked down on the people collecting them, but the Papaseits enjoyed designing new pitxells, objects they, like Ferran Segarra, defined as any piece of pottery with one hole for entering and one for emitting water. For them these pieces offered creative freedom. Indeed they were very proud, and as Dominica put it, '... those collectors enjoy pitxells so much'. They mentioned, as an example of how people could enjoy this pottery, an architect who was supposed to have about a thousand of these objects.[38]

Over the years, they built up a network of pitxells collectors who visited their workshop regularly, buying their pieces. Since the Papaseits were making so many different types of pitxells (they guessed about 70 or 80), they kept a list of the collectors and purchases. On each visit, they could check what each customer already had and thus could prevent duplication. In addition, they took orders for certain types of pitxell, and when they produced a series of that type, they called collectors to come pick them up. They produced new types of pitxells in small numbers. Depending on the response of the collectors, they continued or ceased producing new types. Often it turned out that they had produced a newly designed pitxell only for those collectors, which was never marketed to wholesalers or shop-owners. For instance, there was the 'Greek pitxell' – a copy from a pitxell brought from Greece: not aesthetically appealing but technically curious. They never sold this type to wholesalers or shop-owners, only to the more dedicated collectors who had to come to their workshop to purchase it.[39]

I almost always interviewed Pepe Papaseit and his wife Dominica along with their son Josep. It was their son who had introduced a novel understanding of visitors and of his family's craft. Josep, who at the time of the fieldwork was in his mid twenties, had attended the art school in Tarragona where he learned techniques uncommon to potters of Miravet and gained a perspective on pottery different from that of his elders and of Miravet potters in general. Like Ferran Segarra, he also collected books on pottery, made visits to potter workshops in other parts of the country, collected old pottery and wrote articles on pottery in a local magazine. He also became very interested in ceramics as an art form and knew several ceramists personally. In between the regular periods of production, he was working on reproducing sophisticated ceramics like pharmacy pots – never made in Miravet – and creating artistic vases and objects. Though his father also designed pitxells, Josep became the one who created new designs. Josep also created art ceramics, pieces not departing from the local pottery tradition but from artistic concerns.[40]

* * *

Concomitant with an understanding of their visitors' appreciation and consumption of pottery, potters interacted differently with visitors, and therefore specified the production and marketing of pottery accordingly. To analyse their differences, Walter Benjamin's conception of 'aura' is appropriate. In his article 'The Work of Art in the Age of Mechanical Reproduction' (Benjamin 1968), Benjamin juxtaposes a mechanically produced work of art with a manually produced one. The first is a reproduction of another object (i.e. an original) and is anonymous; the second has an 'aura', a value based upon its uniqueness in time and space, and which is its own particular social history. This conception of aura is illuminating, since by recognition of an object being handmade and hence having an aura, an object is identified with its producer, though this object might be exchanged as a commodity and thus in other aspects be alienated, in the Marxian sense, from its producer.

All potters were aware that certain visitors consumed their products as handmade objects and thus identified them with their person – these potters were flattered by it, but Vicen Sales and most other potters ultimately understood this consumption and identification as a market force no different from the consumption by other visitors or, for that matter, consumers of pottery in general. As people making a living, they saw these visitors as a commercial opportunity and offered them any merchandise they could. These potters perceived a difference between a mechanically and a manually produced object in market terms, they produced and marketed whatever sold best, but they did not identify with their products, i.e. recognize their pottery's aura.

In contrast, the Segarras and the Papaseits understood that certain visitors linked them with their products, that for these visitors their pottery had an aura: that their pottery was unique by being made in Miravet for a long time, by particular people for a particular use by other people. They too were people making a living, but interpreted visitors not only as a commercial opportunity, but also as people identifying traditional local potters with pottery, and in response they identified themselves self-consciously as traditional potters. In addition, for Josep Papaseit, pottery was a medium to express one's individuality, the individuality of the artist as well as of the traditional craftsman.

* * *

Except for Ferran Segarra, the exchanges in the potter workshops had little economic significance, but they did have social ones. Although potters exchanged pottery in their workshops, sometimes even the exact same types of pottery, the social significance of these exchanges differed greatly due to the types of pottery exchanged, the value of the exchanged pottery and the practices and customs around these exchanges. Thus these exchanges encompassed production and consumption. In the exchanges with the locals, potters were as producers of utilitarian wares integrated into the village or local community. In the exchanges with visitors, potters were integrated into the wider society as artisans producing local traditional wares. But from the perspective of the potters, these exchanges with visitors could be encompassed in very different forms of production and consumption. Most potters understood these visitors as an economic opportunity, and had specified their production and marketing accordingly. The Segarras and Papaseits understood the visitors as identifying them with their products, and as a result had specified their production and marketing in ways that emphasized their identification with their pottery.

---

1 Villagers expressed and strengthened their social relationships by these exchanges of objects, but also by two other forms of exchange which contrast sharply with the one analysed above.

The first of these was represented by the presentation of a gift that one was expected to make on occasions such as weddings and First Communions, and that gift was expected to be something which was bought. I asked Pepe Papaseit why at a wedding, for example, he could not give something he himself had produced. Certainly newlywed couples might need things like vessels and planters. He replied that some would look down upon such a gift saying 'they themselves have made it!'. Irrespective of the market value of the piece of pottery given, such a gift would be an embarrassment. It was more appropriate to give items like glasses, though invariably a newlywed couple ended up with too many of them. Hence, when one was obliged to make a gift for such an occasion, the gift needed to have an exchange-value established in a market relationship. This only very recently changed. Upon one return visit in 1994, Josep Papaseit told me he had made dining ware as a wedding gift for friends in Miravet.

The second form of exchange was the gift of a piece of pottery explicitly made by a potter for a certain person to express their relationship. For instance, when Pepe Papaseit was courting his wife Dominica, he made a watering can for her since she loved flowers so much. Potters never made these things and he copied it from a metal watering can. Or, Ferran Segarra made a plate with a drawing of Miravet and a dedication for friends who owned his favourite restaurant. (Similarly, he made two canisters for me with engraved dedications.) These gifts were rare: potters hardly produced anything unique and outside the usual. Tomaset Avante, an uncle of Joan, was an exception in that he was, as Joan called him, *artístic*: he was also the one who designed the gargola de moxó. At the time of the fieldwork, he had long since passed away, but I hade seen a few singular pieces made by him, invariably not for the market but to be given away. These gifts were personal creations intended to express a personal relationship – something inherent to the medium of pottery.

2 Note that on another occasion both Josep and Antoni Pedrola told me that they charged foreign wholesalers the same prices as Spanish ones: they considered it fair.

3 When Pepe Jornet had designed a new planter, he was very proud of it, and placed one in the hallway of his house. Though he did not sell this newly designed planter painted, he painted this pot with latex paint just as he had painted pottery in the 1960s.

4 In a town at the Costa Brava, this has been going on since the end of the twentieth century when people in this town moved to Barcelona during the economic bloom at that time (Pi-Sunyer 1973: 6–7). Noyes describes this as a general Catalan pattern (Noyes 1992: 53), and Barrett for an Aragonese village (Barrett 1974: 83).

5 The latter were a group loosely organized around a local bi-monthly magazine, *la Sanaqueta*, in which they published writings on local events and local history, on feminism, environmentalism, medicinal herbs and, as Ferran Segarra wrote, on pottery. This network was relatively open and I spent a good deal of time with them. Apart from the social network they formed, these people were not clearly distinguishable from other villagers. In general, they were people in their late twenties, thirties, or early forties. Some were professionals, some had rural jobs – though none was a farmer – some were what was locally called *hippie* – yes, men with long hair and untucked shirts! The attitude of these people was similar to one observed by Maddox (Maddox 1993: esp 198).

6 This antique pottery was invariably waxed, just as one would wax wooden furniture. Although the patina of unglazed surfaces of this pottery was appreciated, these people considered these surfaces 'dirty' and too 'coarse', and smoothed them by waxing. Note that this preference for waxed surfaces parallels the preference of the other villagers for fine pottery in their interiors.

7 These texts differ from each other in several dimensions. First, they differ with respect to the broadness of their topic, ranging from all potters and pottery of the Iberian Peninsula to specifically the potters and pottery of Miravet (even, at times, also including all handicrafts). Second, these texts differ with respect to their audiences in complicated ways. For instance, some are in Castillian and address all Spaniards, others are in Catalan and specifically address Catalans. Third, they differ with respect to their 'authors': scholars – indeed also anthropologists – the Generalitat, the municipality of Miravet and mail-order merchants. And fourth, these texts differ with respect to their medium. There are books – ranging from cheap editions to colourful coffee table books – television programmes and exhibitions.

8 A mechanization that was partly turned back at the end of this decade when potters stopped using jiggers.

9 See also the footnote on publications on Spanish pottery in the first chapter.

10 Sempere combined the writing of this book with dealing wholesale in 'traditional' pottery: all participants knew him as an author and many as a trade partner.

11 My translation of: 'Entedemos que la alfarería ha sobrevivido a estos embates porque esos nuevos productos no han sido capaces de reemplazar a los productos tradicionales derivados de la tierra y de las manos hábiles de los artesoans que cumplen una función en las sociedad.'

12 'Hemos tenido que llegar a su casi total desaparación, para percatarnos de la nobleza de la materia, y ha sido precisa una degradación de la calidad de vida, para que sintiéramos la imperiosa necesidad de recuperar los productos naturales.' (Sempere 1982: 11).

13 'La artesanía ha contribuído desde siempre a equilibrar el progreso de una manera armónica, sin destruir el entorno natural, aprovechando las materias primas conocidas y perfeccionadas por el hombre durante milenios.' (Sempere 1982: 11).

14 'La alfarería tiene a su alcance infinidad de posibilidades con tan sólo tres elementos: *tierra, agua y fuego*. Elementos que hombres y mujeres alafareros vienen empleando desde tiempos inmemoriales.' (Sempere 1982: 11, italics in original).

15 'Gracias a estos artesanos que han vivido prácticamente aislados del mundo y cuya produción ha sido insustible hace poco tiempo, la tradición ha sobrevivido y se ha conservado a través de varias generaciones.' (Sempere 1982: 12).

16 'En los utensilios, aún vivos, que hallaremos en cada alfar, prodremos establecer un diálogo evidente con civilizaciones y culturas que durante siglos fueren imprimiendo su inflencia en la muestra.' (Sempere 1982: 12).

17 'Esta labor de los alfareros que fuera parte principal y necesidad físico-funcional imprescindible durante milenios, contenía la auténtica relación uso-acción-forma. Su lógica poseía una lucha común con la estética: nacía "constreñida desde dentro."' (Sempere 1982: 12).

18 The other books, as well as the many published later, are similar in their approach to pottery. Other publications on potters and pottery include magazines and posters. For instance, in 1984 the magazine *SOM, Publicació de cultura popular Catalana*, 'We Are, Publication on the Catalan Popular Culture', had an article meaningfully called 'Miravet, Fang i tradició', 'Miravet, Clay and Tradition' (Ventura 1984). In the local magazine *La Sanaqueta* appeared another article with an almost identical title (without the name of Miravet) by a village author (Alfonso 1983).

When I presented myself as someone who was to write a dissertation on the potters and pottery of Miravet, the participants presumed that I was going to write something similar to these books and guides. When I asked about information that is not offered in these books and guides, for instance about kinship relationships, they often replied, 'what has that to do with pottery?'

The potters and pottery of Miravet were mentioned in two publications by the municipality of Miravet. In the early 1970s the municipality of Miravet began to promote tourism by publishing a handout (Anon., n.d., *Miravet*). This handout contained a brief 'historical-geographical' description of Miravet (especially of its Templar castle), four photographs (two of them of potters who were throwing), a map of the village, and only, under the list of economic activities, a mention of the fact that pottery was produced (Anon., n.d., *Miravet* (Tarragona) Breve Resumen).

In 1982 a tourist fair, *Rodatour*, was held in Barcelona and the Municipality of Miravet occupied a stand. At this fair, the potters demonstrated the throwing of pottery. Indeed, they had been very proud of the attention they received: even the President of the Generalitat, Jordi Pujol, and the mayor of Barcelona, Joan Maragall, came to have a look. For this fair the municipality published another handout, this time twelve pages. Now the pottery production was very prominently present: of the ten pages with text, five were devoted to the pottery workshops. Again, the pottery was divided in *tradicional,* 'traditional' pottery, but also in another, unspecified, category. For instance, on the Pedrola workshop it said '... perhaps the pottery workshop with the largest production, above all vessels in large sizes, 'solseres" to be used as planters, cossis, miniatures, and kitchen utensils as well as pieces from the most pure Miravetan tradition'." (Miravet n.d.: no page numbering) My translation of: '... es potser de les cantereries de més producció, sobre tot en gerres de força magnitut, 'solseres' per jardineria, cossis, pitxells, miniatures i estris cassolans així com peces de la més pura tradiccó miravetana' (Miravet n.d.: no page numbering).

19 Spanish banks are by law obliged to spend a part of their profits to sponsor cultural events.

20 My translation of: 'La tradició terrissaire continua. Els temps es modernitzen, però els artesans de Miravet continuen amb les seves rodes terrisseres i la fermesa d'artistes del fang. Fan honor a un barri anomenat "Las alfarerias"' (Anon., *Art Popular Català, Terrisses de Miravet,* published by Obra Cultural de la Caixa de Pensions per a la vellesa i d'estalvis 1978: no page numbering).

21 On 15 May, 1990 I talked with Mr Joan Ferrer, director of the *Subdirecció General d'Artesania,* an organization founded in 1986 by the Generalitat in order to promote handicrafts. This organization tried to achieve this by providing assistance with the training of handicraftsmen, with the commercialization of handicraft products and with the investment in production equipment. According to the director, the reason it did so was the 'individuality' of handicrafts as opposed to uniform industrial products: this individuality needed to be preserved and disseminated. Mr Ferrer explicitly mentioned that this dids not mean that handicraft production should be denied technical development; on the contrary, it was important that the handicraftsman adapted himself to the changing times. Moreover, he emphasized that the Generalitat was the first to perceive the necessity of this kind of promotion as the central Spanish government had neglected the craftsmen. The latter statement has to be interpreted within the framework of centre-periphery politics. The Franco government had organized handicraftsmen in a 'guild', but its impact was so slight that participants hardly remembered it. Apart from the subsidies Ferran Segarra had received from this organization, the potters of Miravet had not noticed much of this new support. But perhaps potters might find it more helpful in the future.

22 See also Giner (1980: 10): 'This search [of 'stateless nations' like Catalonia] for a common identity and strength results also in the conscious participation of the people in many symbolic acts of ethnico-cultural affirmation.' He then gives the examples of the sardana and the *castells,* the human towers.

23 My translation of 'XARXA és una organizació difusora de la cultura catalana que tots els catalans hauríem de conèixer.'

24 My translation of 'Així XARXA participa també en la gran campanya popular per conquerir un autèntic Estatut.'

25 The full quote is: 'As art, searching for new formulas, had arrived at a certain disorientation, and as artists had distanced themselves from the people by their intellectual elaborations, we rediscover folk art with all its beauty of simple things and our sensibility to appreciate the forms which, during centuries, have been stripped of everything superfluous. To know

the folk art of our people gives us aesthetic pleasure, but in addition helps us to understand history, to deepen our roots, and to shape our personality in the most genuine sense of this word.' (n.d.: 4)

My translation of: 'Quan l'art, a la recerca de noves fórmules, ha arribat a una certa desorientació, quan els artistes s'han allunyat del poble amb llurs elucubracions intel.lectuals, redescrobrim l'art popular amb tota la bellesa de les coses senzilles i la nostra sensibilitat sap apreciar les formes que, al llarg dels segles, eliminaren tot allò que li era superflu. Conèixer l'art popular del nostre poble, a més de produir-nos un plaer estètic ens ajuda a conèixer la història, a profundir en les nostres pròpies arrels i a configurar la nostra personalitat en el sentit més genuí de la paraula.' (Xarxa n.d.: 4, quoting Violant, *L'art popular a Catalunya*, no publication date or page numbers are given).

26 The history of Catalanism and its relation with class politics is much more complex than can be contained within this study. For excellent histories of these complexities, see Johnston (1991) and DiGiacomo (1985).

27 All these texts address an audience interested in visiting potter workshops and in buying and collecting pottery. Pottery is popular as no other Spanish handicraft because of its physical characteristics and its specific history of production and consumption. Like the potters of Miravet, Spanish potters in general experienced a crisis only as late as the 1960s with respect to the marketing and consumption of their products. Hence, when these texts were published, a relatively large number of potters were still working. Therefore, pottery remained relatively cheap and affordable by many, was still produced in a large variety, and potters' workshops could visited often and easily. Pottery, especially the smaller types, easily fits into a modern urban lifestyle.

In contrast, other Spanish handicraftsmen and their products have not had the good fortune of attracting so much attention. For instance, glass is at present manually produced only in the Balearic Islands. Many kinds of glass are no longer produced and are therefore relatively expensive, and the glass that is still produced is not as varied. Hence, glass is not as appropriate an item for collecting as pottery. Moreover, glass does not have a history of production and consumption interwoven into rural life such as pottery has. Or, for another example, baskets are rarely produced in Spain: the Spanish baskets have been pushed from the market by Asian and Moroccan imports. Basket weavers work in rooms not publicly accessible like potters' workshops. The Spanish baskets are relatively big and therefore not as easily adaptable to modern urban life. Probably as a consequence of all these reasons, there are few books published on these other handicraftsmen and their products.

28 Moeran's study of Japanese potters (1984) focuses on a similar appreciation of handicrafts (perhaps with the same intellectual roots in the work of John Ruskin and William Morris) and how this appreciation has affected the production, marketing and consumption of the Onta pottery. As such both our ethnographies overlap and show interesting similarities, like an explicit link between the pottery and its producers as essential to a certain appreciation of pottery. There are also interesting contrasts, like the technical and aesthetic development of pottery and how this development has taken place.

29 In the interviews, these potters often used *tradició*, 'tradition' to explain their understandings. Several social scientists have argued for an irony in the term tradition: with tradition people claim a continuity with the past, while ironically to call a practice or thing tradition constitutes a break with the past (see Handler and Linnekin 1984 and Hobsbawm and Ranger 1992). The irony consists in the juxtaposition of these people's understanding with the analyst's understanding. In opposition to this, I attempt with my analysis to understand how the participants make sense out of their experiences by using 'tradition', especially how different participants use this term differently (see also Sahlins 1993). In the following analysis, I use the term tradition as the participants did and, indeed, without cynicism.

30 As do all of the participants, Vicen Sales relates 'handmade' to that which is thrown on a potters' wheel. Other parts of the production, like the preparation of clay, could be mechanized without affecting the handmade character of the pottery. All potters were keenly aware that their pottery was attractive because it was handmade, which meant thrown by hand. This was highlighted on an occasion in 1991 when the brothers Pedrola ordered an advanced machine for throwing planters. Though it was a large investment, this machine would throw much

faster than they could by hand. Moreover, it would free them from throwing so they could spend more time and attention on other aspects of the production. When they told me about this machine, I asked if that would make their products no longer handmade. They, in turn, asked why it needed to be 'handmade'. What difference would it make if the pottery would look the same? Still they said that they would *pasar*, 'pass', a piece thrown with this machine briefly on the wheel, just move their hands quickly over the surface while revolving, in order to give it a 'handmade' look. So for them it was 'handmade': though they did not understand why customers wanted 'handmade', they acknowledged that that was what the customer wanted.

31 He normally did not ask them how they were going to use a particular piece. I had observed him enthusiastically answering visitors' questions about pottery, but never asking them questions about it.

32 Since the words 'to invent' and 'invention' have become so charged in this context (*The Invention of Tradition*, Hobsbawm and Ranger, 1983) it is interesting to note Ferran Segarra's use of it in this context. Talking about the decorative pitxells he was producing, I asked him if these were his *inventos*, 'inventions'. He replied 'Inventions, copies?' and proceeded by explaining that he '... had adapted the idea of ceràmica popular with a mixture of new shapes ... I don't know if they are mine ... no one invents anything ... Picasso and those people do ... but I grab a piece and give it my shape.' In this explanation, Ferran Segarra used 'to invent' in reference to something like an artistic creation and did not wish to claim to have 'invented' any pitxell. Asking him if he did not like to produce new models, he gave an answer in which 'to invent' had negative connotations, something he emphasized by referring to the then current Catalan craze of commercializing anything by making it 'Olympic'. In this latter use, 'to invent' had the sense of breaking with the past, i.e. what Ferran saw as 'ceràmica popular'.

33 In reference to this, but on another occasion, his wife Teresa told me that as artisans they needed to have a given model that they could copy perfectly. She added that they could not or were not good at designing models from scratch, something better left for artists.

34 There was an overlap between such kinds of visitors and the wholesalers and shop-owners. Milio Sempere was a wholesaler with the same interest in pottery, an interest that he also expressed as an author. He was the only wholesaler with this interest, but there were several shop-owners with the same interest. I interviewed one client of Ferran Segarra. He fitted into Ferran Segarra's description, being of a particular generation, and he described his clientele similarly to Ferran. The Papaseits also described such a similarity between distributors and customers of distributors.

35 Previously, and most potters still did at the time of the fieldwork, one attempted to make all pieces of a certain type to look the same as much as possible.

36 The whole text is: 'In Miravet, pottery, the fundamental support of the antique cultures, is very much represented: Iberian finds, stamped terracotta and remains of Moorish pottery compose the deposits which tell us of the evolution of these handicrafts in the course of the centuries until the present Miravetan pottery. Today, the most traditional pieces are combined with new ornamental forms required by a changing society, which help to keep this craft alive, made by hand, and by affection and love for delicate things.'

This is my translation of: 'A Miravet; la terrissa, suport fonamental de les cultures antigues, es trova molt representada, troballes IBERIQUES, terracotes SIGILATES i restes de terrises musulmanes aporten els sediments que ens parlen de l'evolució d'aquestes artesanies al llarg dels segles fins a l'actual terrisa miravetana. Avui, les peces més tradicionals es mesclen amb noves formes ornamentals demanades per una societat canviant, que ajuden a mantindre viu aquest treball artesà fet amb les mans i amb el caliu i l'estima de les coses fràgils.'

37 As we come to see, Josep Papaseit had a similar collection of books. Other potters too had books but they had always been given these books by the authors. For instance, Milio Treig Jt said that he had all the books written by Emili Sempere, who as a wholesaler in pottery was also a client of his. But when I asked him if he had ever bought a book on pottery, he laughed. It seemed to be a silly question. Similarly Pepe Jornet was acquainted with authors and had received books from them, but he told me that he had only read what was written on Miravet.

38 The Papaseits prided themselves in the production of, in their words, *coses rares*, 'strange things', for example a ceramic frog and owl. These animals had been ordered as a wedding gift: the animals referred to a play on the names of the bridal couple. They were the only potter workshop where people could come with such specialized requests.

39 Milio Treig Jr. also produced and designed new types of pitxells, though did not have this network of collectors. However, he himself was a collector of pitxells: in his house he had shelves with specimens of every pitxell he had ever produced.

40 Joan Avante had two sons. One who had studied natural sciences for a short while in Barcelona and another who went to the same art school as Josep Papaseit. Both had returned to join their father in his workshop. Though the one who had attended the art school expressed to me the wish to 'revive old techniques and glazes and develop them within the context of popular pottery', I never saw the pieces by which he was attempting this. Nor did he collect old pottery or books on pottery, or visit other potter workshops. Although a gifted thrower and very knowledgeable on glazes and other academic techniques, he did not seem to have had such an influence on his workshop as Josep Papaseit had. Joan Avante also had a daughter who went to the same art school in Tarragona but studied sculpture. Though she sometimes worked in the workshop, mainly on her own sculptures, she seemed not to have any influence on their workshop.

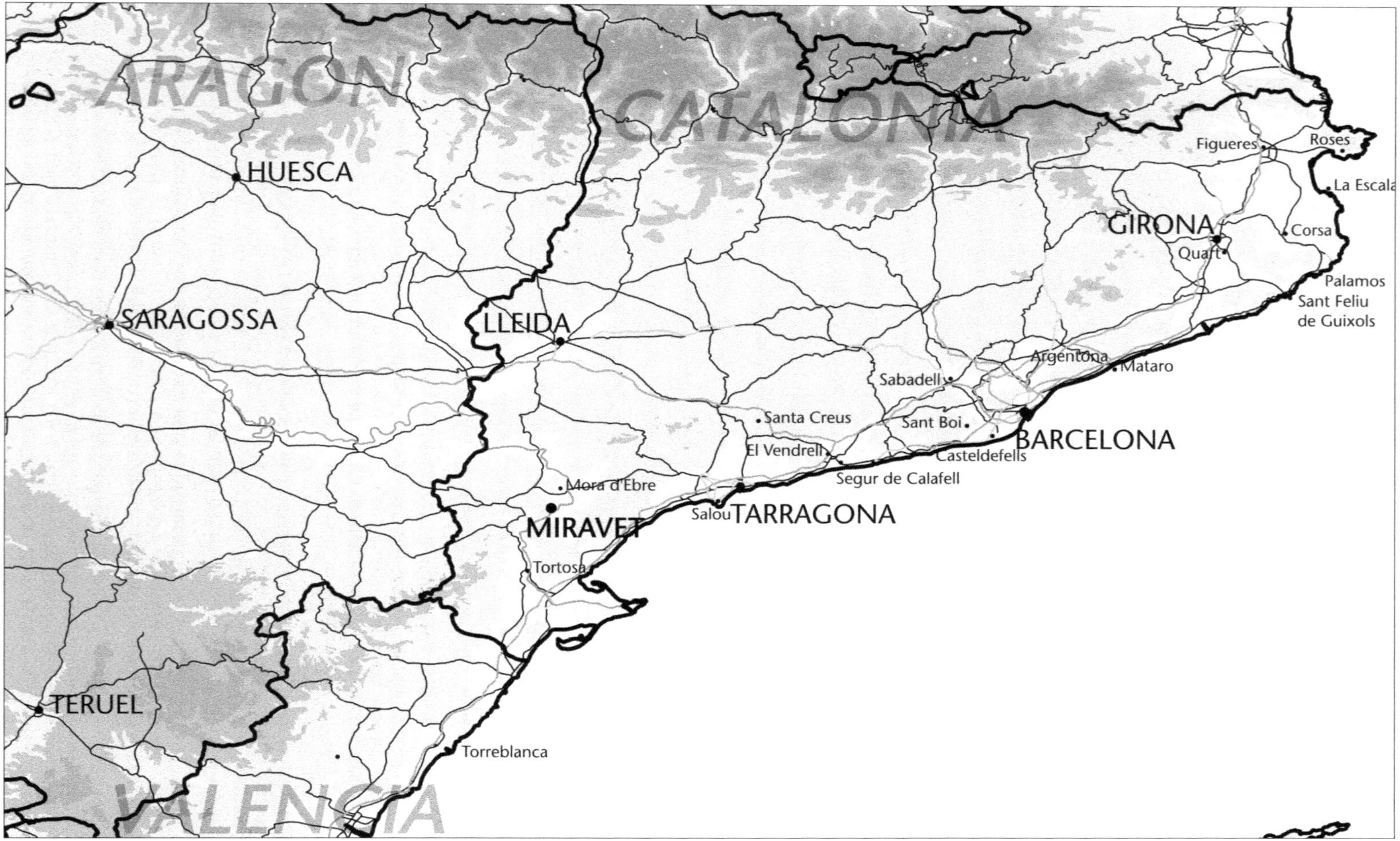

Map 6 *Map to Chapter 8.*

Chapter 8

Potters and consumers in the early 1990s

This chapter parallels the chapter on potters and consumers during the period of the muleteers. In Chapter 3 the channels through which potters learned about the consumers and forms of consumption of their pottery were described. At that time, potters differed in the level of detail in this knowledge, and their overlapping, collective knowledge as a 'geography of consumption' was presented. As this knowledge is a function of the manner of production, of marketing and self-identification, and as these have become individualized, especially during the 1970s and 1980s, potters in the 1990s did not differ quantitatively in their knowledge of consumers, but qualitatively. There are plural geographies of consumption, albeit individual geographies overlap. The relationships they have with consumers through their consumption of pottery are exemplified with three different geographies of consumption provided by three different pottery workshops.

Vicen Sales on consumers and consumption

Since the 1970s the Fabregat-Sales workshop changed very little with regards to investments in equipment and workspace. Especially during the 1980s, Vicen Sales – the only potter left in this workshop – specialized in a repertoire of large pottery.

On several occasions I talked with Vicen Sales about the consumers and consumption. From about 1970 on, the Fabregat-Sales workshop had acquired a number of clients, mostly wholesalers, in Vicen Sales's words, de la costa, *'at the coast', i.e. the Catalan and northern Valencian coast. He listed the most important ones: Castelló, Torreblanca, El Vendrell, Castelldefels, Sant Boi, Sant Feliú de Guixols, Palamós, Corsá, La Escala, Roses and Figueres. While we discussed these shops, he referred to them with different labels, one of these being* botiges de jardineria, *'planters shops', since he supplied them mostly with planters. Besides planters, he also supplied them with other types of pottery, though in small and declining numbers and not regularly. For instance, in a truckload of planters he included at the most 80 pitxells, but usually just half a dozen. He sent clients who wanted more pitxells to the Treig and Papaseit workshops. Next to pitxells, there were bowls, some wholly green glazed, bedwarmers and gargoles. He regularly* supplied a few of these clients with vessels, but in very small quantities; only the client in Castelló ordered small vessels, and two or three other clients ordered large vessels, but at most they ordered ten per year.

From the beginning, Vicen Sales noticed a difference between the orders of these coastal clients and those of the inland ones he was then still supplying. The Aragonese clients ordered the same types of pottery from year to year, with only slight variations in quantities, while the coastal orders varied from one order to the next, and even ordered in one shipment variations on one type. For instance, Vicen always sold the cossis with smooth surfaces and flat rims as they were used as wash-tubs to Aragon, but to the coastal clients he sold cossis with rounded rims to be used as planters, with decorations of lines and bordons, 'ribs'. One order might include the planters with a wavy line decoration as well as planters with rib decoration, the next order with a flower line decoration, or none at all of this type. Vicen Sales interpreted this variation within and between orders simply as moda, 'fashion'.

Although previously he had delivered to these coastal clients, at the time of the fieldwork he only delivered to a few. He described their shops and their customers. These shops sold pottery to be used decoratively and these customers were the same as the visitors to his workshop: people who bought pottery as caprices (he also referred to these shops as de capritxes). Therefore, these clients ordered mostly decorative types of pottery, but Vicen was quite sure that other pottery, like the bowls, bed warmers and gargoles, were also going to be used decoratively. For instance, Vicen had been selling a large quantity of large bowls to his client in Figueres. These were exactly the same as the ones which had been used for pig slaughter, but a client had once told him that he sold them to people who would drill holes underneath the rim and make hanging planters out of them.

That these shops catered to people with disposable income was also evidenced by their locations in tourist resorts or on the main road to France, close to the border, i.e. places where tourists passed. Vicen Sales also referred to these shops as de tourista or de souvenir. While unloading a truck, he had often seen tourists buying his products at these shops. He had also seen his planters and other similar ones used to decorate xalets or torres. These latter were the whitewashed houses with terraces and balconies and gardens in which potted plants were cultivated. This architectural type had become popular during the 1960s and had increased thereafte. Tourist resorts (like the one where my parents live) and the more affluent suburbs consisted of this type. It was an upscale style of house. In contrast to the regular village houses it did not contain any workspace or significant storage space. It was only intended for living and its appeal was enhanced by decorative use of windows, roofs, arches, balconies and terraces.

In those coastal shops, Vicen had seen that planters were also sold which were produced in other places, like Quart (Girona). He explained this: 'in Quart they are making a planter which is red, while we make it white. Sometimes people are looking for white ones and therefore we have to make planters for these clients. If that was not the case, these clients would be better off buying planters in Quart ... that would be cheaper because there would be less cost of transportation since Quart is much nearer to the northern coastal places than Miravet.' He elaborated on this by explaining that these customers wanted a choice. That was also why orders contained variations, so they could pick the planter they liked best. The white, *or rather blond, pottery was also additional evidence that these shops catered to foreigners: Dutch and Belgian exporters had been explicitly interested in this colour, hence, in the potter's estimation, foreign consumers liked it.*

Vicen Sales also supplied a few of these coastal clients with vessels, though in very small quantities. These vessels were intended to be used for the storage of olive oil and the preservation of food, depending on their sizes. Since Vicen Sales sold planters in the shape of the vessel to the same clients (that is, vessels not being glazed inside, and having a hole for draining water and a wide mouth), he was sure that these vessels were used for their original purpose. Vicen Sales: 'Perhaps someone who has a house in the mountains buys a vessel at those shops and stores olive oil in it at home.' With this description, Vicen distinguished the person who used this vessel in this way from regular customers of these coastal shops.[1]

Vicen Sales also described these coastal shops. For instance, the one owned by a client in El Vendrell was '... like a supermarket, very large ... everything they carried was pottery and ceramics, all kinds of ceramics, and no hardware as previous clients had had.' He described this merchandise as decorative, or as primarily aesthetically appealing. It carried anything from fina, *'fine', ceramics, like painted plates from Talavera de la Reina, to very* basta, *'coarse' pottery, like Vicen's, but also fountains, benches, statues and other garden decorations made of* piedra artificial, *'artificial stone', a decorative cement. He said that I would probably know this kind of shop since they were everywhere at the coast, including the resort where my parents were living.*

These shops were located just outside the built-up areas of villages or towns and along the major roads. Between the roads and the lots, they all had parking spaces. The longest sides of their lots and their buildings ran parallel with the roads, so that these shops looked as impressive and large as possible. The lots were fenced in with wide-mesh wirenetting which did not hinder the view at the lots and the buildings. The buildings sat at the back of the lots, taking up a small proportion of the area, while the rest of the lots were large open areas covered with gravel. In these open areas planters were piled up, organized by type and size, and in many cases 'artificial stone' garden decorations were displayed. The buildings had glass walls facing the roads so that one could easily see what they offered. Inside there were metal shelves with the finer ceramics, decorative plates and so on, and often a few decorative articles of basketry, glass and alabaster: all kinds of merchandise and not necessarily limited to handicrafts. In addition there were racks with postcards and souvenirs, like key chains. To further promote their merchandise, shop-owners put up signs advertizing 'souvenirs', 'handicrafts', 'pottery' and/or 'ceramics', usually in several languages such as Castillian, French, German, English and of course, since Franco's death, Catalan. These shops were often decorated with flags of the countries of their clientele, and with attention grabbers such as huge (some up to two metres tall) pots.

Next to these coastal clients, Vicen Sales still had clients with shops de payes, *'of the farmer'. One important long-term client had a shop in Manresa, and he always ordered vessels, especially the larger ones. Previously these vessels had been peddled by muleteers in that area, and this client dated from the period of the shift to marketing to wholesalers in the 1950s. For quite some time, this client sent an employee to pick up the orders and Vicen Sales had not seen this shop lately nor*

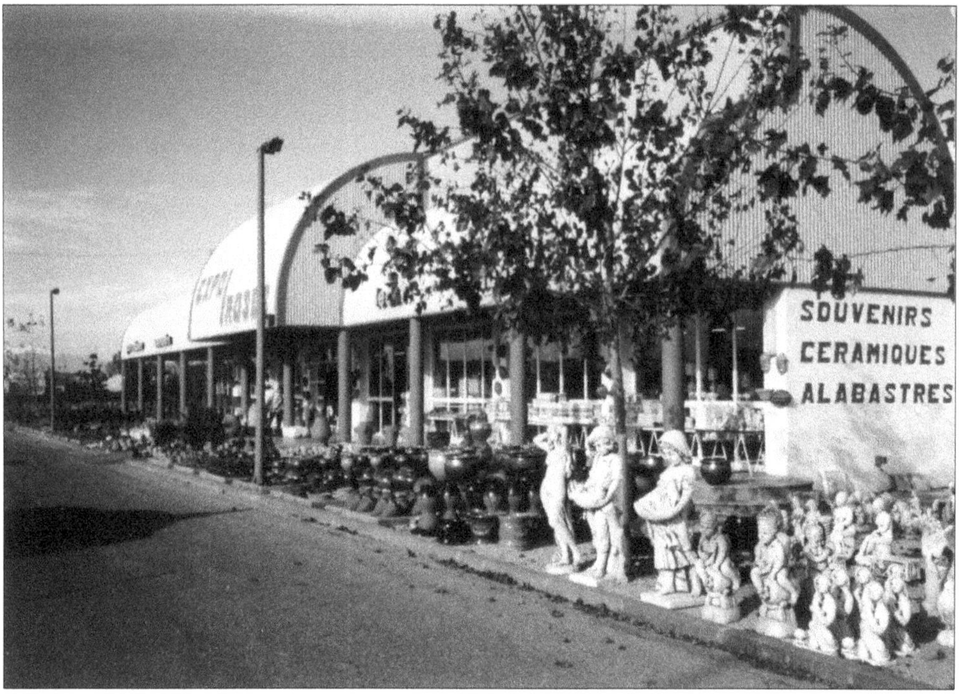

Plate 8.1 ***Shop in Roses (Girona):*** one of the shops described in a coastal tourist resort that carries pottery of Miravet. Author photo, 1997.

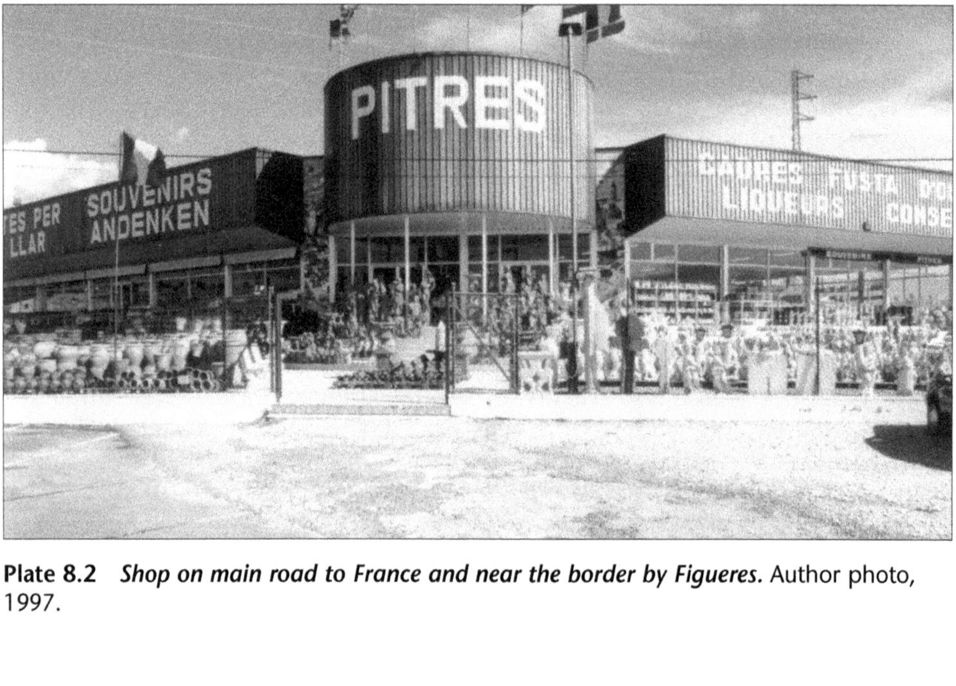

Plate 8.2 ***Shop on main road to France and near the border by Figueres.*** Author photo, 1997.

had he been to that area. He was sure that it was still the same kind of shop as before, one selling hardware, kitchen utensils and 'things for the countryside'. Vicen Sales: 'In Manresa there is much agriculture, and there are many olives but there are also many mountains and many small villages throughout the region. People come down to Manresa to shop and they carry the vessel back up to their farmsteads.' Another such shop, not an important client though, was a shop in Mora d'Ebre. Vicen Sales supplied it with gargoles. He contrasted the consumption of this type of pottery by customers of this shop with the consumption of the gargola by customers of the coastal shops. Customers of the shop in Mora d'Ebre still used the gargola for carrying water when working the land; customers of the coastal shops would place it where it looked pretty. (1990 was the last year he supplied any client in Saragossa: the last orders had also consisted of vessels. He had not been there for a long time so did not know what had replaced his vessels, but he was quite sure that there too, people were still storing olive oil in large vessels, though not on such a large scale as previously.)

Besides these domestic clients, the Fabregat-Sales workshop also acquired foreign exporters. Vicen Sales supplied his first foreign client in the mid 1970s through a representative in Barcelona. Apart from an occasional foreign exporter who placed a single order, in 1987, he began to supply a Belgian exporter directly. This client became a regular with yearly orders and a very important one: Vicen Sales once received an order from him for almost five kiln loads, while on average he made 10–12 kiln loads a year. These orders consisted mostly of planters, though also, in small numbers, of vessels made for the storage of olive oil. These orders were picked up by huge trucks with trailers which simultaneously picked up pottery at the Pedrola and Jornet workshops and old pottery at Ferran Segarra's.

In addition to knowing these clients through their orders, Vicen Sales knew them through their visits. The Belgian client often placed orders by mail or fax, occasionally visited the Fabregat-Sales workshop personally but had his orders picked up by a transport company. This Belgian exporter was a woman who always came with her husband and often the business partner of her husband, or sent her husband with his partner. Her husband and his partner dealt in second-hand and vintage cars and seemed to be in a better position to travel, or perhaps needed to for their car trade. Vicen Sales knew this because one of the men spoke a bit of Castillian. He knew too that she had a daughter, since once she had been vacationing at the coast and brought her daughter with her on a visit to his workshop. This was about all he knew of her: due to the lack of a shared language, his communication with her was limited to business essentials.

Of course Vicen Sales had never delivered to his exporter's warehouse, nor had he travelled abroad to get an idea of the shops, or subsequent consumption, in which his products were offered to the foreign consumers. Yet he had an idea of the countries to which his products were exported and merged this with the information he had through the orders. On one of my first visits to his workshop, Vicen Sales pointed at vessels, regular ones to be used for the storage of olive oil. These were part of an order for the Belgian exporter. He asked me if I, being Dutch, had any idea how these were going to be used. These vessels were glazed on the inside, something which was not necessary (undesirable in fact) if these were used as planters, and in addition to this made these vessels more expensive. (The rest of the order consisted of unglazed planters.) He knew that Belgium and the Netherlands were, as he said,

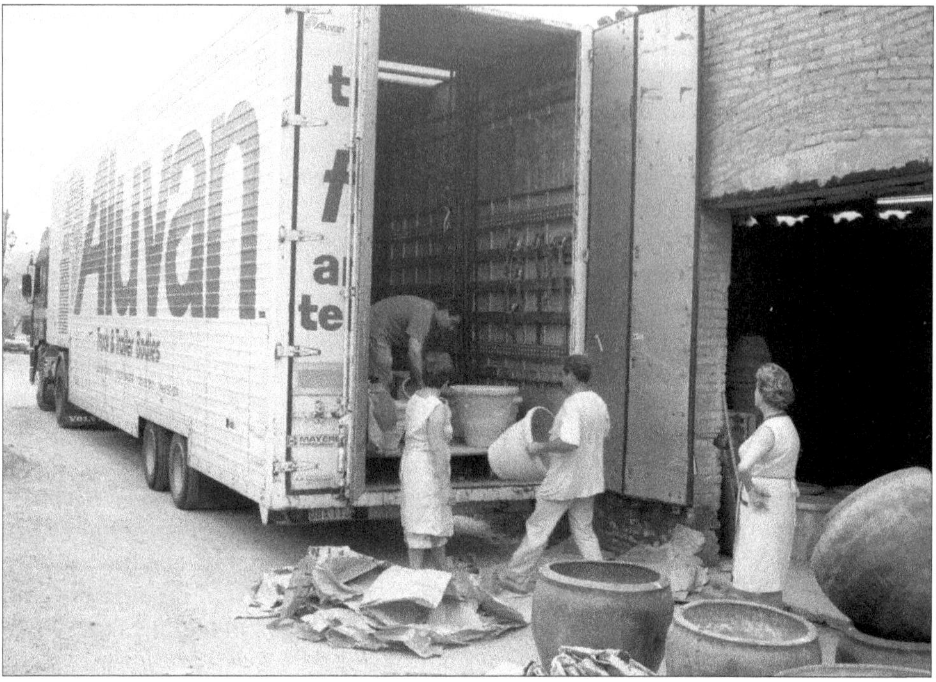

Plate 8.3 ***Loading a truck.*** This is actually one of the smaller trucks. Author photo, 1991.

'countries of butter', so these were not going to be used for the storage of olive oil, but he was puzzled how they were going to be used. On another occasion he expressed his concern that his planters were not resistant to the colder climate in the northern countries. But when we began talking about it, he admitted that he had no idea if his products were going to be used indoors or outdoors. He knew that his products were appreciated by the foreigners because of their colours and by virtue of being handmade, otherwise these exporters could buy the industrially made pots which were usually red and much cheaper. When probed on this, he could not offer any explanation other than that foreign consumers obviously liked his products. Within Spain, he knew that his planters were more expensive than the mechanically produced ones, and he realized that abroad the additional costs of transportation made them even more costly, yet he also had the impression that foreigners had more buying power: they came to Spain to vacation in hotels and often in second homes.

With respect to the consumers and consumption of pottery, Vicen Sales categorized pottery applying three criteria: first the physical characteristics of pottery; second the geographic region where the pottery was sold; and finally the people who consumed it. These criteria were interdependent: some types of pottery, like the planter, were sold in one specific region, in one specific way and to one specific kind of consumer. Other types like the gargola were sold in two different regions in two different ways and hence to two different kinds of consumers. Lastly, some types of pottery, when sold in Spain (regardless of the manner and place) were consumed by one type of person. But when these types were sold abroad, their consumption was puzzling to Vicen.

He further differentiated pottery in three categories. The first, which consisted of planters and small pottery, he sold to clients located on the Catalan and the northern Valencian coasts. This pottery was offered in ever changing variations and in a context of other merchandise used for home and garden decoration. It was offered to people, tourists and xalet owners, people who have the economic means to buy this pottery as a caprice. The second category, which consists of vessels and gargoles, he sold to clients in Manresa and Mora d'Ebre, places in the countryside with mixed agriculture and industry. This pottery was offered in a context of other 'utilitarian' merchandise to people who 'needed' this pottery. For instance, gargoles were used for drinking water and not for decoration, as they were when sold in coastal shops. The third category, planters, he sold to northern European countries. He only knew that these planters were expensive, hence that these people also had the financial means to buy them, but had no further idea how these planters were offered to consumers or how they were consumed.

Milio Treig on consumers and consumption

The Treig workshop specialized in the production of small pottery, especially since 1991 when they shifted from using a large wood fired kiln to a small gas kiln. Milio Treig Senr. threw the types of pottery that had always been produced in Miravet, while his son, Milio Treig Jr., expanded their repertoire of small pottery. He then attempted to limit this repertoire, but still continued designing new types of pitxells as clients were interested in these. They had about 15 clients, two of them small wholesalers.

I interviewed Milio Treig Jr. about these clients and the subsequent marketing to consumers of his products. He did not deliver his products on a regular basis, but there were exceptions. For example, a client had come to pick up an order which was not yet finished and he felt obliged to deliver the remainder of it. From these occasional deliveries and from general knowledge about different areas where his clients were located, he got some impression of his clients' shops. He mentioned that coastal places like Argentona, Salou, Segur de Calafell and Tarragona ordered pitxells in three sizes and in addition some other small pottery, like the fregalls or 'pan-scrubs', troughs for chickens and rabbits, and canisters with 'pasta' written on them. Sometimes they ordered pitxells with texts like 'Camping Playa', the name of a campsite, or 'Segur de Calafell', the name of a town. He called these shops de verano, *'of the summer'. He mentioned one in Segur de Calafell where he had recently made a delivery. It had been a shop near the beach where 'they sold pottery from Miravet and other places, and in addition inflatable animals and beach chairs'. I understood that he referred to the same kind of shop as did Vicen Sales, but located in the centres of villages and towns, and therefore they were smaller and without the open area with planters and garden decorations, focusing more on beach tourist articles. He had not noticed any differences in content between the orders from these different areas. The largest were those of his client in Santa Creus, which is not on the coast. It is a Catalan Cistercian monaster from the 12th and*

13th century and attracts tourists year round and could therefore turn over more merchandise.

He also had two wholesale clients located in Barcelona. One of them ordered the same pottery as those owning coastal shops; this client drove a van saying ferretería, *'hardware store', but apart from that Milio Treig had no idea what kind of wholesaler he was. The other was Milio Sempere, the author of the guide* Las Rutas a los Alfares. *This client ordered only the 'typical' pitxells of Miravet. When probed which ones these were, he mentioned of course de gargola de moxó, de 'pitxell of the barrel',*[2] *the 'pitxell of the trunk', but also the* doll *which was previously made in the Empordá.*[3] *He clarified this by saying that he supplied this wholesaler with the less* treballat, *'worked', pieces. He also produced the cántír for this client: Treig immediately mentioned that it was for collectors, i.e. not for fetching water at a fountain or river as previously. He knew that this second wholesaler traded in 'typical' Spanish pottery. He presumed that these wholesalers catered to the same kind of shops he supplied directly.*

In addition to these clients, the Treig workshop had clients who '... order pitxells and cosas utiles, *'useful things', just a few clients, not located at the coast but inland.' Two of these clients were located in industrial towns around Barcelona, in Sabadell and Mataro, the other one in Valls. Since their workshop had been supplying the client in Sabadell for at least 40 years, Milio Treig Jr. had been delivering to this client's warehouse. This client was a wholesaler and owner of two shops, one selling leather wares, the other shop selling goods such as dining ware, coffee sets and some pottery from Miravet. Previously this client had ordered vessels, but in the early 1990s mainly troughs and oval-shaped money banks: these latter were sold to kids to paint faces on them. He ordered less and less and Milio conjectured that this client wanted to redirect his trade. He was sure that the troughs sold to this client, as well as to the other inland clients, were going to be used for feeding animals. This was in contrast to the troughs he sold (in small quantities) to the coastal shops. There they were going to be used as ashtrays or cactus planters. To the client in Mataro he sold a small vessel, approximately 40 cm high, but with an exceptionally wide mouth. This client had specifically ordered this shape of vessel because many Andalucian immigrants lived in Mataro and they used such vessels for preserving cheese in olive oil, hence the wide mouth to pass the cheeses through. He also sold large vessels to these clients, especially the one in Valls. He was not sure how these were going to be used: these vessels were made for storing olive oil but not many people still did this. On the other hand, to use them as planters did not make much sense, as pottery without glaze inside is cheaper and better for that use. To the client in Valls, he also sold small quantities of the latest and most elaborately designed pitxells, which were definitely sold to collectors.*

The Treig workshop had two clients who were not trading pottery. One was a neighbourhood group in Tarragona that every year ordered a number of pitxells in three newly designed models for its annual festival. These pitxells were sold at this festival as keepsakes. In Saragossa they had a producer of honey who sold his products in pottery jars: these looked pretty in a kitchen or on a table. The selling of honey in pottery jars had been popular in the 1970s and 1980s, but was then waning: pottery was considered a bit too coarse for the kitchen or on the table.

Although Treig mainly produced small pottery, he applied similar interdependent criteria to categorize pottery. Therefore, his geography of consumption greatly overlapped with the first two categories Vicen Sales had recognized with regard to the kind of consumers and of consumption. The first category consisting of pitxells and small pottery was very similar to Vicen Sales' first category of pottery catering to tourists and xalet owners. It was offered in similar kinds of shops as Vicen Sales' products, though smaller ones located within villages and towns and more oriented towards beach tourist articles. In addition, he had one client who had such a shop, but inland near a tourist attraction rather than located at the coast, and he had one wholesaler catering to these tourist shops. The second category was roughly equivalent to the Vicen Sales' one of 'utilitarian' merchandise catering to people who 'needed' this pottery, in similar geographical locations. Since the Treig workshop specialized in the production of small pottery, it did not export, therefore Milio Treig Jr. did not recognize Vicen Sales' third category of planters marketed abroad.

The Papaseits on consumers and consumption

The Papaseits directed their production towards a repertoire of small pieces, reviving old pieces and designing new ones, especially pitxells. Technically they were not equipped for large pieces, but they still needed to produce a few for their method of packing the kiln.

In a series of interviews, I discussed this topic with Pepe Papaseit, his wife Dominica and their son Josep. In the early 1990s the Papaseits had two wholesale clients. One of the wholesalers, located in Tortosa, was contacted by Pepe Papaseit at the end of the 1950s (see Chapter 3). He had been an important client, especially during the 1960s and 1970s, but later he did not order much. Now the Papaseits sold him gargoles and pitxells, specified by Pepe Papaseit as 'for drinking water' and as he said '... still for use, not for tourists, never used as decoration' (i.e. not the newly designed pitxells). In their workshop they had a publicity calendar from this client in which he advertised himself as trading in 'articles of glass, of ceramics, for cleaning, and of palm fibre'. Previously Pepe Papaseit had been delivering pottery to this client's warehouse and he added that this client was also selling a lot of plastic articles. This client offered a range of articles geared towards small shops in villages, and covered the area of southern Catalonia, from Baix Aragon up to the city of Castelló. These were small village shops selling everything from food to household utensils and perhaps a bit of hardware. Josep Papaseit related an anecdote, that once on a trip while he was in such a shop he had seen a piece of pottery he had made. When he mentioned this to the owners they were very surprised. They had in fact bought it from the wholesaler in Tortosa and did not know who had produced it, nor did they seem to be very interested.

The other wholesale client was located in Breda, a centre of pottery casserole and pan production, but also of pottery and ceramics trade. Years ago, Pepe Papaseit had been delivering to this client's warehouse and could describe his merchandise. The client was a wholesaler trading in all kinds of Spanish ceramics and pottery, in alabaster, in garden decorations made of 'artificial stone', and in furniture. (This wholesaler obviously catered to the type of shop Vicen Sales described, the decorative

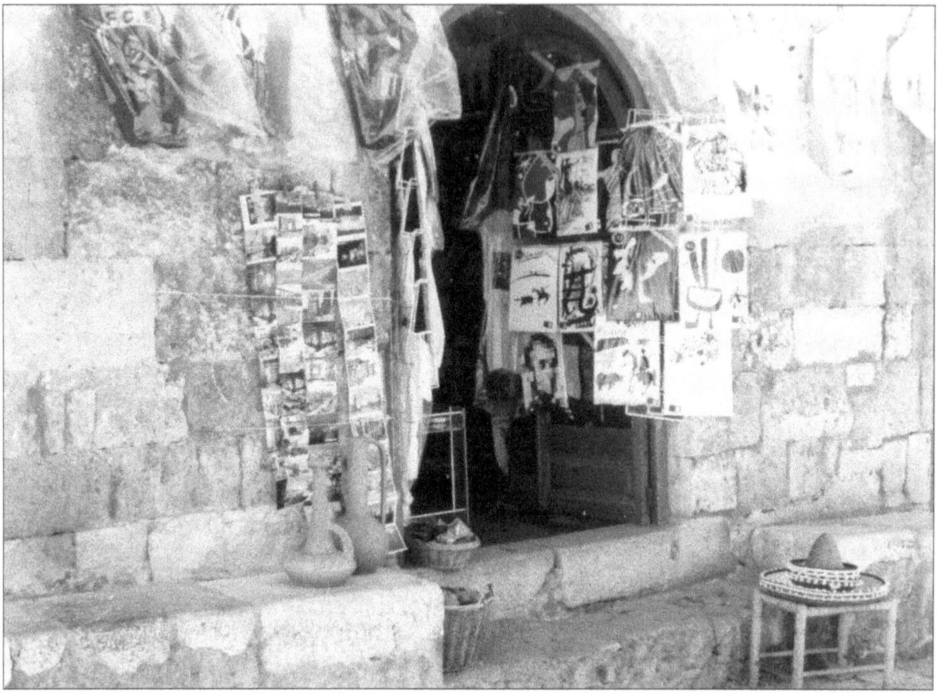

Plate 8.4 ***Shop in Tarragona.*** This shop is located near the cathedral. The two pitchers sitting on the left were produced by the Avante workshop. Author photo, 1997.

pottery stores.). Pepe Papaseit had been contacted by the client in the 1960s and he became a valued customer. This client began ordering vessels and planters and some small decorative pottery. Later, by the end of the 1970s when pitxells became popular as collectibles, he ordered these in large quantities and greatly influenced the direction Papaseit's production took. Gradually, the orders of this wholesaler declined as his clients began to buy directly from Pepe Papaseit.

The Papaseits began accumulating many shop-owners as clients in the 1980s. These clients had all contacted the Papaseits by visiting their workshop and consequently the Papaseits began to know these people but not their shops. In a shop in Barcelona I recognized Papaseit's pottery. I asked the Papaseits about this shop but I had forgotten the name of it, and I could only describe and vaguely locate it. However, they could not identify it: a description and imprecise location of a shop was not enough information to trigger any recognition. They had several clients in Barcelona, and it might be that the owner was a client of theirs, but it could also have come from one of the wholesalers in Breda. The Papaseits knew their clients' shops in Tarragona and at the coast, because they visited these places though never delivered there.

The Papaseits combined their knowledge of shops with the orders they received. Though they could not distinguish rigidly between the kinds of shop on the basis of orders, they did perceive different emphases among shops. Clients with shops in the cities of Barcelona, Tarragona and Madrid ordered mostly a wide variety of different pitxells, especially the newest model pitxells and 'traditional' pottery. They catered to a discriminating domestic clientele looking for gifts or collecting pottery. Such shops specialized in ceramics and pottery, had perhaps a bit of glass and other handicrafts on the side, but specialized in handmade products. They carried ceramics and pottery from all over Spain and usually displayed it by place of origin indicated by labels. They catered to the same kind of people who visited the Papaseit workshop collecting 'traditional' and 'typical' pottery and pitxells.[4]

In addition, the Papaseits had clients with shops in the coastal tourist resorts, mostly the southern Catalan coast, which were the kind of shop described by Vicen Sales and Milio Treig Jr. These shop-owners ordered pottery that was 'pretty', sold well and was not too expensive. They only ordered a small variety of the newest pitxells. In addition, they were always anxious to receive the order before the summer, as these shops catered primarily to tourists. Josep Papaseit: 'these shops sell pottery and at the same time a piece of wood with "souvenir" on it'.

Shops in Tarragona were between these two kinds of shops as Tarragona was a large city with a market of Catalans interested in collecting pottery, but at the same time it was also a tourist city. For example, the Papaseits mentioned a client of theirs who had a shop close to the cathedral, the main tourist attraction. This shop clearly catered to both kinds of customers, and though its sales peaked in summer, it sold year round. It was a shop like this that made the Papaseits apprehensive about categorizing their clients. If shop-owners asked their advice for what to order, they were reluctant to give any.

As the Papaseits knew their clients from their visits, they had noticed a slight difference in the knowledge of pottery between owners of shops primarily catering to domestic customers and those catering to tourists. The shop-owner catering to domestic customers knew the pottery as visitors interested in pottery also knew it, and they knew what constituted good quality pottery; the second kind had generally less understanding of pottery.

With respect to consumers and consumption, the Papaseits categorized pottery in ways structurally similar to Vicen Sales and Milio Treig Jr.. For instance, a pitxell could be sold in an inland village shop to be used for drinking water, or to a large city shop to be consumed as a collectible, or to a coastal resort shop to be consumed as a souvenir. The shops in Tarragona took an ambiguous position due to their location and therefore these potters could not categorize the pottery sold to these clients with regard to how it was going to be consumed, regardless of its physical characteristics.

Though the Papaseits categorized pottery in structurally similar ways, they applied different criteria. Like Vicen Sales and Milio Treig Jr., they recognized a category of pottery sold in the inland area consisting of gargoles and pitxells. This pottery was offered in a similar context to that which Vicen Sales and Milio Treig Jr. had described. But whereas Vicen Sales and Milio Treig Jr. only distinguished one other category of pottery marketed in Spain, the Papaseits distinguished two. They recognized a category sold to clients located in the large cities of Barcelona, Madrid and Tarragona, to whom they sold 'traditional' and 'typical' pottery and the newest pitxells. Such pottery was offered to customers in a context of other Spanish pottery, ceramics and glass, and with explicit indications of the places of production. All merchandise was sold under the category of handicrafts. These shops catered to customers who purchased pottery as gifts or as collectors: eventually it was used decoratively. The other category of pottery marketed in Spain was that sold to clients at the coast, which consisted of all kinds of small pottery. The primary purchase factor for these clients was price. This pottery was sold in shops catering to tourists, probably shops similar to those described by Vicen Sales and Milio Treig Jr. The Papaseits could make this latter differentiation between the large city and the coastal resort shops because they recognized in the customers of these shops similar motivations to those they recognized in the visitors to their workshop. They could make this distinction because they self-identified as traditional and as creative potters.

* * *

Potters categorized their pottery in relation to its consumption and its consumers. As they had differentiated their production, repertoires of products and clients, they also determined these categorizations differently. Vicen Sales produced planters which were exported: this pottery constituted a category which Milio Treig Jr. did not recognize, but then the latter produced small pottery marketed only in Spain. Potters also differed in their self-identification and concomitantly in their understanding of consumers of their products. Vicen Sales and Milio Treig Jr. understood their craft as a way of making a living, while the Papaseits also understood it as a traditional and creative craft. Therefore, the latter could differentiate pottery on the basis of marketing to customers as traditional pottery and creative expressions, or simply as souvenirs.

Temporal dimensions

Just as in the period when pottery was marketed by muleteers, production, marketing and consumption by the early 1990s also constituted temporal dimensions.

Previously, potters produced pottery in a cycle of throwing, glazing, packing the kiln, firing and exchanging to muleteers. From the 1960s onwards, potters extended their workspace and they glazed the pottery whenever it was dry enough to do so. Previously the glazing, packing and firing of a kiln required so much labour force that potters needed to exchange labour between households. With the glazing spread out over time, and with the new kilns, potters no longer needed to exchange labour and the packing and firing of a kiln became less important markers of their work cycles. Some potters purchased gas or gas oil kilns which they fired while continuing with other tasks. Some potters used more than one kiln and thus disrupted the cyclic rhythm even more. Because of all these changes in the production, potters organized the different phases in the production of pottery less cyclically.

Previously, upon unpacking the kiln and the conclusion of the production process, potters sold their products to muleteers. When potters delivered their own products, they did this following the unpacking of the kiln. Later, when clients were required to pick up their orders, potters needed to store the pottery. Eventually, potters primarily worked on orders, some potters produced types of pottery which they knew were in demand and stocked these. Moreover, foreign clients often ordered quantities which encompassed several kiln loads. Because of these changes in the marketing, potters exchanged their products with clients independently of the phases of the production process.

In the consumption of pottery, potters continued to link different types of pottery to different seasons. The planters were a product for spring when people potted plants. In summer other pottery was sold to tourists during tourist season. The vessels which were used for storing olive oil and food were, as they had been always, products for autumn. But potters did not perceive any link to the seasons with the small pottery sold to Catalan and Spanish customers and which was consumed by them for decoration.

Although some types of pottery were linked to specific seasons, potters no longer produced them according to season. With seasonally specific pottery, some clients were eager to get their orders in before a particular season. After the 1970s potters hardly met the demand for pottery with a short lead time, so clients had to accept waiting periods, sometimes up to a year. Even clients who had to have their orders in by a certain time began to accept delays. In autumn 1990, while Pepe Jornet was loading a truck with planters, I asked him who needed planters this time of year. He replied that his client would have liked to receive this order earlier and more in season, but he had had such a workload that he could not produce it on time.

The cyclical dimensions of the production, marketing and consumption became replaced by a linear one.[5]

Over the years, potters had always changed their production. For instance, they clearly remembered how a new kiln type was introduced shortly after the Civil War, and many had vague memories of even earlier experiments with changes in the preparation of clay. Since the 1960s, as potters began to invest more and more in production, in new equipment and work space, these changes accelerated and each change succeeded the previous one at a faster pace. All the potters, except for the ones who did not have successors, planned further changes.

During the period of the muleteers, the exchange of pottery already had a linear aspect. For instance, potters and muleteers in long-term trade relationships blurred the strict rhythm of delivery and payment. With respect to Catalan and Spanish clients, potters still had long-term relationships in which their clients annually placed an order, but with foreign clients they had shorter relationships, sometimes just a single order. The foreign wholesalers were especially unstable. For example, a foreign wholesaler who had ordered the largest part of Pepe Jornet's production in 1989, 1990 and 1991, hardly ordered anything in 1992. Clearly, although foreign demand overall was growing, it was irregular with regards to specific clients and countries. The distance foreign wholesalers came also grew. The first foreigners were located in the south of France, but gradually clients in northern parts of France came, and then those from Belgium, the Netherlands and Switzerland. The Pedrolas had a Canadian client, but only briefly, yet in 1995 they told me that they had been contacted by a commercial agent in Chicago: perhaps they would export to North America.

Similarly, the repertoire of products was never static. For instance, around 1930 the Catalan bowl began being produced by the potters. Since the 1960s, potters produced many new types of pottery and have since evolved their repertoires by adapting old types to new uses, like the planters developed from vessels and cossis, by copying types produced at other places, and by designing new types. As they said, pottery became a fashion phenomenon.

The temporal dimensions were never wholly cyclical – there had always been linear aspects to them. However, after the period of the muleteers, the cyclical aspects became less and less pronounced and the linear ones increasingly more so.

* * *

Potters continued to relate to consumers through their knowledge of consumption and consumers, which they acquired through several channels. These channels had become the direct personal contact with consumers as visitors to their workshops, and the contacts with distributors in the form of orders and deliveries to warehouses and shops. As these channels had changed, the knowledge transmitted through them had also changed. With respect to this knowledge, potters no longer varied in detail, that is, quantitatively in the sense that one potter knew more of particular forms of consumption and consumers than another potter. Eventually each potter had qualitatively different knowledge of consumption and consumers, knowledge that differed in relation to production, marketing and self-identification of a particular potter. The knowledge of Vicen Sales, Milio Treig Jr. and the Papaseits represented three qualitatively different forms of this knowledge. These potters categorized pottery on the basis of its consumers and consumption in a structurally similar way, but applied different criteria. By the synergistic effect of changes in production, marketing and consumption, potters constituted a complexity of relationships with distributors and consumers of their products in temporal dimensions that were no longer predominantly cyclical, but linear.

---

1 When talking about the period of the muleteers, he had referred to mountains as areas where people predominantly engaged in agriculture, i.e. were more likely to produce their own olive oil, and where they had larger houses and therefore tended to store olive oil in large vessels: he appeared to see in this consumption a historical continuity.

2 Since this client ordered the 'typical' pottery I asked if he produced this piece for him without glaze as was previously done. He did not: '... I glaze because people, women, don't want it unglazed. It's cleaner glazed.'

3 Note that he did not use this term, but pointed the type out. *Doll* seems to be the local Empordanees expression, though often mentioned in the literature on Catalan pottery: Ferran Segarra and Josep Papaseit knew this term, probably from their readings, Milio Treig Jr. did not.

4 In such shops, there was pottery distributed by Emili Sempere: he sold it with a small card guaranteeing its authenticity. Clearly, Milio Treig Jr. was not aware of this division of the indigenous market.

5 The only potter who clearly produced and marketed in an annual cycle was Ferran Segarra. He marketed his whole production to tourist visitors during the summer months. In addition to producing pottery, he also traded in old pottery, doing so year round.

Chapter 9

The central themes

I have built this book around three central themes. The first central theme was my approach to the meaning of pottery as constituted by internal differentiation and categorization (as the category of pottery became differentiated into subcategories like types and prices of pottery) and external differentiation and categorization (as pottery was categorized together with other objects, like the contexts in which it was marketed). I related these differentiations and categorizations always to physical aspects of the pottery (like size, material, colour and shape), and to human interaction with pottery, (i.e. the production, exchange and consumption of pottery).[1] Thus, by analysing these differentiations and categorizations in the history of the pottery, that is, the history of the physical appearance of these objects, I showed how the physical or material history complements the potters' social history. I studied how physical transformations in the pottery resulted from and affected its role in the social relations people formed as they produced, marketed and consumed this pottery.

In the first part of the twentieth century, potters marketed their pottery to muleteers. Potters and muleteers reached agreements upon the expression of the pottery's value in money, and thus categorized the pottery as a commodity, and they agreed upon the differentiation of types and sizes of pottery. Yet potters and muleteers often contested aspects of the price calculation, like the differentiation in good or flawed ware.

Through their marketing to muleteers, potters learned about the consumption of their products. The potters' knowledge about the consumption and the consumers formed a 'geography of consumption'. This semiotic space was largely structured by the way in which pottery, as specified by its physical characteristics, was marketed in different regions and in different contexts, by the consumption of a certain piece of pottery, and by consumers having certain social characteristics. These dimensions were all linked with actual space and entailed differentiations and categorizations of the pottery.

Potters contested the differentiation and categorization of pottery in their trade with muleteers, and thus were motivated to make the shift to marketing to wholesalers and shop-owners. With these new clients, potters calculated prices based on the invested labour and raw materials. Another result of these new marketing relationships was that the pottery became standardized in its actual measurements. As the marketing relations shifted, the relationships among the potters shifted as well: see the turbulent history of the potters' Co-operative. In the marketing to muleteers, pottery was differentiated as good or flawed. But an important factor in the demise of this Co-operative was the lack of this differentiation as the Co-operative accepted all pottery produced by its members as of equal quality. The shift to a new marketing practice did not affect the consumption of pottery, but it affected marketing and the knowledge potters acquired of the consumption of their pottery, especially of the contexts formed by other products in which the pottery was marketed to consumers.

During the pivotal 1960s changes in the differentiations and categorizations of the pottery related to physical changes, as well as to new production and marketing relationships. The physical changes ranged from developing existing shapes into new ones, to designing totally new shapes. This pottery was mainly differentiated into planters and small pottery: potters specialized in producing one or the other, and each category had different marketing practices and networks. These pivotal changes were further developed during the subsequent decades of the 1970s and 1980s.

Exchanges at the workshops resulted in two very distinct differentiations and categorizations of pottery, internally as well as externally. In short, when potters exchanged pottery with locals, it was categorized in relation to other local, agricultural products. When, starting in the 1970s, potters sold pottery to visitors, it was categorized as a typical local product and as a handicraft.

In my description of the potters' relationships with consumers in the early 1990s, I applied the same analysis as I did when describing how they marketed their products to muleteers in the first half of the twentieth century. By analyzing two different periods in the same manner, I highlighted the different differentiations and categorizations of the pottery during these different periods.

* * *

The second central theme was the mutual encompassment of production, exchange and consumption, thus covering all social relationships constituted and articulated through objects, and these social relationships as interconnected with each other.

In the synchronic analysis of the period of the muleteers, this mutual encompassment was evident in the relationships between production, exchange and consumption:

1) the relationship between the production of pottery and the productive consumption it involved (in comparison with and complementing the production of fellow-villagers – i.e. agriculturists and especially muleteers);
2) the relationship between the production of pottery and the exchanges between potters and muleteers;

3) the relationship between production and consumption in the form of the potters' knowledge, gained through exchanges with the muleteers, of the use-values of their products acquired in the consumption of it;

4) the relationship between this knowledge of the consumption of pottery and the exchanges between potters and muleteers in the form of the potters' contestations of the price calculations.

In the subsequent diachronic analyses of the 1940s and 1950s, I approached the mutual encompassment by analysing how production, exchange and consumption affected each other as changes in these occurred. Because of the increase in demand for pottery (i.e. a change in consumption), the tension in the marketing or exchange relationships between potters and muleteers rose. Some potters explored new exchange relationships with new partners; other potters had to follow later when agricultural production changed and muleteers disappeared. This change in the exchange relationships also effected a change in the potters' knowledge of consumption of their products.

Next followed the analysis of the synergy of effects of production, exchange and consumption with each other during the 1960s, 1970s and 1980s. This synergistic complex could be characterized thus:

1) the demand for old products declined while a demand for new products developed;

2) the productive consumption implied in pottery production changed as the context formed by other kinds of production (like the modernization of Spain) changed and therefore potters shifted from labour intensive to capital intensive forms of production;

3) the new forms of production and new products caused potters to look for new clients;

4) new clients ordered new types of products;

5) because of their increased productivity, potters changed to their advantage the power balance in their marketing to clients;

6) the consumption of pottery as luxuries caused potters to perfect their products;

7) with higher quality products, potters charged higher prices.

The combined result of these changes was that the complex formed by production, exchange and consumption of pottery changed completely during the period studied.

Potters exchanged their products with locals, for agricultural products. I analysed these exchanges with locals by comparing the production and consumption of pottery with the agricultural production and consumption of goods. I hypothesized about possible shifts in these exchanges due to shifts in these forms of between these exchanges and the consumption of pottery and of other exchange goods.

I analysed the exchanges with visitors in relationship to changes in general Spanish production and to changes in repertoires of types of pottery and the marketing of these types. On the one hand, Spanish production in general changed, causing the manual production of pottery to become exceptional within the context of dominant industrial production: this was one of the factors by which this pottery became consumed in new ways. On the other hand, because of this new form of consumption, potters changed their repertoire of products to meet demand and began to market their products in their workshops.

* * *

The third theme built on the previous ones. In the first I studied the relationship between the physical characteristics of pottery and the human interaction with pottery. In the second I approached this human interaction as the mutual encompassment of production, exchange (or marketing) and consumption. The third theme consisted of the different social levels – ranging from the individual to the international – on which this production, exchange and consumption occurred, with all these levels interacting dialectically with each other. By juxtaposing the individual potters and the local, micro-history of the small group of potters with macro histories, I presented a local perspective of producers as affected by and acting upon global developments, ultimately localizing the European transition to one single integrated market economy.

I began this study by a synchronic analysis of the dialectic relationships integrating several social levels. The potters produced and marketed their pottery as embedded in the local political economy of the village based on the mixed market-oriented and subsistence agriculture of a relatively isolated Spain within a more industrialized European context. In this context, potters were in a weaker position against muleteers and, therefore, to strengthen their positions potters assessed collective prices for their pottery, i.e. potters dealt not only on the individual level with muleteers, but also as a group. The pottery was consumed within the Spanish market, a market that varied enormously with respect to the integration of consumers within it: for some consumers this pottery was one of the few goods acquired from outside their village, other consumers were purchasing goods produced abroad.

For the balance of my study, the dialectic relationships between the different social levels consisted of the effects on, and responses to, each other.[2] After the Civil War the Franco government was internationally ostracized and pursued a policy of autarky. This development on the national and international level sharpened the tension on the village level between potters and muleteers. Within the network of potters and muleteers, two potters, Josep Fabregat and Pepe Jornet, took individual positions in their local networks that incited them to explore new marketing practices by using the new national transportation network. These individual explorations affected the mutual relationships between the potters in the village: the potters founded a co-operative to explore these new practices collectively. As the Potters Co-operative failed, potters marketed their products individually and eventually gave up collective prices. Due to a worsening national social and economic situation after 1950, the Franco government was forced to reach out to other countries to procure international economic aid. With this aid Spanish agriculture and transportation modernized, and industry developed. But still the Spanish economy deteriorated. In 1958, the Franco government gave up its policy of autarky and together with international organizations developed an economic plan that triggered off emigration and migration. In the village of Miravet, many men who had peddled migrated to the city. All potters were now forced to market their products to wholesalers and shop-owners.

During the 1960s Spain became increasingly integrated into the European economy and the Spanish economy experienced an unprecedented growth. As Western Europe in general experienced an economic boom, foreign tourism in Spain exploded. One result of these developments was a change in the consumption of pottery. Other results included the rise of labour costs in Miravet, and the increasing availability of money for local investments. Potters shifted from labour intensive to capital intensive production techniques: Josep Fabregat as the economically strongest potter was the first one to buy a jigger, crucial equipment in this shift. Potters differed individually, as not all potters made the shift at the same pace and in the same manner. As a result, potters differentiated their production, repertoires, clients and knowledge of consumption from each other, and potters ceased to exchange labour amongst themselves. During the 1970s and 1980s Spain continued to develop economically and to become integrated into the European economy. Due to their increasing productivity and the increasing demand for pottery, potters improved their economic positions within the village community, now a community with an entirely market-oriented economy. Because of Spain's integration into the international economy, foreign wholesalers began to export the pottery of Miravet. The end result of these developments was that potters developed almost individual forms of production, repertoires, marketing and knowledge of consumers and consumption of their products. I exemplified this by describing the productions and repertoires of the Pedrolas and the Papaseits, and by describing the knowledge of consumers and consumption of Vicen Sales, Milio Treig Jr. and the Papaseits.

Potters exchanged products with locals. These exchanges changed due to more general developments. Locals changed their consumption of pottery as functionally superior and/or cheaper industrial goods produced elsewhere became available to them. These exchanges also changed as the local agriculture shifted to production for the market, first the national one and later international as well.

When, during the 1960s, artisanal production transformed from the prevalent form of Spanish production to an exceptional one (even within the international context), pottery became appreciated by Spaniards and foreigners alike as handmade and traditional. When Spain transformed from a rural country into an urban one, rural life became valued positively and this pottery was appreciated as an aspect of it. Furthermore, these two developments towards new appreciations were strengthened by the development of tourism, largely international tourism. Franco died in 1975 and Catalonia was granted a limited autonomy. Catalans were free to celebrate their culture and delight in everything Catalan, including pottery. Potters responded to the international, national and regional developments by producing types of pottery that appealed to these appreciations and that could be viewed as 'Catalan', and by catering to people who, motivated by these appreciations, visited their workshops. Tomaset Avante had designed the gargola de moxó to be given to a particular person: a unique gift for an individual. In response to the newly emerged appreciation, all potters included this piece in their repertoires and the gargola de moxó became the most typical piece of Miravet. On an individual level, potters differed in their understanding of the new appreciation of their pottery, and to exemplify this I compared and contrasted the understanding of Vicen Sales, the Segarras and the Papaseits with each other.

* * *

The three central themes of this study were: the meanings of objects, as they were differentiated and categorized in human interaction; this human interaction with objects as consisting of the mutual encompassment of production, exchange and consumption; and the human interaction with objects as occurring on several dialectically interrelated social levels. Through these themes, I have represented the relationships between the potters and their pottery as these potters produced and exchanged this pottery, and learned of its consumption.

---

1 The differentiation and categorization of physical characteristics of objects as an aspect of a culture is the main topic of ethnoscience. Anthropologists as cognitive scientists working in this tradition share with each other a methodology and theoretical assumptions. Their methodology consists of:

1) the singling out from a language a set of expressions which are semantically related and form a cluster;

2) the singling out a part of the total of the phenomena using the same description or definition as for singling out that cluster and then analysing this part in its physical characteristics in order to construct an 'etic grid' (Kay 1969: 79) or a 'reference scheme' (Colby 1966: 7);

3) the study of how the relatively great number of characteristics of the etic grid is decreased to the relatively small number of words by differentiating and categorizing these characteristics. The way this decrease is achieved is considered characteristic of a culture in the ethnoscience approach.

One of their theoretical assumptions is that one can analyse physical characteristics unmediated by a specific culture, and therefore they apply in their studies positivistic natural science. The other assumption is based on Charles Morris' conception of semiosis that makes it possible to devise 'semantical rules' which link a sign to something outside of the world of signs without taking the pragmatics of sign use into account. I have, on the contrary, approached in a previous publication and in this study the differentiation and categorization of physical characteristics as only existing with respect to human interaction with the material world and this human interaction an aspect of social relationships – 'cultural' and 'social' are almost inseparable in my analysis (see van Veggel 1991).

2 Note that in the first two chapters, I used biographies of individual potters and muleteers as examples of a class. In the later chapters I used biographies as the most individual social level to demonstrate how persons were affected and responded in their unique ways to changes on higher levels.

Appendix A

Composition of population

The *Rectificación el Censo electoral corespondiente al año 1934* (Archive of Instituto Nacional de Estadistica, Tarragona) is a list of the people registered to vote in the year 1934. Next to their names, addresses, ages and their capacities to read and write, it mentions their 'professions' or 'offices'. Miravet was divided in two districts, one consisting of the village itself, the other of the Raval. The list of the village itself is complete, the one of the Raval, the neighbourhood where the potters lived, unfortunately is not.

Men	Village	Raval
alfarero ('potter')		6
agricultor ('farmer')	2	
jornalero ('day labourer')	40	1
labrador ('farmer')	143	13
lortelano ('vegetable gardener')	1	
(subtotal in agriculture)	**(194)**	**(14)**
obrero ('worker')	5	
pastor ('shepherd')	3	
albañil ('bricklayer')	5	
arriero ('muleteer')	1	
barbero ('barber')	3	
barquero ('bargeman')	9	
cafetero ('cafe-owner')	1	
camarero ('waiter')	1	
carpentero ('carpenter')	1	
carrero ('cart-driver')	6	2
comercio ('trade')	12	
comerciante ('trader')	2	
estudiante ('student')	1	
esquilador	2	
guarnicero	1	
inútilla ('useless person')	1	
maestro ('school teacher')	2	
peluquero ('hairdresser')	1	
sacerdote ('priest')	1	
sastre ('tailor')	1	
sillero (someone making chairs)	2	
not mentioned	2	
Total	**249**	**22**

Women	Village	Raval
hortelana ('vegetable gardener')	3	
inútila ('useless person')	1	
jornalera ('day labourer')	7	
maestra ('elementary school teacher')	2	
modista ('dressmaker')	1	
labradora ('farmer')	6	
labores ('works')	4	1
pantalonera ('trouser maker')	1	
obrera ('worker')	2	
sirvienta ('domestic servant')	1	
su sexo ('her sex')	199	21
not mentioned/unclear	10	
Total	**237**	**22**

Padrón Municipal

(Municipal Archive of Miravet)	1950	1960	1970
Men			
alfarero ('potters')	**37**	**21**	**22**
agricultor ('farmer')	234	82	
peón agricola ('day labourer')			128
labrador ('farmer')	286		
ganadero ('cattle farmer')		3	4
pastor ('shepherd')	11	3	
tractorista ('tractor driver')			2
(subtotal in agriculture)	**(297)**	**(240)**	**(216)**
albañil ('bricklayer')	7	7	9
peón albañil ('bricklayer's helper')			7
barbero ('barber')	4	1	1
abogado ('advocate')	1		
alguacil ('messenger')	1	1	1
arriero ('muleteer')	21		
ayudante ('helper')			1
barquero ('bargeman')	13	2	
calafate ('caulker')	2	2	1
camarero ('waiter')		2	8
canicero ('butcher')		1	
carpentero ('carpenter')	6	7	6
carretero ('carter')	1	1	1
arg. chófer ('agricultural driver')		4	
chófer ('driver')	4	8	9
chófer/ayudante ('driver/helper')			1
commercio ('trade')	26	23	8
commerciante ('trader')			1

(Municipal Archive of Miravet)	1950	1960	1970
Men			
arg. commercio ('agricultural trade')		1	
cons. carro (?)	1		
constable ('constable')			1
cts. carros (?)		1	
cura ('priest')			1
ebanista ('cabinetmaker')			1
electrista ('electrician')	2	3	3
electrónico ('electronic')			1
empresador ('entrepreneur')			1
escribiente ('clerk')	1		
escolar ('student')	78	65	41
esquilador ('sheep-shearer')	2		
estudiante ('student')	5	12	27
farmaceútico ('pharmacist')	1		
fotógrafo ('photographer')	1		
herrero ('smith')	4	2	1
hojalatero ('tinsmith')	1		
guerda m. ('watchman')	1		
lampista ('plumber')	2	1	
maestro ('schoolteacher')	1	2	2
marino ('sailor')	1		
agr. mecánico ('agr. mechanic')		1	
mecánico ('mechanic')	1	2	7
médico ('physician')	2	1	1
militar ('soldier')	1		
minero ('miner')		2	
oficiante ('officiant')	1		
panadero ('baker)	6	4	1
paralítico (paralysed)	1		
peluquero ('hairdresser')			1
pintor ('painter')	1	1	3
planchista ('ironer')			1
repartido coreos ('postman')			1
sacerdote ('pastor')		1	
sastre ('tailor')		1	
secretario ('secretary')	1		1
sereno mol. ('night-watchman')		1	
transporte ('transporter')	2		
tratante ('trader')	3	1	1
propietario ('owner')	1	2	
rentista ('person of private means')		1	
sondeos (?)		1	2
viajante ('travelling salesman')		1	
zapatero ('cobbler')	1		

(Municipal Archive of Miravet)	1950	1960	1970
Inactive			
not mentioned (infants)	39	27	29
impedido ('handicapped')		2	
inválido ('disabled person')		1	1
jubilado ('pensioner')		34	68
not mentioned		1	
unclear handwriting	3	7	
Women			
agricultor ('farmer')		7	
bachilerato ('secondary education')	1		
carnicera ('butcher')	1	2	4
comadrona ('midwife')	1		
commercio ('trade')	1	6	2
e.hogar ('visiting nurse'?)			2
enfermera ('nurse')	1		1
escolar ('school girl')	87	73	53
estudiante ('student')	1	6	11
inválida ('disabled person')			1
jubilada ('pensioner')		5	47
labores ('housewife')	528		384
sus labores ('housewife')		437	
labrador ('farmer')	2		
maestra ('school teacher')	4	3	4
minera ('miner')		1	
modista ('seamstress')	9	4	
oficinista ('office clerk')	1		3
panadera ('baker')		1	
peluquera ('hair dresser')	3		3
practicante ('apothecary's ass.')			1
prof corte (?)	1		
sirvienta ('domestic servant')	1	2	
telefonista ('telephonist')	1		
teléfonos ('at telephone company')		2	1
venta ('sales person')			1
venta? (unclear)			1
not mentioned (infants)	34	19	18
unclear handwriting			4

Appendix B

Population of Miravet

1900	1868 inhabitants
1910	1566 inhabitants
1920	1647 inhabitants
1930	1391 inhabitants
1940	1218 inhabitants
1945	1283 inhabitants
1950	1234 inhabitants
1955	1157 inhabitants
1960	1019 inhabitants
1965	959 inhabitants
1970	1014 inhabitants
1975	846 inhabitants
1980	837 inhabitants
1985	838 inhabitants
1990	785 inhabitants

Data of 1900–86: Anon., *La Ribera d'Ebre* (1989: 46–7). Data of 1991: Anon., *Censo de Población de 1991* (not dated).

Appendix C

Land use

	1900	1948	1957	1965	1971	1985
Irrigated land	*45*	*165*	*163*	*162*	*159*	*454**
irrigated land		165		131	128	
irrigated cereals	25		131			
irrigated fruit			32	31	31	263
vegetables	20					100
legumes						10
tubercles						59
fodder						19
citrus						8
Dry land	*1,115*	*1,173*	*1,183*	*1,187*	*1,049*	*1,166*
cereal	298	61	75	75	74	18
wine	90	239	324	324	194	673
almonds and olives		469				
almonds	405					145
fruit			284	284	281	229
olives	267	348	479	483	479	308
carobs	55	56	21	21	21	0
vegetables						4
Not cultivated						
pines		374	259	259	259	
low scrub		1,007	833	832	736	
'erial rasos'		499	421			
pastures				421	421	
reed		12	7	7	6	
river bank trees		28	56	56	56	
wood	1,408					
useless and not cult.	355					

Data for 1900 from Anon., *La Ribera d'Ebre* (1989: 360).

Data for 1948 based on Anon., *Amillaramientos de Rustica of the Municipality of Miravet* (Arxiu Historic, Tarragona).

Data for 1957 based on Anon., 'Ministerio de Hacienda, Servicio de Catastro de la Riqueza Rustica' (Municipal Archive of Miravet).

Data for 1965 based on Anon., 'Ministerio de Hacienda, Servicio Tecnico Facultado para la Aplicacion de los Tributos' (Muncipal Archive of Miravet).

Data for 1971 based on Anon., 'Ministerio de Trabajo, Mutualidad Agraria de la Seguridad Social' (Municipal Archive of Miravet).

Data for 1985 from Anon., *La Ribera d'Ebre* (1989: 93, 98).

* Please note that the specified uses of irrigated land do not add up mathematically to the total.

Appendix D

Livestock

	1948	1952
horses	17	18
mules	125	116
donkeys	19	14
sheep	429	168
goats	350	336
pigs	75	80
beehives	90	17

Data for 1948 based on Anon., Amillaramientos de Rustica of the municipality of Miravet found (Arxiu Historic, Tarragona).

Data for 1952 based on Anon., Expediente of de *riqueza pecuaria* (Municipal Archive of Miravet).

Appendix E

Prices of pottery

	1946	1951	1967	1971	1979
vessels					
gerrot	45			410	
gerrot gran		75	300		1600
gerrot petit		65	290		1540
gerreta					
solsera gran	27	42.5	160	210	
solsera petita	22	32.5	140	200	
tafol	8				
tafol gran		20	65	95	468
tafol petit		15	50	85	457
tofulet				75	
solseró	7	10	30	35	
solseronet			26	28	132
petxin	6	5.5		23	
petxinet				19	
saboner	2	2.5	9	12	
bowls					
mandonguer	15	20	70		
mitjaner	10	10	35	65	
two in one				40	
sangant	6	5.5		30	
three in one				20	
quatreun	4	2.75		15	
grella	4				
pitança	2	1.75			
sangant verde		6			
quatreun verde		2.75			
tubs					
cossi per passar baguda gran	20				
cossi petit per baguda	10				
cossi calandrin		15			
cuenco 1			60		165
cuenco 2			50		154
cuenco 3			40		116
cuenco 4			35		
cuenco 5					72
cuenco palmera 1				50	
cuenco palmera 2				60	

	1946	1951	1967	1971	1979
cántirs					
cántir gran del cul estret		5		30	
cántir gran de cul ample		5		42	
cantir petit de cul estret		3.5		25	
cántir petit de cul ample		3.5			
cántir de cul ample (children)		1.5			
cántir de cul estret (children)		1.5			
gargola gran pel camp		6			
gargòla		3		20	
cántir comuner (serveix de comuna detràs de l'estable)		10			
Cántir comuner per transportar la al camp		8			
other					
Ribellet de pentinar	2				
Morter	2				
setrill gran	5				
Setrill mitja	3				
setrill per endur-se oli	2				
pitxell (modern) per aigua5	5				
plateret per aguantar el pitxell	2				
cántir per mesura vi	10				
mixòla per deixarhi la mesura de vi sota	8				
porró tipo antic vi	2				
1/2 porró per vi	1				
fidrigòla (guardiola)	4				
cau de conills	6				
cadúfo	2				
mostra -pixador	3				
bassí per fer les feines	10				

Data for 1946: List of acquisitions made on 8 July 1946 from the potter Francesc Sarroca Pedrola (of the Petit household) by the Museo del Poble Espanyol, Barcelona. (Note that these pieces were individually purchased, something exceptional for this period. Usually pottery was sold in 'pieces' to muleteers.)

Data for 1951: Remainder of the bookkeeping of the Cooperativa Alfarería de Miravet, "Albaran de entrega, Relación de los géneros que en esta fecha se han vendido a D....". Prices of vessels and bowls from 3 September 1951; of green bowls from 19 July 1951; of cántirs from 25 January 1950 (one dozen for 50 Pts, in this bookkeeping only once specified with respect to bottom, and very rarely as 'gran'); of cantirs petit from 3 January 1950 (usually as such specified or with diminutive canteret); and of cossi calandrin from 28 January 1950. (Pedrola family archive).

Data for 1971: from bookkeeping of Pepe Papaseit (remaining part is from 16 June 1971 until 14 July 1972). Most prices of vessels from 11 December 1971: on that slip appear the old labels used for several types of vessels together with the then usual numbering. In the remaining part of the bookkeeping, the prices of the vessels vary a little (other prices are from 2 October 1971 and 4 December 1971). Prices of vessels are from 7 August 1971, 27 May 1972 and 8 July 1972. Prices of cántirs are from March 11, 1972 (called 'grande' and agueder') and 8 July 1971 (Papaseit family archive).

Data for 1967 and 1979: from bookkeeping of Pepe Jornet (remaining part goes from 22 June 1967 until 22 July 1979 with several breaks) are from slip of 22 June 1967 and of 22 July 1979 (Jornet family archive).

Appendix F

The Papaseit bill books of 1971 and 1972

The Papaseits showed me a carbon copy bill book from a part of 1971 and a part of 1972. Together we leafed through it and discussed both the bills and the clients, clients from Benicarló, Vinaroz, Ampolla, Huesca, Tarragona, Reus, Sabadell, Tortosa and Breda.

The clients in Benicarló and Vinaroz were still the same shop-owners Pepe had contacted in 1957 or 1958 when he had returned from military service. The one in Ampolla was contacted somewhat later. Still this happened under similar circumstances and conditions as the others in that region. In 1971 and 1972 these clients bought the following pottery: vessels in the sizes from saboner up to solsera gran; 'painted pottery' (this was specified by codes from the catalogue from which this pottery was copied, and specified by labels such as 'umbrella stands' and 'large jugs'.); flowerpots (hand thrown; some were specified as *grabadas,*' with a textured surface'); planters (some were specified as *cuenco palmera*, 'palm-tub'. Dominica explained to me that they, the Papaseits, had given this name to a bowl-shaped tub that was nicked with a kind of cogwheel underneath the rim. Pepe added that they did not sell many planters to these clients.); saucers for underneath planters (they added that they hardly made this type.); troughs for rabbits in several sizes (as Pepe said, the tourist market had begun to develop, but these troughs were still for household use); bowls (only the grelas, used for serving food on the table, were mentioned.); pitxells (if they were specified, they were only specified as small and large: these must have been the simple pitxell.); colanders in two sizes; *Gazpacheros* (this was a jug for serving *gaspatxo*, 'chilled tomato soup'); Sangría pitchers in two sizes (when Dominica heard me reading this, she exclaimed: 'Those pitchers, my husband made a *barbaridad* [an excess, of those] for there, for Benicarló.'); mortars in two sizes; and ashtrays with *recuerdo de Benicarló*, 'souvenir of Benicarló'.

Pepe Papaseit placed these bills in their historical context. When he began to supply these clients, he sold mainly vessels, bowls from the mitjaner down, and canterets. Gradually the pottery he sold became more decorative since this was an economically strong region where there was money for such things, but also especially since tourism was on the rise. These bills were from that transitory period. At that time he still was visiting these clients to see if they would place an order, but as he was loosing clients due to their retirements, and as clients were coming to Miravet, he stopped visiting these clients. One of these clients eventually bought a van and came to Miravet to order and to pick up, but with this one he has also lost contact. He conjectured that this region was no longer a good market for his products. Indeed, Peníscola, immediately south of this region, is a big tourist attraction with its papal castle. Shops there were selling pottery made in Andalucia that was a lot cheaper than his.

The client in Huesca was billed only for vessels, from the smallest up to the tafol gran. Pepe explained that they had contacted this client through a befriended muleteer. This client had been buying from this muleteer and when the muleteer no longer supplied him, he had written a letter to this muleteer with an order and had requested to forward that order to a potter. Pepe had supplied him several times, always sending the pottery by train and always getting orders and checks by mail: he had never met that client. From the location of this client and his orders, Pepe figured that this pottery was consumed as it always had been there.

The clients in Tarragona and Reus had ordered large vessels, tafols and solseres. The one in Reus had ordered ten: he was an insignificant client and they could not remember him. They did remember the one in Tarragona: they pointed out where, in the inner part of the city, his shop was located. It had been a shop selling fodder, and, at that time, vessels as well. By now it had become a pet shop.

The client in Sabadell was billed for vessels in all sizes from the solsera gran down, and for sets of flowerpots. This client had been supplied by a muleteer, a relative of theirs, and when this muleteer ceased doing this, this client had come to Miravet and was presented to them by the muleteer. He had a large shop with two floors. On the first floor he sold 'coarse' merchandise like their pottery, on the second 'fine' merchandise like fine ceramics from Manises and many toys. Though Pepe delivered to his shop, this client and his wife visited Miravet every year on their way to an annual fair in Valencia. Through these visits, they developed a hearty friendship: Dominica remembered that this client once made her daughter a present of a beautiful doll's pram. They stopped supplying this client when he closed his shop.

Two clients in Tortosa were billed. One was a shop-owner with whom Pepe had the same contact as with the ones on the coast (i.e. in Benicarló, Vinaroz and Ampolla) but whom he was supplying with other pottery. This client in Tortosa had ordered vessels in all sizes, troughs for rabbits and chickens, pitxells, colanders, cántirs and canterets. His order differed from the orders placed by the clients on the coast in that it did not include the decorative pottery, but instead included pottery consumed in a rural context. Pepe figured that though this shop was located in the city of Tortosa, it must have been selling this pottery to villagers visiting Tortosa. (Note that in the same period, Pepe Jornet sold 'painted pottery' to a florist client in Tortosa.)

The other client in Tortosa was a wholesaler, also one contacted after Pepe had returned from military service, and one Pepe was still supplying at the time of the fieldwork. This order contained almost the same pottery as the other order from Tortosa, but in addition it contained grelas, small bowls from the quatreun down, and planters. This was pottery to be used in a utilitarian manner, with the exception of the planters, which were ordered only in small quantities. The Papaseits displayed a calendar given by this wholesaler hanging in their workshop. On this calendar, the wholesaler advertised himself as trading in 'articles of glass, of ceramics, for cleaning, [and] of palm fiber'. Initially Pepe Papaseit had been delivering his products to the warehouse of this client and he could add that this client was also selling a lot of plastic articles. This client offered a range of articles geared towards small shops in villages and covered the area from Baix Aragon down to the city of Castelló. Because he offered all these articles, it was convenient for these shops to obtain all their merchandise from this wholesaler. There were, for instance, villages closer to Miravet but shop-owners from these villages did not come to Miravet just to obtain pottery: it was easier to order pottery together with other merchandise with this wholesaler and have it delivered to their shops. The Papaseits told me several anecdotes of how they had come across their products that were marketed through this wholesaler. They found it ironic that those shop-owners did not know that these products were made in Miravet, even when Miravet was so nearby. The strongest case was the shop-owner in Miravet who was selling colanders made by the Papaseits without knowing it.

The client in Breda, a wholesaler, was by far the most important client. His orders had contained vessels in all sizes specified with numbers. On the other bills vessels were specified with both numbers and their labels, but as Pepe explained to me, those other clients knew the labels which potters and muleteers had been using but this client did not. As a matter of fact, he had given the Papaseits stamps to number their vessels. The vessels were also specified with 'glazed rim' or on other bills for the same client 'flowers'. These were vessels to be used as planters, something that was new for that time and hence had to be specified with these predicates. Another vessel that only appeared on these bills was one specified as 'mini'. This was a vessel in the shape of the regular vessel, but much smaller, smaller in fact than the saboner. It was used for the storage of honey and during four or five years the Papaseits produced an enormous quantity of these: Dominica even claimed they had produced 200,000 pieces of these! She added that this piece had only been popular for a short period and that nowadays people prefer wholly glazed and finer pieces for honey, generally for use in the kitchen. The last type of vessel that only appeared on these bills was one specified as 'super extra'. These were old vessels, the same vessels Ferran Segarra would later trade on a large scale. At that time Pepe obtained these vessels by exchanging these with new ones. The other types of pottery that appeared on these bills were planters, the cuenco palmera and colanders. At the bottom of most of these bills, also in contrast to the other bills, appeared as an item viruta: this was the saw dust used as packing material for which this client was charged. He was also charged the costs of transportation but these did not appear on the bill: that was a matter between the truck driver/owner and this client.

This wholesaler had come to Miravet and contacted Pepe's father, probably sometime in the 1970s, and they had been supplying him ever since. He was located in Breda. He was trading in all kinds of ceramics and pottery from all over Spain, in alabaster, garden decoration made of piedra artificial, and furniture. He offered a range of articles geared towards home and garden decoration shops in tourists resorts.

When Pepe had still been accompanying the truck driver on the deliveries to this client, he had seen that the vessels and planters were exported. Pepe estimated that for every vessel or planter that was sold in Spain, 50 were exported. This client had been varying his orders of planters without, according to Pepe, any apparent method.

Bibliography

Anon. (1936/37), *Anuario Comercial de España, Tomo II*. No place given: Guia Pascual.

Anon. (1963), *Censo y Classificación de Vehiculos de Motos y Bicicletas* (for the fourth trimester 1963), Municipal archive of Miravet.

Anon. (1978), *Art Popular Català, Terrisses de Miravet*. No place given: Obra Cultural de la Caixa de Pensions per la Vellesa i d'estavis.

Anon. (1989), *La Ribera d'Ebre, transformacions sòcio-econòmiques i perspectives de futur.* Col.lecció Catalunya comarcal. Barcelona: Caixa d'estalvis de Catalunya.

Anon. (n.d.), *Censo de la Población de 1991*. Madrid: Instituto Nacional de Estadistica.

Anon. (n.d.), *Miravet*. No place of publication given: Municipality of Miravet.

Anon. (n.d.), *Miravet (Tarragona), Breve Resumen historico-geográfico de la villa*. No place given: Municipality of Miravet (though not mentioned on this handout).

Anon. (n.d.), *Xarxa* (catalogue of mail order company). Barcelona: Xarxa Cultural.

Aceves, Josep B. (1971), *Social Change in a Spanish Village*. Cambridge, Mass: Schenkman Publ.

Aceves, Josep B (1987), The urbanization of 'El Pinar', *Anthropological Quartely* 60(4): 189–93.

Alfonso, Santi (1983), 'Fang i tradició', *La Sanaqueta* 15: 5–6.

Alvaro Zamora, María Isabel (1980), *Alfareria Popular Aragonesa*. Zaragosa: Libros Porticos.

Anderson, Charles W. (1970), *The Political Economy of Modern Spain, Policy Making in an Authoritarian System*. Madison: University of Wisconsin Press.

Appadurai, Arjun (1986), 'Introduction: commodities and the politics of value'. In Appadurai, Arjun (ed.), *The Social Life of Things*, pp.3–63. Cambridge: Cambridge University Press.

Arnhold, Anthony G. (1981), 'Conditions of existence in a Castilian village,' *Anthropology Papers* 26, Winnipeg, Canada: University of Manitoba.

Bailey, Anne M. (1990), 'Rural revolution in Catalonia during the Spanish Civil War, the meaning of multiple sovereignty'. In Family, *Class and Nation in Catalonia*. Llobera, Josep R. (ed.), Critique of Anthropology X 2 and 3: 179–96.

Barrera González, Andrés (1990), *Casa, Herencia y Familia en la Cataluña Rural, Lógica de la Razón Doméstica*. Madrid: Alianza Editorial.

Barrett, Richard A. (1974), *Benabarre: the Modernization of a Spanish Village*. New York: Holt, Rinehart & Winston.

Baudrillard, Jean (1981[1969]), *For a Critique of the Political Economy of the Sign*. St Louis: Telos Press.

Benedict, Ruth (1989[1934]), *Patterns of Culture*. Boston: Houghton Mifflin Company.

Benjamin, Walter (1968[1955]), 'The work of art in the age of mechanical

reproduction'. In Arendt, Hannah (ed.), *Illuminations*. pp. 219–53. New York: Harcourt, Brace & World inc.
Bierhenke, Wilhelm (1932), 'Ländliche Gewerbe der Sierra de Gata, Sach und Wördhandliche Untersuchungen', *Hamburger Studien zu Volkstum und Kultur der Romanen* 20. Hamburg: Seminar für Romanische Sprache und Kultur.
Bladé Desumvila, A. (1966), *El Castell de Miravet*. Barcelona: Rafael Dalmua.
Bourdieu, Pierre (1984[1979]), *Distinction, A Social Critique of the Judgement of Taste*. Cambridge, MA: Harvard University Press.
Brandes, Stanley H. (1975), *Migration, Kinship and Community, Tradition and Transition in a Spanish Village*. New York: Academic Press.
Brandes, Stanley H. (1990), 'The Sardana, Catalan dance and Catalan national identity', *Journal of American Folklore* 102(407): 24–41.
Breen, T. H. (1986), 'An empire of goods: the anglicization of colonial America, 1690–1776', *Journal of British Studies* 25: 467–99.
Brenan, Gerald. (1950), *The Face of Spain*. London: Penguin.
Cabezón Cuélar, Miguel, Catelló Puig, Ana and Ramón Oliván, Tirso (1984), *La Alfarería en Huesca, Descripción y Localización*. Huesca: Instituto Aragonés de Antropologia.
Calvo, Lluis (1995), 'L'antropologia com a disciplina acadèmica (del 1968 fins ara)'. In Calvo, Lluis and Mañà, Josep (eds), *De l'Ahir i de l'Avui, El Patrimoni Etnològic de Catalunya.*. Barcelona, Generalitat de Catalunya, Departament de Cultura.
Carr, Raymond (1987), 'Introduction: the Spanish transition to democracy in historical perspective'. In Clark, Robert P. and Haltzel, Michael (eds), *Spain in the 1980s: the Democratic Transition and a New International Role*, pp. 1–13. Cambridge, MA: Ballinger Publishing Company.
Carr, Raymond (1982[1966]), *Spain 1801–1975*. Oxford, UK: Clarendon Press.
Carr, Raymond and Fusi, Juan Pablo (1979), *Spain, Dictatorship to Democracy*. London: Unwin Hyman.
Carretero, Andrés, Fernández, Matilda, Limón, Antonio Morcillo, Angela, Ortíz, Carmen, Paoletti, Celsa and Pérez, Angel (1984), *Ceramica Popular de Andalucia*. Madrid: Editora Nacional.
Carrion Garzaran, Angel (1976), 'La población española y su territorio'. In Gonzales Seara, Luis (ed.), *Fundación Foessa, Estudios Sociológicos sobre la situación social de España*. Madrid: Euramerica.
Colby, B.N. (1966), 'Ethnographic semantics: a preliminary survey', *Current Anthropology* 7 (1): 3–17.
Conversi, Daniele (1993), 'Domino effect or internal developments? The influences of international events and political ideologies on Catalan and Basque nationalism', *West European Politics*, .16 no. 3: 245–70.
Curtis, Freddie (1962), 'The utility pottery industry of Bailén, southern Spain', *American Anthropologist* 64: 486–503.
Del Arco Alvarez, José Luis (1971), *Analisis economico y Sociológico del Cooperativismo Agricola*. Madrid: Confederación Española de Cajas de Ahorro.
Del Campo, Salustiano (1975), *La Población de España*. Paris: CIRED.
De Miguel, Amando (1972), *Sintesis del Informe Sociològico sobre la Situaciòn Social de España*. Madrid: La editorial católica, S.A.
De Miguel, Amando (1985), 'The Politics of Identity: Nationalism in Catalonia'. Ph.D. Dissertation, Department of Anthropology University of Massachusetts.
DiGiacomo, Susan M. (1987), '"La caseta i l'hortet": rural imagery in Catalan urban politics', *Anthropological Quartely* 60(4):160–6.
DiGiacomo, Susan M. (1986), 'Images of class and ethnicity in Catalan politics,

1977–1980'. In McDonogh, Gary W. (ed.), *Conflict in Catalonia, Images of an Urban Society*, pp. 72–91. Gainesville: University Presses of Florida.

Fabian, Johannes (1983), *Time and the Other, How Anthropology Makes its Object*. New York: Columbia University Press.

Feito, José Manuel (1985), *Ceramica Tradicional Asturiana*. Madrid: Editora Nacional.

Fraser, Ronald (1973), *Tajos, The Story of a Village on the Costa del Sol*. New York: Pantheon Books.

Freeman, Susan Tax (1968), 'Corporate village organization in the Sierra Ministra: An Iberian Structural Type', *Man* (NS) 3(1): 477–84.

Freeman, Susan Tax (1970), *Neighbors, The Social Contract in a Castillian Hamlet*. Chicago, IL: University of Chicago Press.

Frigolé, Juan (1975), 'Creación y evolución de una cooperativa agrícola en al vega alta del Segura desde 1962 a 1974', *Revista de Estudios Sociales* 14–17: 167–200.

García Alén, Luciano (1983), *La Alfareria de Galicia, 2 volumes*. La Coruña: Fundación Pecho.

Gasòliba i Böhm, Carles (1991), 'Catalonia in the Context of EC Integration in 1992'. In Azevedo, Milton M. (ed.), *Contemporary Catalonia in Spain and Europe*, pp. 21–9. Berkely, CA: University of California at Berkeley.

Giner, Salvador (1980), *The Social Structure of Catalonia. The Anglo-Catalan Society Occasional Publications 1*. Sheffield: University of Sheffield.

González Antón, Rafael (1987), *La Alfarería popular en Canarias*. Santa Cruz de Tenerife: Aula de Cultura de Tenerife.

Guerrero Martín, José (1988), *Alfares y alfareros de España*. Barcelona: Ediciones del Serbal.

Handler, Richard and Linnekin, Jocelyn (1984), 'Tradition, genuine or spurious', *Journal of American Folklore* 97, 385: 273–90.

Hansen Edward C. (1977), *Rural Catalonia under the Franco Regime, the Fate of Regional Culture Since the Spanish Civil War*. Cambridge: Cambridge University Press.

Harding, Susan Friend (1984), *Remaking Ibieca, Rural Life in Aragon under Franco*. Chapel Hill & London: University of North Carolina Press.

Harrison, Joseph (1978), *An Economic History of Modern Spain*. New York: Holmes & Meier Publishers.

Harrison, Joseph (1985), *The Spanish Economy in the Twentieth Century*. New York: St Martin's Press.

Herr, Richard (1971), *An Historical Essay on Modern Spain*. Berkeley, CA: University of California Press.

Hobsbawm, Eric (1992[1983]), 'Introduction: inventing traditions'. In Hobsbawm, Eric and Ranger, Terence (eds), *The Invention of Tradition*. Cambridge: Cambridge University Press.

Jackson, Peter, and Thrift, Nigel (1995), 'Geographies of consumption'. In Miller, Daniel (ed.), *Acknowledging Consumption, a review of new studies*, pp. 204–37. London: Routledge.

Jociles Rubio, Maria Isabel (1989), *La Casa en la Catalunya Nova*. Madrid: Ministerio de Cultura.

Johnston, Hank (1991), *Tales of Nationalism, Catalonia, 1939–1979*. Brunswick, NJ: Rutgers University.

Kay, Paul (1969), 'Comments on Colby'. In Tyler, Stephen (ed.), *Cognitive Anthopology*, pp. 78–90. New York: Holt, Rinehart and Winston.

Köpke, Wulf (1974), 'Frauentöpferei in Spanien', *Baesler Archiv Neue Folge* Band XXII: 335–441.

Köpke, Wulf (1985), *Töpferöfen, Die Brennenanlagen der Tradionellen Töpferein Spaniens*. Bonn: Dr Rudolf Habelt GMBH.

Köpke, Wulf (1986), 'Töpfer, Ton and Teufel, Über Religion and Magie der Spanishen Töpfer', *Tribus band* 35: 41–55.

Laitin, David D. (1987), *Linguistic Conflict in Catalonia. Language Problems and Language Planning* 11(2): 129–47.

Martínez Alier, Joan (1971), *Laborers and Landowners in Southern Spain*. London: George Allen and Unwin.

Lison-Tolosana, Carmelo (1983[1966]), *Belmonte de los Caballeros, Anthropology and History in an Aragonese Community*. Princeton, NJ: Princeton University press.

Llorens Artigas, J. and Corredor Matheos, J. (1970), *Cerámica Popular Española*. Barcelona: Blume.

Maddox, Richard (1993), *El Castillo, The Politics of Tradition in an Andalucian Town*. Urbana & Chicago: University of Illinois Press.

Marx, Karl (1964[1932]), *Economic and Philosophical Manuscripts of 1844*. New York, NY: International Publishers.

Marx, Karl (1978[1972]), 'The Grundrisse'. In Tucker, Robert C. (ed.), *The Marx-Engels Reader*, pp. 221–465. New York: Norton & co.

Marx, Karl (1981 [1964), *A Contribution to the Critique of Political Economy*. New York, NY: International Publishers.

Marx, Karl (1984[1867]), *Capital*. New York, NY: International Publishers.

Mauss, Marcel (1967), *The Gift, The Forms and Function of Exchange in Archaic Societies* (translation by Ian Cunnison). New York: The Norton Library.

Mauss, Marcel (1990), *The Gift, the Form and Reason for Exchange in Archaic Societies* (translation by W.D. Halls),. London: Routledge.

Miller, Daniel (1987), *Material Culture and Mass Consumption*. Oxford: Basil Blackwell Ltd.

Miller, Daniel (1995a), 'Consumption as the vanguard of history. A polemic by way of an introduction'. In Miller, Daniel (ed.), *Acknowledging Consumption, a Review of New Studies*, pp. 1–57. London: Routledge.

Miller, Daniel (1995b), 'Consumption studies as the transformation of anthropology'. In Miller, Daniel (ed.), *Acknowledging Consumption*, pp. 264–98. London: Routledge.

Miller, Daniel (1995c), 'Introduction: Anthropology, modernity and consumption'. In Miller, Daniel (ed.), *Worlds Apart*, pp. 1–22. London: Routledge.

Miller, Daniel (1995d), 'Consumption and commodities', *Annual Review of Anthropology* 24: 141–61.

Moeran, Brian (1984), *Lost Innocence, Folk Craft Potters of Onta, Japan*. Berkeley, University of California Press.

Moncada, Jesus (1994), *The Tow Path*. London: Harvill.

Munn, Nancy D. (1977), 'The spatiotemporal transformation of Gawan canoes', *Journal de la Societe des Oceanistes* 33(4): 39–53.

Munn, Nancy D. (1983), 'Gawan kula: Spatiotemporal control and the symbolism of influence'. In Leach, Jerry W. and Leach, Edmund (eds), *The Kula. New Perspectives on Massim Exchange*, pp. 277–308. Cambridge, MA: Cambridge University Press.

Munn, Nancy D. (1986), *The Fame of Gawa, A Symbolic Study of Value Transformation in a Massim (Papua New Guinea), Society*. Durham and London: Duke University Press.

Munn, Nancy D. (1990), 'Constructing regional worlds in experience: Kula exchange, witchcraft and Gawan local events', *Man* (NS) 25: 1–17.

Nadal, Jordi (1984), *La Población Española, Siglos XVI a VV*. Barcelona: Ariel.

Noyes, Dorothy (1992), 'Tradition in action: the uses of folk culture in contemporary Catalonia'. In Auladell, Joaquim (ed.), *Catalonia 1992: A*

European Nation, pp. 36–66. Barcelona: Llibres de l'Index.
Ortega, Pacqual (1986), 'Alguns aspectes de la repoblació de Miravet (Ribera d'Ebre), després de l'expulsió dels moriscos (1610–1623)'. Pedralbes, *Revista d' History moderna*. Departament de Geografia i Història, Universitat de Barcelona 16: 36–70.
Pérez Baró, Albert (1989), *Història de les Cooperatives a Catalunya*. Barcelona: Editorial Crítica.
Pérez-Díaz, Victor (1991), *Structure and Change of Castilian Peasant Communities, a Sociological Inquiry into Rural Castille 1550–1990*. New York: Garland Publishing.
Pérez Vidal, José (1983), *La Cerámica Popular Española, Zona Norte*. Barcelos: Câmara Municipal de Barcelos.
Pi-Sunyer, Oriol (1973), 'Tourism and its discontents: the impact of a new industry on Catalan community', *Studies in European Society* 1: 1–20.
Pitt-Rivers, Julian A. (1971[1954]), *The People of the Sierra*. Chicago: University of Chicago Press.
Prat, Joan (1983), 'Reflexiones criticas sobre la cultura popular'. In *Actas del II Congreso Iberoamericano de Antropología 1983*, pp. 11–15. Cabildo Insular de Gran Canaria.
Prat Joan (coordinador) (1985–86), 'Trenta ans de literatura antropològica sobre espanya', *Arxiu d' Etnografia de Catalunya*. 4–5.
Ramos Pérez, Herminio (1976), *La Cerámica Popular de Zamora Ceramicas Vivas*. Zamora: Herminio Ramoz Pérez.
Ringrose, David R. (1996), *Spain, Europe, and the "Spanish Miracle" 1700 – 1900'*. Cambridge: the Cambridge University Press.
Roseberry, William (1994), 'Marxism and culture'. In *Anthropologies and Histories: Essays in Culture, History, and Political Economy*. Roseberry, William, pp. 30–54. New Brunswick: Rutgers University Press.
Saez, Josep Maria (1992), *La Terrissa Miravetana*. Flix: Centre d'Estudis de la Ribera d'Ebre.
Sahlins, Marshall (1972), *Stone Age Economics*. New York: Aldine Publishing Company.
Sahlins, Marshall (1981), *Historical Metaphors and Mythical Realities, Structure in the Early History of the Sandwich Islands Kingdom*. Ann Arbor, MI: The University of Michigan Press.
Sahlins, Marshall (1991), 'The return of the event, again; With reflections on the beginnings of the great Fijian war of 1843 to 1855 between the kingdoms of Bau and Rewa'. In Biersack, Aletta (ed.), *Clio in Oceania: Towards a Historical Anthropology*, pp. 37–99. Washington: Smithsonian Institution Press.
Sahlins, Marshall (1993), 'Goodbye Tristes Tropes: ethnography in the context of modern world history', *Journal of Modern History* 65: 1–25.
Schütz, Ilse, n.d. 'La alfarería tradicional de Agost'. *Naria* 37–38: 16–25.
Segarra, Ferran (1990a), 'Història de la ceràmica (I)', *La Sanaqueta* 41: 5.
Segarra, Ferran (1990b), 'Història de la ceràmica (II)'. *La Sanaqueta* 42: 5.
Segarra, Ferran (1990c), 'Història de la ceràmica (III)'. *La Sanaqueta* 43: 5.
Segarra, Ferran (1993), 'Canters i canterers', In *La Sanaqueta* 54: 8.
Sempere, Emili (1982a), *Rutas a los Alfares*. Barcelona: les Punxes
Sempere, Emili (1982b), *La Terrissa de las Terres de l'Ebre*. Barcelona: les Punxes.
Sempere, Emili (1985), *La Terrissa Catalana*. Col. Gent Nostra. Barcelona: Ediciones de Nueve Arte Thor.
Seseña, Natacha (1976), *Barros y Lozas de España*. Madrid: Editorial Prensa Española.
Shubert, Adrian (1990), *A Social History of Spain*. London: Unwin Hyman.
Silverman, Sydel F. (1968), 'Agricultural organization, social structure, and values in

Italy: Amoral Familism Reconsidered', *American Anthropologist* 70: 1–20.
Subirats, Santíago Martín (1988), *Mostra de Terrissa Popular Catalan*. Barcelona: Generalitat de Catalunya, Department de Cultura.
Tamames, Ramón (1986), *The Spanish Economy, An Introduction*. New York: St. Martin's Press.
Torres, Pablo (1982), *Cantáros Españolas*. Madrid: Editorial Arteras
van Veggel, Rob J.F.M. (1991), 'Peirce's semiois, Morris's semiosis, and studies of reference', *Semiotica* 87(1/2): 95–117.
Ventura, Josep (1984), 'Postals de Catalunya: Miravet, fang i tradició'. *Som, Publicació de cultura popular catalan* 83: 22–5.
Vossen, R. (1972a), 'Systematische Bestandsaufname der Töpfereitradition in Spanien', *Mitteilungen aus den Museum für Volkenkunde, Hamburg Neue Folge* Band 2: 51.
Vossen, R. (1972b), *Töpferei in Spanien*. Hamburg: Hamburgisches Museum für Volkenkunde und Vorgeschichte.
Vossen, R. (1974), 'Prinzipien der Benennung sowie Bewertungssysteme Spanischen Gebraucheskeramik', *Mitteilungen aus dem Museum für Volkenkunde, Hamburg. Neue Folge* Band 4: 119–68.
Vossen, R. (1984), 'Towards building models of traditional trade in ceramics'. In Van der Leeuw (ed.), *The Many Dimensions of Pottery*, pp. 339–406. Amsterdam: Universiteit van Amsterdam.
Vossen, R., Seseña, N and Köpke, W. (1980[1975]), *Guia de los Alfares de España*. Madrid: Editoria Nacional.
Vilanova, N., Font, N., Davies, R. Udina, T. Jenè, A. (1988), *Vox Diccionari, Anglès-Català, Català-Anglès*. Barcelona: Biblograf.
Whyte, William Foote and Whyte, Kathleen King (1988), *Making Mondragon, The Growth and Dynamics of the Worker Cooperative Complex*. New York: State School of Industrial and Labor Relations.
Woolard, Kathryn A. (1989), *Double Talk, Bilingualism and the Politics of Ethnicity in Catalonia*. Stanford, CA: Stanford University Press.
Ysàs, Pere (1994), 'Les condicions de vida durant la postguerra'. In Anonymous, *Els anys de la postguerra a Catalunya (1939–1959)*, pp. 7–22. Barcelona: Publicacions de l'abadia de Montserrat.

General index

Author index

www.ingramcontent.com/pod-product-compliance
Ingram Content Group UK Ltd.
Pitfield, Milton Keynes, MK11 3LW, UK
UKHW020801290926
15470UKWH00003B/891

9 780955 640025